D0805827

AROMATIC NUCLEOPHILIC SUBSTITUTION

REACTION MECHANISMS IN ORGANIC CHEMISTRY

A SERIES OF MONOGRAPHS EDITED BY

C. EABORN
Professor of Chemistry,
University of Sussex, Brighton, Great Britain

AND

N. B. CHAPMAN
Professor of Chemistry,
University of Hull, Great Britain

MONOGRAPH 8

Other titles in this series

1 Nucleophilic Substitution at a Saturated Carbon Atom *by* C. A. BUNTON

2 Elimination Reactions *by* D. V. BANTHORPE

3 Electrophilic Substitution in Benzenoid Compounds *by* R. O. C. NORMAN and R. TAYLOR

4 Electrophilic Additions to Unsaturated Systems *by* P. B. D. DE LA MARE and R. BOLTON

5 The Organic Chemistry of Phosphorus *by* A. J. KIRBY and S. G. WARREN

6 Aromatic Rearrangements *by* H. J. SHINE

7 Steroid Reaction Mechanisms *by* D. N. KIRK and M. P. HARTHSORN

In preparation

Free-Radical Polymerisation *by* A. D. JENKINS

AROMATIC NUCLEOPHILIC SUBSTITUTION

J. MILLER

Superintendent, Organic Chemistry Division,
Defence Standards Laboratories,
Melbourne (Australia)

ELSEVIER PUBLISHING COMPANY

AMSTERDAM/LONDON/NEW YORK

1968

QD476
.R4
v. 8

ELSEVIER PUBLISHING COMPANY
335 JAN VAN GALENSTRAAT
P.O. BOX 211, AMSTERDAM, THE NETHERLANDS

ELSEVIER PUBLISHING CO. LTD.
BARKING, ESSEX, ENGLAND

AMERICAN ELSEVIER PUBLISHING COMPANY, INC.
52 VANDERBILT AVENUE
NEW YORK, NEW YORK 10017

INDIANA
UNIVERSITY
LIBRARY

NORTHWEST

LIBRARY OF CONGRESS CARD NUMBER: 68-20648

STANDARD BOOK NUMBER 444-40683-2

WITH 125 FIGURES AND 74 TABLES.

COPYRIGHT © 1968 BY ELSEVIER PUBLISHING COMPANY, AMSTERDAM

ALL RIGHTS RESERVED. THIS BOOK OR ANY PART THEREOF MUST NOT BE REPRODUCED IN ANY FORM WITHOUT THE WRITTEN PERMISSION OF THE PUBLISHER, ELSEVIER PUBLISHING COMPANY, AMSTERDAM, THE NETHERLANDS.

Printed in Great Britain

"what good shall I do you, unless what I say contains something by way of revelation, or enlightenment, or prophecy, or instruction?"

I Corinthians 14, v.6
(The New English Bible)

PREFACE

The above quotation succinctly summarises what I have sought to achieve in this book. If I have succeeded the credit is due to the early training and life-long stimulus received from Sir Christopher Ingold, the late Professor E. D. Hughes, and Dr. O. L. Brady, among other distinguished figures, at University College London, which its graduates must surely regard as not so much a place as a way of life. I would also express my real appreciation of the valuable comments of the two General Editors of the series, namely, Professor N. B. Chapman and Professor C. Eaborn, and of Professor Sir Christopher Ingold. I thank Dr. M. E. C. Biffin and Dr. D. B. Paul for their valuable assistance in checking proofs.

There is all too much danger in the context of the "information explosion" that a scientific book may be obsolescent by the time it is in print. I hope I have avoided this, not only by the obvious method of trying to ensure that it is up to date when it goes to the publisher, and by references to still unpublished work, but also by a willingness to supply my own "revelation and prophecy" even at the risk of being proven wrong at a later date.

I have not attempted to make the book an exhaustive review, but I have tried to cover all areas of major interest. I trust that in doing so, I have given due credit to the many workers in the field, whose manifest achievements have made the subject of Aromatic Nucleophilic Substitution so rewarding and intellectually satisfying, and so illuminating of other areas of chemistry and scientific thought.

I hope that there is something of value in the book for a wide spectrum of readers from undergraduate and graduate students at Universities to those concerned in production in chemical industry.

Like many other authors I have found the task of scientific writing a lengthy one. It was almost entirely carried out while serving as Head of the Chemistry Department of the University of Hong Kong.

The co-operative and understanding attitude of that University, and more recently of the Australian Defence Scientific Service deserve my thanks and appreciation; as does the forbearance of the publishers to its long-overdue author, and of my wife for the long hours away from home.

J.M.

CONTENTS

Chapter 1

OCCURRENCE AND MECHANISMS OF AROMATIC NUCLEOPHILIC SUBSTITUTION

1. Introduction

Formally nucleophilic substitution at an aromatic carbon atom resembles other nucleophilic substitutions on carbon in that a bond to the carbon at the reaction centre is formed by a reagent Y, and a group X is correspondingly displaced with its bonding electrons. This is illustrated in Fig. 1, which excludes consideration of reversibility, relative timing of bond formation and rupture, charge, and any rearrangement which may occur. In such reactions, which include isotopic substitution and solvolytic reactions, Y becomes one unit more positive and X one more unit more negative.

$$\ddot{Y} + Ar \vdots X \longrightarrow Y{-}Ar + \ddot{X}$$

Fig. 1. Generalised representation of aromatic S_N reactions.

In many simple Ar–X compounds such as halogeno-benzenes and -toluenes, nucleophilic substitution reactions require either vigorous or else very basic conditions, whereas S_N reactions of halogenoalkanes are relatively facile. Such aromatic reactions are nevertheless included among industrial reactions of major importance, as well as those on the laboratory scale. For example, phenol and aniline, which are among the most important synthetic aromatic compounds in scale of production and use, are made largely by nucleophilic substitution reactions. Most phenol is made by hydrolysis of chlorobenzene with caustic soda (370°) or water and a catalyst (425°), and the chlorobenzene may be made separately, or the reaction carried out as the first of a two-step process from benzene. A major process for the manufacture of aniline is the catalysed ammonolysis of chlorobenzene

References p. 26

(190–210°) in water. At a much lower temperature but in a much more basic medium, aniline is formed by reaction of chlorobenzene with amide ion in liquid ammonia at –33°. By isotopic tagging, this reaction has been shown to produce aniline in which the amino group has entered at the position *ortho* to the point at which chlorine was originally attached, as well as the directly substituted product. High temperature alkaline hydrolysis of halogenotoluenes gives correspondingly a mixture of rearranged and unrearranged cresols.

Many reactions of aromatic diazonium ions involve displacement of the diazonium group by a nucleophile. They present special and unusual features, and preparatively, catalysis by cuprous salts is common.

With suitable substituents, replacement of chlorine and other typical leaving groups becomes facile. As long ago as 1854, Pisani *[1]* reported the preparation of picramide and picric acid from picryl chloride (1-chloro-2,4,6-trinitrobenzene) under very mild conditions. In the form of cotton, cellulose, a weak nucleophilic reagent when acting through its hydroxyl group, reacts readily with "reactive dyes", *e.g.* cyanuric chloride (2,4,6-trichloro-1,3,5-triazabenzene) derivatives. Activation in the aromatic system in this case is by heterocyclic nitrogen instead of nitro groups. Like picryl chloride, 1-chloro-2,4-dinitrobenzene reacts readily with a wide range of nucleophilic reagents. As a cheap and readily available material it constitutes a most important chemical intermediate. 1-Fluoro-2,4-dinitrobenzene is much more reactive with some reagents and much less with others than the 1-chloro compound. Very high reactivity is involved for example in the displacement of fluorine by amino or peptide nitrogen, utilised in studies of amino-acids, polypeptides and proteins. Reaction with water and alcohols is also facile, especially in alkaline conditions, but the substrate is very *un*reactive with most heavy nucleophiles, *i.e.* reagents in which the atom forming the bond to aromatic carbon is an atom of the second or higher row of the Periodic Table.

The nitro groups in *m*-dinitrobenzene are difficult to displace with nucleophilic reagents, but those in *o*- and *p*-dinitrobenzene are much more reactive, though reaction normally proceeds only to the displacement of one nitro group in each.

The reactions cited exemplify industrial and laboratory scale aromatic S_N reactions, which proceed by a range of reaction mechanisms now well understood. Known mechanisms are discussed in detail below and in subsequent chapters, and the most important and fundamental are compared in this chapter.

2. Multiplicity of Mechanisms

(*a*) *General*

There is a substantial and valuable body of early experimental work on aromatic S_N reactions with important implications as to mechanism, as for example the series of papers by Dutch research workers *[2–5]*. However the most active research has been carried out since about 1950. In 1951 a comprehensive review was published by Bunnett and Zahler *[6a]* and a shorter review by Miller *[7]*. More recent reviews include those of Bunnett *[6b]* and Ross *[8]*.

The mechanistic forms of nucleophilic substitution in saturated aliphatic compounds have been discussed at length by Ingold, and by Bunton *[9a,10]*. The main forms (*cf.* *[11]*) are the unimolecular (S_N1) mechanism involving a rate-limiting heterolysis of the bond to the displaced group followed by reaction with the nucleophile; and the bimolecular (S_N2) mechanism involving both bonding by the nucleophile and heterolysis of the bond to the displaced group in forming the transition state of the rate-limiting step. Important variations are mechanism S_N1cB in which the rate-limiting step is the heterolysis of the conjugate base of the substrate; and mechanism S_N2C^+ in which the rate-limiting step is the reaction of the nucleophile with a carbonium ion in equilibrium with the reactants. It should be mentioned that there are corresponding elimination reactions, which are discussed fully by Banthorpe *[12]*.

Whereas nucleophilic substitution is the characteristic substitution reaction at carbon in saturated aliphatic compounds other than hydrocarbons and those which contain metal–carbon bonds, electrophilic substitution is the characteristic reaction in aromatic compounds with their π-electron system in molecular orbitals external to the main σ-bond framework. Further, many of the groups typically displaceable by nucleophiles have unshared electrons which are conjugated with the ring, thus rendering their displacement with their

References p. 26

bonding electrons still less facile. With appropriate substituent groups, however, nucleophilic substitutions can occur readily and include many important reactions. An example already given is the widespread and large-scale use of 1-chloro-2,4-dinitrobenzene as an industrial chemical intermediate, because of the facile replacement of its chlorine atom.

There are parallels with aliphatic S_N reactions in the nature of nucleophilic reagents and displaced groups. There are also some mechanistic resemblances but these are much less close. The great majority of aromatic nucleophilic substitution reactions follow the bimolecular addition–elimination mechanism—the typical activated aromatic S_N2 reaction. An elimination–addition S_N2 mechanism is well characterised but much less common. There are also some S_N1 reactions and there are some uncommon special reactions, differing from any of the above, which are discussed in Chapter 9.

Both aromatic and aliphatic S_N reactions are heterolytic reactions. As such they are very largely confined to the condensed phase and commonly encountered as homogeneous reactions in solution.

(*b*) *Unimolecular mechanism*, S_N1; *and* S_N1-*like reactions*

In nucleophilic substitution at a saturated aliphatic carbon atom a rate-limiting heterolysis of the C–X bond may occur, especially when electron-releasing groups stabilise the carbonium ion formed. This electron release is most effective if acting by the conjugative mechanism ($+T$ groups)*, for there is then delocalisation of charge resulting also in a decrease in charge on carbon. There is also an increase in bonding in the system, including the point of separation of the leaving group, to compensate for the heterolysis. Such a heterolysis is however rarely facile except in media where there is substantial ion solvation. It is exemplified in Fig. 2 by a side-chain aryl system, for a more illustrative comparison with substitution at aromatic carbon.

In comparison even with the S_N1 reactions of less reactive aliphatic substrates, the simple S_N1 mechanism is much less probable in aromatic substrates because (*i*) internal conjugation to stabilise an aryl

* Ingold now uses the symbol *K* for conjugation.

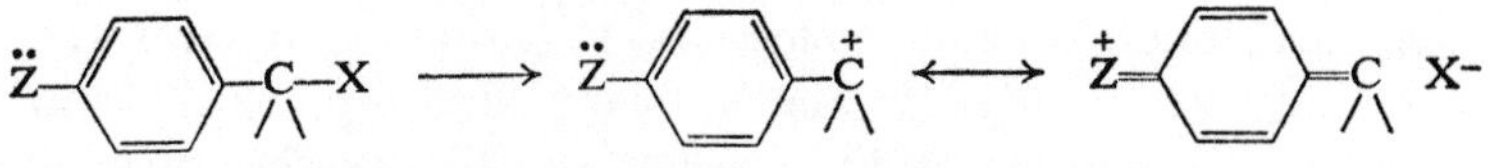

Fig. 2. Aliphatic S_N1 heterolysis in a side-chain aryl system.

cation Ar^+ formed by a heterolysis of Ar–X is not possible, as the axis of the vacant orbital is at right angles to those of the ring pπ-orbitals, and the inductive effects available are much less efficacious; (*ii*) there is no increase in bonding in the rest of the system to compensate for the bond heterolysed; (*iii*) the positive charge of an aryl cation (Ar^+) would be much less exposed to solvation than in an alkyl or side-chain aryl cation; (*iv*) a decrease in steric requirements in forming Ar^+ is unlikely to be kinetically important, as it sometimes is in forming the cation $Alph^+$ in aliphatic S_N1 reactions; (*v*) bonds to common leaving groups are stronger in Ar–X than Alph–X compounds, associated usually with conjugation of unshared electrons on X with the ring, and this is unfavourable to the required heterolysis. Taft *[13]* has suggested that some stabilisation would result if such a cation were a triplet-state ion-radical (see Chapter 2, pp. 29, 30).

The meagreness of evidence in favour of aromatic S_N1 reactions despite extensive mechanistic studies, shows that there must be important adverse factors such as have been mentioned. Nucleophilic displacements of the diazonium group of aromatic diazonium salts discussed in Chapter 2 possess many special features, but can formally be regarded as S_N1 reactions. They are clearly very different from normal bimolecular addition–elimination and elimination–addition reactions. Copper-catalysed reactions of unreactive halogen compounds, which are discussed in Chapter 8 are also S_N1-like in character, but there is no reliable evidence to suggest that any uncatalysed aromatic nucleophilic substitution takes place by a mechanism involving heterolysis of an Ar–X bond to form an intermediate aryl cation, which then reacts with a nucleophile.

(*c*) *Bimolecular mechanisms*, S_N2

(*i*) *Introduction*

Aromatic compounds consist of a σ-bonded framework sandwiched between the aromatic π-electron system in molecular orbitals. This is illustrated, for example, in Ingold's text *[9b]*. Repulsive electronic and steric interactions between the unshared electrons which confer nucleophilic character on the reagent, and the π-electrons of the aromatic system, render attack by a nucleophilic reagent on aromatic carbon unfavourable. Structural changes in the aromatic system which result in one or more carbon atoms of the ring becoming electron-deficient, and stabilise cyclohexadienide (benzenide) transition states (see below), result in marked enhancement of reactivity.

It should be recalled that there is always an ambiguity in consideration of substituent effects in bimolecular reactions *[9a]*. For an S_N2 reaction, electron-withdrawal from the reaction centre favours bond-formation by the nucleophile, but hinders rupture of the bond to the leaving (displaced) group, and *vice versa* for electron-release. Even in saturated aliphatic S_N2 reactions the normal pattern is for rates of reactions to be reduced by electron-release, implying the greater importance of bond-formation in the transition state. It can be otherwise only in S_N1-like reactions with bond-breaking considerably advanced relative to bond-formation. As regards aromatic systems reasons have already been adduced to explain why an S_N1 mechanism involving heterolysis of the Ar–X bond to form an aryl cation (Ar^+) is not favoured. The special S_N1 reactions of diazonium ions and S_N1-like copper-catalysed reactions of aryl halides are the only reactions reliably known to proceed by other than S_N2 mechanisms. In these few reactions rather special substituent effects are encountered (see Chapter 2, pp. 31–34; and Chapter 8, p. 342), so that there is no reaction series with a clear pattern indicating favourable kinetic effects of electron-release in aromatic S_N reactions.

It is difficult to visualise an aliphatic type S_N2 mechanism, *i.e.* one with concerted bond-formation and -rupture at the point of substitution, which can retain the full aromatic delocalisation of the π-electron system. If this is not retained then an S_N2 mechanism with addition as a key step preceding elimination (see below) is energetically preferred and sterically favoured. There are in any case numerous features of aromatic S_N reactions which are inexplicable on the basis of a synchronous (aliphatic type) mechanism. These include examples of base and electrophilic catalysis, and the general phenomenon of facility of group displacement especially of halogens, which are discussed in Section 5 and in later chapters.

As the main mechanism one is therefore left with an S_N2 mechanism, with addition as the key step, favoured by electron-withdrawing substituents. This is discussed below in some detail, and in later chapters.

There is however another well-known, though much less common S_N2 mechanism. In this the key step is abstraction by bases of sufficient strength of hydrogen attached to the ring. These are commonly nucleophiles towards aromatic carbon also, so that such reactions are to be sought, and are found, with substrates of low reactivity in the normal S_N2 mechanism. This alternative mechanism is the benzyne mechanism.

(ii) The benzyne or elimination–addition mechanism

It is clear from the preliminary discussion above that most simple aromatic compounds, even with attached groups typically displaceable by nucleophiles, are not readily susceptible to attack by nucleophilic reagents. In many of the cases where such compounds can be made to react, it is with strongly basic reagents by an elimination–addition mechanism known as the benzyne mechanism, illustrated in Fig. 3. These reactions are detectable characteristically by the

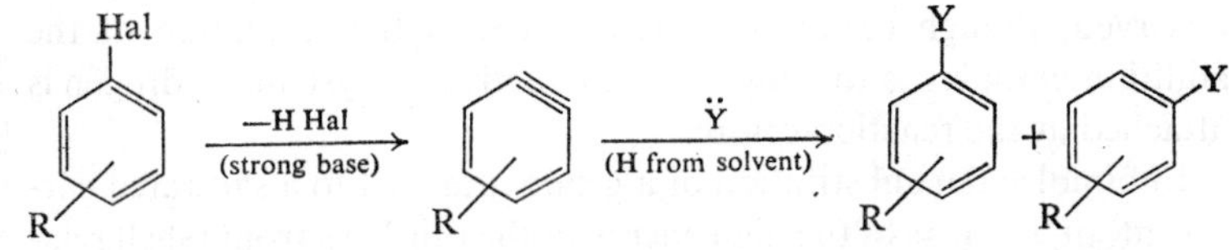

Hal, R; —H Hal (strong base); R; $\ddot{Y}$ (H from solvent); Y, R; +; Y, R

Fig. 3. A general form of the elimination–addition or benzyne S_N2 mechanism.

occurrence of vicinal *cine*-substitution in which the entering group occupies a position neighbouring that of the displaced group. This is readily recognised by production of rearranged as well as unrearranged products when a substituent is present, otherwise it may be shown by isotopic labelling of ring atoms. Roberts and his co-workers *[14]* showed for example, that chlorobenzene-1-^{14}C reacts with amide ion in liquid ammonia to give almost equal amounts of aniline-1-^{14}C and aniline-2-^{14}C; and that *p*-iodotoluene reacts with aqueous caustic soda to give almost equal amounts of *m*- and *p*-cresol (see Chapter 2, p. 38).

The elimination stage may be stepwise or concerted, and there is evidence for both situations. The stepwise elimination reaction is classifiable as an E1cB, and the concerted reaction as an E2 reaction, and we may assume that reactions between the two extremes also occur *[12]*. The benzyne or elimination–addition mechanism is discussed in detail in Chapter 3.

(*iii*) *The activated or addition–elimination mechanism*

Facile nucleophilic substitution in aromatic systems can only be expected when there is (*i*) a group X at the point of substitution readily replaceable with its bonding electrons, and (*ii*) suitably placed electron-withdrawing substituents or ring hetero atoms present, which can facilitate formation of a bond to a ring carbon atom by a nucleophilic reagent, and stabilise the transition state. It is for this reason that reactions proceeding by this mechanism are often called activated aromatic S_N2 reactions. Such substitutions when uncatalysed will be overall second order reactions and the great majority of aromatic nucleophilic substitutions fall into this category.

In the presence of an oxidising agent, or in otherwise suitable conditions, nucleophilic replacement even of hydrogen may be observed, though cessation of reaction with the formation of the addition complex is the norm when in activated systems hydrogen is attached at the reaction centre.

In bimolecular substitution of a group attached to a saturated carbon atom, because of the high energy of the third electronic shell relative to the two inner occupied shells, a fully 5-covalent intermediate is not possible. Bond formation by the nucleophile is therefore synchronous with the breaking of the bond to the replaceable group. The reaction path of lowest energy then involves a change from sp^3 hybridisation to sp^2 hybridisation with a shared p-orbital and results in stereochemical inversion. This is shown in Fig. 4.

Aromatic carbon atoms are bonded directly to only three other atoms, and in aromatic substitution a relatively stable intermediate, which is isolable in suitable cases, can readily be visualised, involving an increase of one in the number of species fully bonded at the reaction centre, with accompanying change from sp^2 to sp^3 hybridisation *[6–8,15–29]*. The situation is analogous to that in aromatic S_E

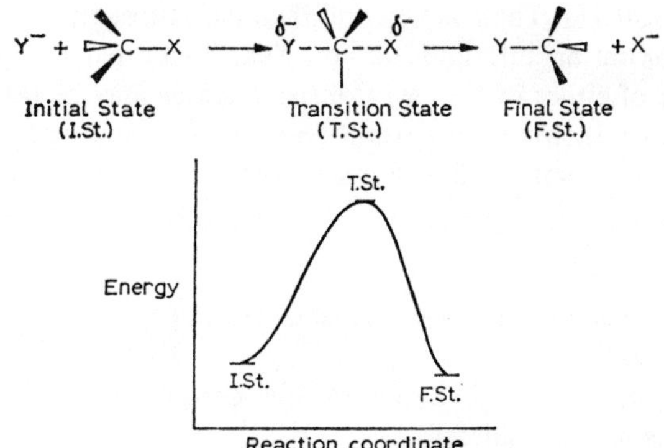

Fig. 4. Representation of bimolecular substitution at a saturated carbon atom. Full bond above plane of paper, ◄; full bond below plane of paper, ◁; full bond in plane of paper, —; partial bond, ---.

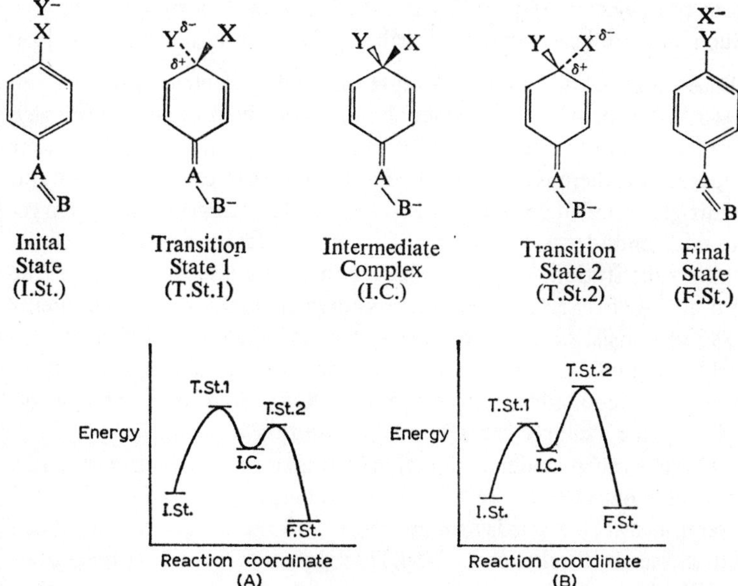

Fig. 5. Representation of addition–elimination (activated) aromatic S_N2 reactions. In (A) formation of the first transition-state is rate-limiting. In (B) formation of the second transition-state is rate-limiting. Electron delocalization in the benzenide (cyclohexadienide) system is not represented.

reactions *[30,31]*. The reaction will then pass through two transition states flanking an intermediate complex, illustrated in Fig. 5, and formation of either of the two transition states may be rate-limiting (see Chapter 5). Many important features of aromatic nucleophilic substitutions depend on this. Some further details are given in Section 6 (p. 22).

3. Charge-Transfer (π) and Covalent (σ) Complexes

The possible role of charge-transfer complexes in aromatic S_N reactions is now considered.

The interaction of benzene and iodine to produce a substance characterised by a new spectroscopic absorption band was reported by Benesi and Hildebrand *[32]* to be due to a 1:1 complex. Mulliken *[33]* worked out a theory for such complexes in which one molecule acts as an electron-donor (D) and the other as electron-acceptor (A). Such a complex may be described *[34a]* as a resonance hybrid ($D–A \leftrightarrow \overset{+}{D}–\overset{-}{A}$) in which $\overset{+}{D}–\overset{-}{A}$ represents a one-electron transfer from D to A. The field has been covered more recently in a book by Briegleb *[35]* and more briefly by Mulliken and Person *[36]*. The latter suggest that there is no sharp distinction between charge-transfer and hydrogen-bonded complexes. Brown *[37]* has suggested that charge-transfer complexes are the main intermediates in aromatic electrophilic substitution, but in a relevant investigation, *inter alia*, of the bromine–benzene (1:1) complex, known from X-ray measurements *[38]* to consist of alternate layers of benzene rings and perpendicularly oriented bromine molecules, Hooper *[39]* has inferred from pure electric quadrupole resonance (PQR) data that there is no evidence for a charge-transfer complex in this system.

There is much evidence that charge-transfer complexes exist which may alternatively be regarded as π-complexes, or fully bonded σ-complexes. As regards aromatic S_E reactions, the evidence for both is discussed by Streitwieser *[34b]*. He suggests that their existence does not implicate them necessarily as intermediates, whereas on the other hand there is a close parallelism between the effects of substituents on formation of isolable proton σ-complexes and on the intermediates in electrophilic substitution reactions. He concludes therefore that the σ-complex is a good model for the transition state. In aromatic S_N reactions the formation of stable complexes has been known for over 60 years *[15,16]*. The formation of covalent adducts by addition of alkoxides to picryl ethers (Meisenheimer complexes) *[15,16]*, more recent work, especially spectrometric, *[17–22,26a]* on such compounds; and recent work on Meisenheimer complexes of dinitro-aromatic ethers *[25a]*, has lent confidence to their structural representation* as in Fig. 6, including the equivalence of the methoxyl groups in this example.

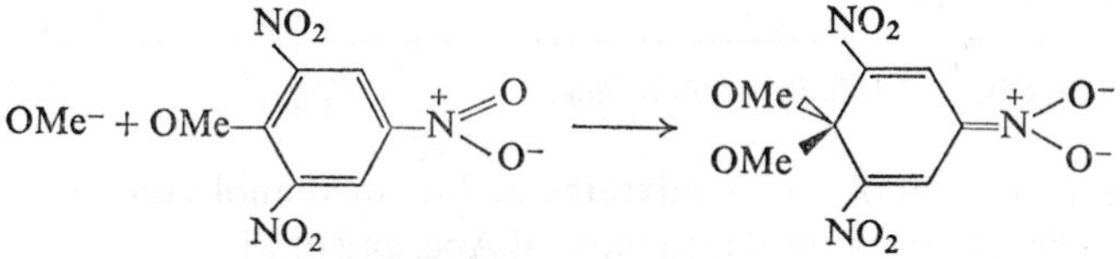

$$OMe^- + OMe\text{–}C_6H_2(NO_2)_2\text{–}\overset{+}{N}(=O)O^- \longrightarrow (OMe)_2C_6H_2(NO_2)_2{=}\overset{+}{N}(O^-)_2$$

Fig. 6. Formation of a symmetrical Meisenheimer complex.

Abe *[23a]* has made quantum mechanical calculations of the kinetic stability of such complexes with one (*p*-nitro), two (2,4-dinitro) and three (2,4,6-trinitro) nitro groups, comparing the energy of the transition states with that of the σ-complexes and initial states. Miller and his co-workers *[24a,b]* have made corresponding calculations by use of bond and solvation energies, electron affinities, and an allowance for electron reorganisation. Their results, which are compared in Table 1, are in good agreement as regards energies of transition states and intermediate complexes, but not initial states. This is because Abe had not considered the effects of the solvent. The results of Miller and his co-workers also correctly relate the intermediates to the initial states and show that in solution†, only the trinitrobenzene complex is thermodynamically more stable than the initial state *[cf. 25]*. Abe and his co-workers *[23b]* have recently produced some kinetic results on the decomposition to 2,4,6-trinitroanisole of the Meisenheimer complex illustrated in Fig. 6. The precision of the results of Abe and his co-workers has been criticised by Murto and Kohvakka *[40]*, but the results of the latter authors, which

* Delocalisation in this benzenide (cyclohexadienide) structure involves all three nitro groups.
† In protic solvents.

References p. 26

TABLE 1

CALCULATED VALUES OF $\Delta E^{\ddagger}$ IN kcal·mole^{-1} FOR THE DECOMPOSITION OF MEISENHEIMER COMPLEXES TO REACTANTS

Method	$\Delta E^{\ddagger}$ values for decomposition of adducts of methoxide ion with		
	p-nitro-anisole	2,4-dinitro-anisole	2,4,6-trinitro-anisole
Quantum mechanical *[23a]*	7.5	11.9	15.7
Thermochemical *[24a,b]*	9.5	12.5	16[a]

[a] And 1-methoxy-2,4-dinitronaphthalene.

relate to methanol–water mixtures of low methanol content, show that their values are similar to those of Abe, and of Miller.

It is of interest that comparison of the delocalisation energy of benzene and naphthalene systems suggests that intermediate complexes, which retain one ring fully benzenoid are relatively more stable in the latter system. More specifically the difference is about equivalent to "one nitro group" (*ca.* 9 kcal·mole^{-1}) so that we would expect for example the stability of Meisenheimer (ether) complexes in 2,4-dinitronaphthalene and 2,4,6-trinitrobenzene systems to be about equal, and have values of $\Delta E^{\ddagger}$ about 14 kcal·mole^{-1} for their formation, and 16 kcal·mole^{-1} for their decomposition in protic solvents. This is confirmed by the experimental results of Fendler *[25b]*. The occurrence of such complexes even when no direct substitution occurs, as is often the case when hydrogen is at the point of attachment, has also been demonstrated *[17,25,26]*.

It should be mentioned here that Servis *[26b]*, by utilising NMR measurements, has studied the addition of methoxide ion to 2,4,6-trinitroanisole in DMSO *inter alia*, and reported that addition is more facile at C-3 than C-1, but the σ-complex for addition at C-1 is thermodynamically the more stable. He regards this as contrary to the results based on Miller's calculations *[24a,b]* but this comparison ignores the fact that the calculations refer to solutions in protic solvents and not DMSO, and further are a comparison of addition to 2,4,6-trinitroanisole and 1,3,5-trinitrobenzene, for he takes no account of the influence of the 1-OMe group on addition at C-3. This includes steric interactions with the 2- and 6-nitro groups, and cross-conjugation with all three nitro groups will affect both kinetic and thermodynamic parameters. While the calculations cannot satisfactorily deal with these, they are important factors which cannot be ignored. Entropy factors may also be important and have not been considered. None of the above necessarily implies that charge-transfer complexes may not also be formed in such systems *[18]*.

Recent work by Ainscough and Caldin *[18]* has been interpreted as demonstrating the existence of both charge-transfer (π) and covalent (σ) complexes in the reactions studied. Both "fast" and "slow" reactions occur in the reaction of ethoxide ion with 2,4,6-trinitroanisole (TNA), which they studied. The reversibly-formed product of the "fast" reaction is readily decomposed to reactants by general acid catalysis, and is regarded as a charge-transfer complex. The product of the "slow" reaction, which is decomposed by dilute sulphuric acid, but not undissociated acids, to 2,4,6-trinitrophenetole, is regarded as the same σ-complex as that prepared by Meisenheimer *[15]* in this way and from 2,4,6-trinitrophenetole with methoxide ion—both visible and infra-red spectra confirming its identity *[21]*. The corresponding compound from TNA and methoxide has been shown recently by NMR measurements *[17d,e,22]* to have both methoxyl groups identical. Ainscough and Caldin were able to separate the kinetic data for the two reactions. At –80° the "fast" reaction (with a large excess of ethoxide ion, so that the reaction is first-order in both directions) has values of $\log_{10} k_1$ (rate constant for forward reaction) E_{act} or $\Delta E^{\ddagger}$, $\log_{10} B$, and $\log K$ (equilibrium constant), shown in Fig. 7. The values for the "slow" reaction are also given.

The relative stability of what is regarded as the charge-transfer complex is noteworthy, and corresponds to the stronger class (lone pair donors), referred to by Mulliken and Person *[33]* as n-donors. It is of interest also to note that whereas 1,3,5-trinitrobenzene is regarded as forming relatively weak charge-transfer complexes with aromatic amines *[41,42]*, more stable products are formed by aliphatic amines and ammonia *[43,45]* and are regarded as n-π charge-transfer complexes *[45,46]*.

On the basis of all this evidence Ainscough and Caldin suggested the reaction scheme shown in Fig. 7, but noted that their evidence did not necessarily exclude the possibility that the 1,1-dialkoxy σ-complex

(TNA, OEt^-)

↕

(TNA^-, OEt)

$\underset{k_{-1}}{\overset{k_1}{\rightleftharpoons}}$ $TNA + OEt^-$ $\underset{k_{-2}}{\overset{k_2}{\rightleftharpoons}}$ [1-OMe-1-OEt-2,4,6-$(NO_2)_3$-cyclohexadienide]$^-$

Charge-transfer or π-complex[a] | Initial state | Covalent or σ-complex

Kinetic data at $-80°$:

π-complex formation $\log_{10} k_1 = 1.09$ l·mole^{-1}·sec^{-1}; $\Delta E^{\ddagger} = 10.4$ kcal·mole^{-1}; $\log_{10} B = 10.7$ l·mole^{-1}·sec^{-1}; $\log_{10} K_1 = 2.07$ l·mole^{-1}.

σ-complex formation $\log_{10} k_2 = -4.8$ l·mole^{-1}·sec^{-1}; $\Delta E^{\ddagger} = 13.7$ kcal·mole^{-1}; $\log_{10} B = 10.9$ l·mole^{-1}·sec^{-1}; $\log_{10} K_2$ (at 0°) = "large".

Fig. 7. Scheme for the charge-transfer (π) and covalent (σ) complex formation in reaction of ethoxide ion with 2,4,6-trinitroanisole (TNA). [a]*Cf.* ref. *26b.*

is formed in a reaction following formation of the charge-transfer complex. It is generally assumed however that aromatic S_N2 reactions do not proceed via formation of π-complexes. It is relevant, in any case, to point out that Servis *[26b]* suggested that the initially-formed complex is *also* a σ-complex, *viz.* that formed by addition of ethoxide ion to C-3.

If there is preliminary formation of a charge-transfer or π-complex en route to the σ-complex, as illustrated in Fig. 8 below (*cf.* Fig. 5), this would not alter the activation or free energies of reaction. In the illustration given the activation energy would be the energy difference between the initial state (A) and the transition state (D) for formation of the σ-complex, as in Fig. 5(A).

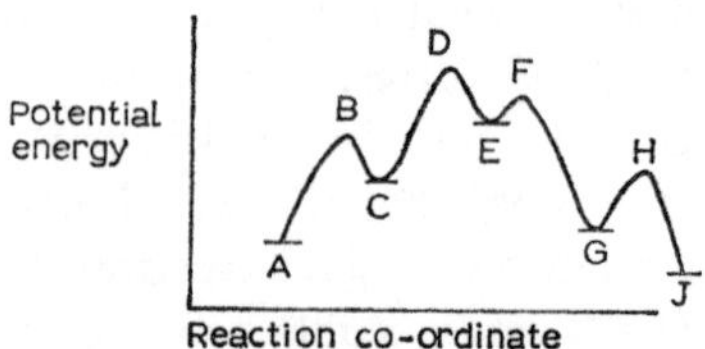

Fig. 8. Representation of aromatic S_N2 reaction proceeding via formation of *both* charge-transfer (π) and covalent (σ) intermediate complexes. A, initial state; B, transition state for formation of π-complex 1; C, π-complex 1; D, transition state for formation of the σ-complex; E, σ-complex; F, transition state for decomposition of the σ-complex; G, π-complex 2; H, transition state for decomposition of π-complex 2; J, final state.

4. The Energetics and Course of the Substitution Reactions

The activation energy (E_{act} or $\Delta E^{\ddagger}$) of a reaction is obtained from experimental rate constants (k) by using the equation

$$k = B\exp(-\Delta E^{\ddagger}/RT) \quad (1)$$

where the pre-exponential term (B) is related to the entropy of activation. This is the integrated form of the alternative expression

$$d\ln k/dT = \Delta E^{\ddagger}/RT^2 \quad (2)$$

In transition state theory the rate constant is given in terms of the product of a universal frequency factor $\kappa T/h$, which varies only with temperature and has the dimensions of reciprocal time, and another term ($K^{\ddagger}$) equivalent to the equilibrium constant for the equilibrium between molecules in the initial state and the activated complex. This is treated as a thermodynamic equilibrium constant and one can then define free energy, heat and entropy of activation ($\Delta G^{\ddagger}$, $\Delta H^{\ddagger}$, $\Delta S^{\ddagger}$). Thus

$$K^{\ddagger} = \exp(-\Delta G^{\ddagger}/RT) = \exp(\Delta S^{\ddagger}/R)\cdot\exp(-\Delta H^{\ddagger}/RT) \quad (3)$$

for a reaction in solution $\Delta E^{\ddagger}$ and $\Delta H^{\ddagger}$ have very similar values. The pre-exponential or frequency factor B is equal to $(\kappa T/h)\exp(\Delta S^{\ddagger}/R)$ so that $\Delta G^{\ddagger}$, $\Delta H^{\ddagger}$ and $\Delta S^{\ddagger}$ can be readily calculated from the Arrhenius parameters measured experimentally.

Transition state theory considers the potential energy surface for reacting molecules. Discussions often deal with the interactions of three atoms in a straight line, such as a reaction, X + Y–Z → X–Y + Z. This leads to a surface which has a saddle-point between two valleys. A potential energy–reaction coordinate plot for an S_N2 reaction at a saturated carbon of the type Y + R–X → Y–R + X resembles that of the lowest energy path across such a potential energy surface. The surface and corresponding path for an aromatic S_N2 reaction proceeding via an intermediate complex is a more complicated version of this, as has been discussed already (p. 9).

Assessment of the facility or otherwise of a reaction depends on the resultant of a number of exothermic and endothermic factors and corresponding changes in entropy. In an S_N1 reaction in a saturated aliphatic system exothermic components are solvation of formed ionic centres; increased conjugation energy; any bond formation to compensate for the heterolysis; and reduction of non-bonding repulsion consequent on reduction in the number of species covalently bonded at the reaction centre. The bond-breaking process is endothermic. Changes in bond energy of other attached groups due to hybridisation changes are small enough to be neglected in comparison. In a corresponding aromatic S_N1 reaction, solvation energy of the aryl cation formed by the heterolysis would be considerably less than that of an aliphatic or side-chain aryl cation; there would be no increased conjugation energy and bond-formation to compensate for the heterolysis, reduction of non-bonding repulsions would probably be negligible; and since Ar–X bonds are stronger than Alph–X bonds this too would militate against the heterolysis. The energetics of an aromatic S_N1 reaction are therefore much less favourable. There is in fact no reliable evidence for an aromatic S_N1 reaction, apart from some reactions of diazonium salts, which form the very stable nitrogen molecule, and do not proceed via a simple aryl cation. There are some copper-catalysed reactions.

Entropy considerations are also important when considering formation and solvation of ions, and the smaller steric requirements in forming an S_N1 transition state. Formation and solvation of ions lower, and reduced steric requirements raises the entropy of activation.

In an S_N2 reaction in a saturated aliphatic system there is an exothermic contribution due to bonding by the nucleophile, but the solvation change in proceeding from the initial to the transition state is not always exothermic (as for example in the common anion–dipole reactions). Non-bonding repulsions tend to increase so that the entropy of activation is generally lower than in S_N1 reactions.

In an addition–elimination aromatic S_N2 reaction*, formation of the first transition state (T.St.1) involves bond formation *without* concurrent bond rupture, but in contrast to the saturated aliphatic reaction there is a change in delocalisation energy in the rest of the molecule, which is probably endothermic except in the most activated systems *[23]*. Also contrasting with the pattern in saturated aliphatic

* Often called activated aromatic S_N2 reactions since they do not proceed readily in simple aromatic systems bearing replaceable groups, but are facile only when activated by electron-withdrawing substituents.

systems is that the typical increase in repulsion between non-bonded atoms may in the aromatic reaction be substantially less, since the increase in number of species at the reaction centre is from 4 to 5 in the one and from 3 to 4 in the other; or may even result in a decrease of repulsion because any substantial repulsions in the planar initial state are likely to be reduced when forming the benzenide (cyclohexadienide type) transition state, with the entering and replaced groups both about 55° out of the ring plane. The first and second transition states are alike in general structure but in the latter case the entering group is fully bonded and rupture of the bond to the leaving (displaced) group is initiated. Conversion of the benzenide intermediate complex to benzenoid initial and final states is a heterolysis at 4-covalent carbon.

Depending on the details of the factors considered above, formation of either of the two transition states (Fig. 5, p. 9) may be rate-limiting. This has been considered in detail by Miller *[24a,b]*, who made calculations for a wide variety of reactions, which suggest that in energy and structure the transition states in activated aromatic S_N2 reactions more nearly resemble the intermediate complex than either initial or final states.

The concept that the bond between nucleophile and ring is largely formed in transition states gains support from the similarity in sensitivity to substituent effects of the ionisation of phenols in protic solvents and the reactivity of phenoxides in activated aromatic nucleophilic substitution in the same solvents *[24d]*.

In activated aromatic S_N2 reactions, typically less sterically demanding than S_N2 reactions in saturated aliphatic systems, the entropy of activation for typical reactions of anions with neutral substrates in protic solvents is close to zero, though Miller has reported *[24a,b]* that when both the entering nucleophile and the leaving group are joined to the ring by bonds between aromatic carbon and heavy atoms the entropy of activation is low (about −15 to −25 e.u., see Chapter 6, p. 190), and he ascribes this to increased steric requirements. It is also well known *[47]* that the entropy of activation is high for reactions involving ions of unlike charge and low for reactions involving ions of like charge. Supporting data have been reported for aromatic S_N reactions. Thus, after extrapolating the results to zero ionic strength for the reaction of methoxide ion with dimethyl-*p*-

nitrophenyl-sulphonium salts *[24c]*, the value of $\log_{10}B$ (which is related to the activation entropy) is 16.8_5, whereas with sodium 4-chloro-3,5-dinitrobenzoate *[48]* the value is 8.1_5.

5. Kinetic Evidence

Many substitution reactions of diazonium ions in aqueous solutions in which the diazonium group is replaced by a nucleophile have been shown to follow first order kinetics, the rates being largely independent of the concentration of nucleophiles, but acceleration by nucleophiles has been shown clearly in some cases. The occurrence of rearrangement of the diazonium group ($\text{Ar–}\overset{+}{\text{N}}{}^{\star}\equiv\text{N} \rightleftharpoons \text{Ar–}\overset{+}{\text{N}}\equiv\text{N}^{\star}$) has been detected by the use of ^{15}N. Study of the product pattern for reaction with aqueous solutions of thiocyanates has been especially fruitful, and it has been analysed in great detail. The latest conclusions are that the reactions proceed via two intermediates, an unselective species of high reactivity resembling an aryl cation but still containing the two nitrogen atoms, and a less reactive and more selective (possibly spirocyclic) species also containing the nitrogen. This is discussed in Chapter 2.

Investigation of the hydrolyses of some halogenobenzenes has shown the occurrence of both elimination–addition (benzyne) and addition–elimination S_N2 mechanisms. Catalysed S_N1-like reactions are also known.

There have been only scanty reports on the kinetics of reactions proceeding by the benzyne mechanism, but there is good evidence for both E1cB and E2 elimination in the primary stage. Overall reactions are however second order. These reactions often but not necessarily occur with rearrangement. The evidence also clearly shows that certain reagents act in different steps as base and nucleophile.

There is ample kinetic support for the normal S_N2 mechanism (activated aromatic nucleophilic substitution) via a relatively stable covalent intermediate or σ-complex. Many workers have reported detailed kinetic results. Reactions usually follow clean second-order kinetics whether the concentrations of nucleophile, substrate or a product are followed. Where there are divergences these can be shown to be due to side-reactions, discussed in Section 6, or to catalysis (Chapter 8).

With many reagents, typically those in which the nucleophilic atoms belong to the first horizontal row of the Periodic Table, the ease of replacement of aromatic halogen is in the order: $F \gg Cl > Br > I$ *[6,7]*, in contrast to the reverse order in *both* S_N1 and S_N2 reactions in saturated aliphatic systems, for in the transition states of reactions proceeding by either of these mechanisms, the bond between carbon and halogen is partly broken. In aromatic S_N reactions giving the contrasting order of reactivity, bond-breaking can have made little or no progress in forming the rate-limiting transition state, and the order $F \gg Cl > Br > I$ has been ascribed *[7;24a,b;49]* to the predominant influence of electronegativity of the group which becomes the leaving group, on bond formation by the nucleophile.

On the other hand, with many reagents, in which typically the nucleophilic atom is from the second or higher row of the Periodic Table, the formation of the second transition state becomes rate-limiting *[24a,b;48b;50]* and correspondingly the order of mobility of the halogens is inverted to that occurring in saturated aliphatic S_N reactions. Miller and his co-workers have recently analysed these patterns in detail *[24a,b;50]*, and also discussed the applicability of group mobility factors in mechanisms of inorganic S_N reactions *[51]*. A similar investigation by Bunnett and his co-workers *[29a,52]* showed that in reactions of some activated aromatic compounds with piperidine, six different substituents differing little in electronegativity, but in all attached by five different atoms, are all displaced at about the same rate.

Attempts to explain halogen mobilities on the basis of the one-stage aliphatic type S_N2 mechanism *[e.g. 53]* are based on the argument that the pattern is a consequence of a spectrum of transition states with a greater or less degree of bond-breaking. In the absence of any special factors this implies corresponding changes in proportions of bond-making, if such transition states are not to be of low stability as in S_N1 reactions of aromatic compounds, or transition states in which the bond order to *both* entering and leaving groups is low.

Differences should clearly be reflected in substituent effects which differ substantially in magnitude and conceivably in direction. Miller and his co-workers *[54]* have measured substituent effects in reactions of azide and thiocyanate ions with activated aromatic substrates in methanol, for which corrected ρ values of 3.77 for

References p. 26

azide and 4.13 for thiocyanate ion at 50° were found. These values are comparable, indicating large substituent effects in both series and the difference is contrary in direction to that required to support the one-stage mechanism. This requires a substantial degree of bond-breaking and correspondingly little bond-making to explain the low reactivity of fluoro compounds with thiocyanate ion and *vice versa* for the high reactivity of fluoro compounds with azide ion. The data are however fully in accord with the two-stage mechanism and its detailed analysis by Miller *[24a,b]*. In this the rate-limiting transition state has the bond to the nucleophile largely made with azide ion as reagent, and fully made with thiocyanate ion as reagent, thus one can predict large substituent effects and large Hammett ρ-values in each, but with somewhat higher values of ρ for the reaction of the thiocyanate ion.

Other investigations, including the effects of base-catalysis, also support the concept of reaction proceeding via an intermediate complex. This work includes an investigation by Bunnett and co-workers *[55]* of the reactions of aniline and *N*-methylaniline with some 1-halogeno-2,4-dinitrobenzenes, the scheme being illustrated in Fig. 9. This follows some work by Hammond and Parks *[56]*, who had reported that in the reactions of 1-halogeno-2,4-dinitrobenzenes with *N*-methylaniline in nitrobenzene and in ethanol, the order of replacement is Br > Cl > F. Bunnett and co-workers *[55]* confirmed this, except for reporting in ethanol the order Cl ≈ F. They suggested that the intermediate complex mechanism was operative in all the reactions, but with $k_{-1} \gg k_2$ for the fluoro compound, and that if this were the case the reaction should then be found to be subject to general base-catalysis since this would increase k_2; and they confirmed this by using acetate ion. Potassium acetate (0.1*M*) has a small positive salt effect similar to that of sodium perchlorate (0.1*M*), but in the reaction of the fluoro compound in the same concentration (0.1*M*) it increases the rate approximately 14-fold, about one hundred times the normal salt effect. They showed separately that this was due neither to replacement of fluorine by acetate, nor to reaction of acetate with ethanol to form ethoxide ion and subsequent reaction of this. From the scheme of Fig. 9, by using the steady state approximation they derived the equation (4).

$$k_{\text{overall}} = \frac{k_1 k_2 + k_1 k_3[\text{B}]}{k_{-1} + k_2 + k_3[\text{B}]} \tag{4}$$

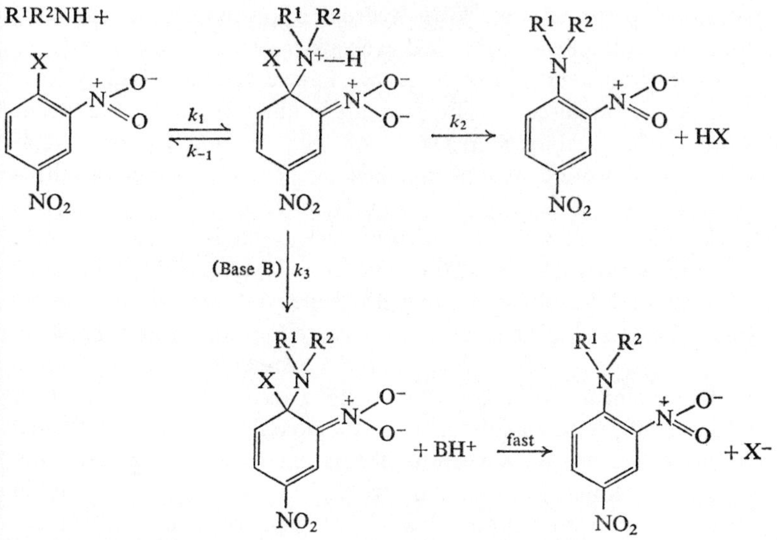

Fig. 9. Paths of reaction of 1-X-2,4-dinitrobenzenes with a primary or secondary amine.

When $k_{-1} \gg (k_2 + k_3[B])$ this becomes

$$k_{\text{overall}} = (k_2 + k_3[B])\,k_1/k_{-1} \tag{5}$$

and the rate coefficient is then linearly dependent on [B]. They then showed experimentally that the catalysis had a linear dependence on acetate ion concentration, and added acetic acid did *not* depress the catalysis by acetate ion. With $k_2 \gg k_{-1}$ equation (4) becomes:

$$k_{\text{overall}} = k_1 \tag{6}$$

and reaction is insensitive to base-catalysis, as for the chloro and bromo compound. They suggest that data obtained by using hydroxide ion as base also support the addition–elimination mechanism. At very high base concentrations or with bases of very high catalytic activity $k_3[B] \gg k_{-1}$ and the rate is again insensitive to base concentration *i.e.* $k_{\text{overall}} = k_1$. The plot of catalysed rate for an active base is then a concave downward curve *[cf. 57]*. In contrast any likely saturated aliphatic S_N2 type reaction would have a simple linear relationship throughout. With hydroxide ion in 60% dioxan–40% water they obtained a concave downward curve.

Some of the details of Bunnett's work have been criticised by Ross *[8]*, but the main conclusions are not in question, and more recent work on base-catalysed reactions of 1-fluoro- and 1-phenoxy-2,4-dinitrobenzene and of *p*-nitrophenyl phosphate and *p*-chloronitrobenzene *[58–60]* lends further support to schemes for reaction via the addition–elimination mechanism, such as that illustrated in Fig. 9.

More recently Lam and Miller *[61]* have investigated the reactions of 1-fluoro-2,4-dinitrobenzene with azide, iodide and thiocyanate ions. The choice of reagents and substrate followed from theoretical considerations *[24a,b;50]*, based on the two-stage addition–elimination mechanism but inconsistent with a one-stage process. These suggested that electrophilic catalysis should only be found in activated aromatic S_N2 reactions when the formation of the second transition state is rate-limiting and that this would be most likely to be found in reactions of fluoro compounds with heavy nucleophiles. Of the three reagents, one is a light and two are heavy nucleophiles; they are related further in that one is a halide and the other two are pseudo-halide ions, and their kinetic nucleophilicities towards a standard aromatic substrate are not too disparate *[24b]*. The preliminary results of Lam and Miller demonstrated large catalytic effects, with rate increases from 10^3 to 10^5 caused by addition of hydrogen and thorium ions, where expected, *viz.*, in iodide and thiocyanate ion reactions. The rates of reactions of azide ion were in no way increased, but on the contrary were decreased by addition of thorium ion.

In his review, Ross *[8]* quotes additional evidence for the formation of intermediate complexes, *e.g.* the formation of charged species on addition of a nucleophile; and protium–deuterium exchange in basic conditions.

6. Side Reactions

(*a*) *Competitive reactions*

In one of the most common, substitutions can occur at more than one point. An example is the replacement of both chlorine and fluorine

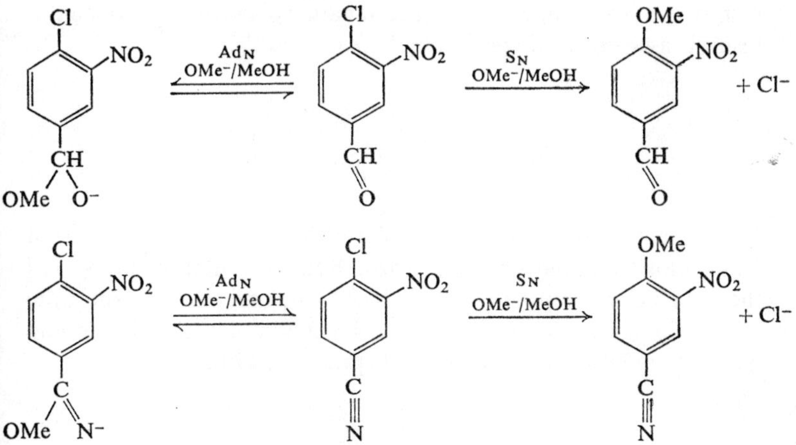

Fig. 10. Concurrent addition and substitution (Ad_N and S_N) reactions of 4-chloro-3-nitro-benzaldehyde and -benzonitrile.

in methanolysis of 1-chloro-4-fluoro-2-nitrobenzene *[62]*. From the above discussion the competitive reaction would not be expected with heavy nucleophiles, and a recent report by Brieux and his co-workers *[63]* mentions this. Another type of competition occurs when the reagent reacts reversibly at an alternative site in the molecule. Reaction can proceed to completion but the kinetics are affected. An example is in the methanolysis of 4-chloro-3-nitrobenzaldehyde and the corresponding nitrile *[64,65]*, illustrated in Fig. 10. In these reactions substitution of chlorine occurs only in the aldehyde and nitrile since the hemiacetal and iminoether conjugate base, formed by the addition, are much less reactive (see Chapter 4, pp. 81, 85).

(*b*) *Alternative ring and side-chain substitution*

Another type of side reaction is exemplified by the reactions of *N,N,N*-trimethyl-*p*-nitroanilinium salts which undergo aromatic (ring) substitution with methoxide ion, giving *p*-nitroanisole *[24d]*, but aliphatic (side chain) substitution with thiocyanate ion, giving *N,N*-dimethyl-*p*-nitroaniline *[66]*. The relative enhancement of reactivity of heavy nucleophiles in aliphatic S_N reactions to which this result is ascribed is discussed in Chapter 6, p. 197.

(c) *Consecutive replacement*

An example of consecutive replacement is the methanolysis of 1,3-dichloro-4,6-dinitrobenzene *[67]*. Both the chloromethoxy and the dimethoxy compound are produced when reagent and substrate are in similar concentrations, since the mobilities of chlorine and methoxyl are not too dissimilar (Chapter 5, p. 170).

(d) *Product decomposition*

Another kind of side reaction occurs as a result of product decomposition forming a new nucleophile which then reacts in parallel with the initially added nucleophile. Thus, for example, it is usual in reactions with thiocyanate ion as nucleophile to isolate diaryl sulphides as well as aryl thiocyanates *[66a]*, due to the reaction of ArS^- formed by decomposition of the products.

(e) *Solvolytic reactions*

Since many solvents are also nucleophiles, solvolysis, which may of course be used as a primary reaction, is sometimes a competitive side reaction. An example is in the uncatalysed reaction of 1-fluoro-2,4-dinitrobenzene with thiocyanate ion in methanol, in which formation of 2,4-dinitroanisole is a major reaction *[68a]*.

An alternative type of solvolysis involves direct interaction of the nucleophile with solvent to form a new reagent. A very well-known example is hydroxide ion in ethanol in which ethoxide ion is the main reactant *[69,70]*. Similarly reaction by fluoride ion in protic solvents commonly results in attack by the lyate ion *[71]*.

(*f*) *Alternative mode of reaction by the reagent*

Nucleophiles may also be reducing agents and instances of this are known in aromatic reactions. Thus in reaction of *p*-chloronitrosobenzene with methoxide ion in methanol, although the chlorine is expected to be more labile than in *p*-chloronitrobenzene, which methoxide converts into *p*-chloroanisole, chlorine substitution is unimportant and the main reaction is reduction to 4,4′-dichloroazoxybenzene *[66b]*. This is because the relative reactivity of the nitroso compared with the nitro group is considerably greater in reduction than when acting as an activating group in the addition–elimination aromatic S_N2 mechanism.

7. Reactions Classified by Charge Types

Bunton in his monograph on aliphatic nucleophilic substitution *[10]* mentions the four possible charge types of reaction for the nucleophile Y and the leaving group X bearing initially zero or unit charge. All these are known also in aromatic nucleophilic substitution. In these, species $\overset{+}{Y}\text{–Ar}$ formed from neutral nucleophiles (Y) commonly have hydrogen attached to Y, which is transferred to another molecule of Y acting as a base.

Type I $\quad Y^- + \text{Ar–X} \longrightarrow \text{Y–Ar} + X^-$

e.g. $\text{OMe}^- + \text{ArHal} \longrightarrow \text{MeOAr} + \text{Hal}^-$

Type II $\quad \ddot{Y} + \text{Ar–X} \longrightarrow Y^+\text{–Ar} + X^-$

e.g. $C_5H_5\ddot{N} + \text{ArHal} \longrightarrow C_5H_5\overset{+}{N}\text{Ar} + \text{Hal}^-$

Type III $\quad Y^- + \text{Ar–X}^+ \longrightarrow \text{Y–Ar} + \ddot{X}$

e.g. $\text{OMe}^- + \text{Ar–NMe}_3^+ \longrightarrow \text{MeOAr} + \ddot{N}\text{Me}_3$

Type IV $\quad \ddot{Y} + \text{Ar–X}^+ \longrightarrow Y^+\text{–Ar} + X$

e.g. $C_5H_{11}\ddot{N}H + \text{Ar–NMe}_3^+ \longrightarrow C_5H_{11}\overset{+}{N}\text{HAr} + \ddot{N}\text{Me}_3$

$$\rightleftharpoons$$

$$C_5H_{11}\ddot{N}\text{Ar} + \text{Me}_3\overset{+}{N}\text{H}$$

These reaction types are commonly considered in relation to the

References p. 26

Hughes–Ingold theory of solvent effects. However the most marked solvent effect in aromatic S_N reactions is probably that demonstrated, for example, by Miller and Parker *[68a]* of marked increase of reactivity of anionic, especially small, nucleophiles in changing from a protic to a dipolar aprotic solvent in which they are less solvated.

REFERENCES

1 F. Pisani, *Compt. rend.*, 39 (1854) 852.

2 C. A. Lobry de Bruyn, *Rec. Trav. Chim.*, 2 (1883) 205 and subsequent papers.

3 P. K. Lulofs, *Rec. Trav. Chim.*, 20 (1901) 292.

4 J. J. Blanksma, *Rec. Trav. Chim.*, 20 (1901) 309 and subsequent papers.

5 A. F. Holleman, *Rec. Trav. Chim.*, 23 (1904) 256 and subsequent papers.

6*a* J. F. Bunnett and R. E. Zahler, *Chem. Rev.*, 49 (1951) 273; *b* J. F. Bunnett, *Quart. Revs.* (*London*), 12 (1958) 1.

7 J. Miller, *Rev. Pure and Appl. Chem.*, 1 (1951) 171.

8 S. D. Ross in S. G. Cohen, A. Streitwieser and R. W. Taft (Eds.), *Progress in Physical Organic Chemistry*, Vol. 1, Interscience, New York, 1963, p. 31 *et seq.*

9 C. K. Ingold, *Structure and Mechanism in Organic Chemistry*, Cornell Univ. Press, Ithaca, N.Y., 1953, (*a*) Chapter VII, (*b*) p. 163.

10 C. A. Bunton, *Nucleophilic Substitution at a Saturated Carbon Atom*, Elsevier, Amsterdam, 1963.

11 M. C. Whiting, *Chem. Brit.*, (1966) 482.

12 D. V. Banthorpe, *Elimination Reactions*, Elsevier, Amsterdam, 1963.

13 R. W. Taft, *J. Am. Chem. Soc.*, 83 (1961) 3350.

14*a* J. D. Roberts, H. E. Simmons, L. A. Carlsmith and C. W. Vaughan, *J. Am. Chem. Soc.*, 75 (1953) 3290; *b* E. F. Jenny and J. D. Roberts, *Helv. Chim. Acta.*, 38 (1955) 1248; *c* J. D. Roberts, D. A. Semenow, H. E. Simmons and L. A. Carlsmith, *J. Am. Chem. Soc.*, 78 (1956) 601, 611; *d* A. T. Bottini and J. D. Roberts, *J. Am. Chem. Soc.*, 79 (1957) 1458.

15 J. Meisenheimer, *Ann.*, 313 (1902) 242.

16*a* C. L. Jackson and F. H. Gazzolo, *Am. Chem. J.*, 23 (1900) 376; *b* C. L. Jackson and R. B. Earle, *Am. Chem. J.*, 29 (1903) 89.

17*a* R. Foster and D. L. Hammick, *J. Chem. Soc.*, (1954) 2154; *b* R. Foster, *Nature*, 176 (1955) 746; 183 (1959) 1042; *c* R. Foster and R. K. Mackie, *J. Chem. Soc.*, (1963) 3796; *d* R. Foster and C. A. Fyfe, *J. Chem. Soc.*, *B*, (1966) 53; *e* R. Foster, C. A. Fyfe, P. H. Emslie and M. I. Foreman, *Tetrahedron*, 23 (1967) 227.

18 J. B. Ainscough and E. F. Caldin, *J. Chem. Soc.*, (1956) 2528, 2540.

19 R. C. Farmer, *J. Chem. Soc.*, (1959) 3425, 3430.

20 L. K. Dyall, *J. Chem. Soc.*, (1960) 5160.

21 R. J. Pollitt and B. C. Saunders, *J. Chem. Soc.*, (1964) 1132.

22 V. Gold and C. H. Rochester, *J. Chem. Soc.*, (1964) 1687.

23*a* T. Abe, *Bull. Chem. Soc., Japan*, 37 (1964) 508; *b* T. Abe, T. Kumai and H. Arai, *Bull. Chem. Soc., Japan*, 38 (1965) 1526.

24*a* J. MILLER, *J. Am. Chem. Soc.*, 85 (1963) 1628; *b* D. L. HILL, K. C. HO AND J. MILLER, *J. Chem. Soc.*, *B*, (1966) 299; *c* B. A. BOLTO AND J. MILLER, *Austral. J. Chem.*, 9 (1956) 74, 304; *d* G. D. LEAHY, M. LIVERIS, J. MILLER AND A. J. PARKER, *Austral. J. Chem.*, 9 (1956) 382.

25*a* W. E. BYRNE, E. J. FENDLER, J. H. FENDLER AND C. E. GRIFFIN, *J. Org. Chem.*, 32 (1967) 2506; *b* J. H. FENDLER, private communication; *c* C. F. BERNASCONI, private communication.

26*a* R. A. ABRAMOVITCH AND G. A. PAULTON, *Chem. Commun.*, (1967) 274; *b* K. L. SERVIS, *J. Am. Chem. Soc.*, 87 (1965) 5495; 89 (1967) 1508.

27 N. MULLER, L. W. PICKETT AND R. S. MULLIKEN, *J. Am. Chem. Soc.*, 76 (1954) 4770.

28 E. J. COREY AND C. K. SAUERS, *J. Am. Chem. Soc.*, 79 (1957) 248.

29*a* J. F. BUNNETT, E. W. GARBISCH AND K. M. PRUITT, *J. Am. Chem. Soc.*, 79 (1957) 385; *b* J. F. BUNNETT AND J. J. RANDALL, *J. Am. Chem. Soc.*, 80 (1958) 6020.

30 L. MELANDER, *Acta Chem. Scand.*, 3 (1949) 95.

31 H. C. BROWN AND J. D. BRADY, *J. Am. Chem. Soc.*, 74 (1952) 3570.

32 H. A. BENESI AND J. H. HILDEBRAND, *J. Am. Chem. Soc.*, 70 (1948) 2832; 71 (1949) 2703.

33 R. S. MULLIKEN, *J. Am. Chem. Soc.*, 72 (1950) 600; 74 (1952) 811; *J. Phys. Chem.*, 56 (1952) 801; *Rec. Trav. Chim.*, 75 (1956) 845.

34 A. STREITWIESER, *Molecular Orbital Theory for Organic Chemists*, Wiley, New York, 1961 (a) p. 199; (b) p. 313–322.

35 G. BRIEGLEB, *Elektronen-Donator–Acceptor-Komplexe*, Springer, Berlin, 1961.

36 R. S. MULLIKEN AND W. B. PERSON, *Ann. Rev. Phys. Chem.*, 13 (1962) 107.

37 R. D. BROWN, *J. Chem. Soc.*, (1959) 2224, 2232.

38 O. HASSEL AND K. O. STROMME, *Acta Chem. Scand.*, 12 (1958) 1146.

39 H. O. HOOPER, *J. Chem. Phys.*, 41 (1964) 599.

40 J. MURTO AND E. KOHVAKKA, *Suomen Kemistilehti*, B39 (1966) 128.

41 R. FOSTER AND D. L. HAMMICK, *J. Chem. Soc.*, (1954) 2685.

42 A. BIER, *Rec. Trav. Chim.*, 75 (1956) 866.

43 R. E. MILLER AND W. F. K. WYNNE-JONES, *J. Chem. Soc.*, (1959) 2375.

44 G. BRIEGLEB, W. LIPTAY AND M. CANTNER, *Z. Physik. Chem. (Frankfurt)*, 26 (1960) 55.

45 R. FOSTER AND R. K. MACKIE, *Tetrahedron*, 16 (1961) 119; 18 (1962) 161.

46 C. R. ALLEN, A. J. BROOK AND E. F. CALDIN, *J. Chem. Soc.*, (1961) 2171.

47 S. GLASSTONE, K. J. LAIDLER AND H. EYRING, *The Theory of Rate Processes*, McGraw Hill, New York, 1941, p. 435.

48*a* J. MILLER AND V. A. WILLIAMS, *J. Chem. Soc.*, (1953) 1475; *b* B. A. BOLTO AND J. MILLER, unpublished work.

49 A. L. BECKWITH, G. D. LEAHY AND J. MILLER, *J. Chem. Soc.*, (1952) 3552.

50*a* J. MILLER AND K. W. WONG, *Austral. J. Chem.* 18 (1965) 117; *b* K. C. HO AND J. MILLER, *J. Chem. Soc.*, *B*, (1966) 310; *c* J. MILLER AND H. W. YEUNG, unpublished work.

51 S. C. CHAN, K. Y. HUI, J. MILLER AND W. S. TSANG, *J. Chem. Soc.*, (1965) 3207.

52 J. F. BUNNETT AND W. D. MERRITT, *J. Am. Chem. Soc.*, 79 (1957) 5967.

53*a* R. E. PARKER AND T. O. READ, *J. Chem. Soc.*, (1962) 12; *b* R. E. PARKER in M. STACEY, J. C. TATLOW AND A. G. SHARPE (Eds.) *Advances in Fluorine Chemistry*, Vol. III, Butterworths, London, 1963, p. 71–75.

54 B. A. BOLTO, J. MILLER AND A. J. PARKER, *J. Am. Chem. Soc.*, 79 (1957) 93.

55*a* J. F. BUNNETT AND K. M. PRUITT, *J. Elisha Mitchell Sci. Soc.*, 73 (1957) 297; *b* J. F. BUNNETT AND J. J. RANDALL, *J. Am. Chem. Soc.*, 80 (1958) 6020.
56 G. S. HAMMOND AND L. R. PARKS, *J. Am. Chem. Soc.*, 77 (1955) 340.
57 H. ZOLLINGER, *Helv. Chim. Acta*, 38 (1955) 1957.
58 C. BERNASCONI AND H. ZOLLINGER, *Tetrahedron Letters*, (1965) 1083.
59 A. J. KIRBY AND W. P. JENCKS, *J. Am. Chem. Soc.*, 87 (1965) 3217.
60*a* J. F. BUNNETT AND R. H. GARST, *J. Am. Chem. Soc.*, 87 (1965) 3875, 3879; *b* J. F. BUNNETT AND C. BERNASCONI, *J. Am. Chem. Soc.*, 87 (1965) 5209.
61 K. B. LAM AND J. MILLER, *Chem. Commun.*, (1966) 642.
62 R. L. HEPPOLETTE AND J. MILLER, *J. Am. Chem. Soc.*, 75 (1953) 4265.
63 A. M. PORTO, L. ALTIERI, A. J. CASTRO AND J. A. BRIEUX, *J. Chem. Soc., B*, (1966) 963.
64 J. MILLER, *J. Am. Chem. Soc.*, 76 (1954) 448.
65 N. S. BAYLISS, R. L. HEPPOLETTE, L. H. LITTLE AND J. MILLER, *J. Am. Chem. Soc.*, 78 (1956) 1978.
66 B. A. BOLTO AND J. MILLER, *J. Org. Chem.*, 20 (1955) 558.
67 M. LIVERIS, P. G. LUTZ AND J. MILLER, *J. Am. Chem. Soc.*, 78 (1956) 3375.
68*a* J. MILLER AND A. J. PARKER, *J. Am. Chem. Soc.*, 83 (1961) 117; *b* J. MILLER AND A. J. PARKER, *Austral. J. Chem.*, 11 (1958) 302.
69 J. F. M. CAUDRI, *Rec. Trav. Chim.*, 47 (1929) 422, 589, 778.
70 J. MURTO, *Suomen Kemistilehti*, B35 (1962) 157.
71*a* R. BOLTON AND J. MILLER, unpublished work; *b* R. BOLTON, B.Sc. Hon. Thesis, University of Western Australia, 1957.

Chapter 2

S_N1 MECHANISMS

1. Introduction

Four general reasons were advanced in Chapter 1 (pp. 4, 5) to explain why an S_N1 mechanism as found in the aliphatic series would occur much less readily as a mechanism involving fission of an Ar–X bond. These are because (*i*) internal conjugation to stabilise Ar^+ formed by heterolysis of Ar–X is not possible, as the axis of the vacant orbital is at right angles to those of the $p\pi$-orbitals, and the inductive effects available are much less efficacious; there can be no increase in bonding in the rest of the system to compensate for the bond heterolysed; (*ii*) the positive charge of an aryl cation (Ar^+) is much less exposed to solvation than that of alkyl or side-chain aryl cations; (*iii*) a decrease in steric requirements in forming Ar^+ is unlikely to be kinetically important, as it sometimes is in aliphatic S_N1 reactions; (*iv*) bonds to common replaced groups are stronger in Ar–X than in Alph–X compounds, associated usually with conjugation of unshared electrons on X with the ring, and this is unfavourable to the required heterolysis.

There is nevertheless an additional factor which could favour an S_N1 mechanism. Taft *[1]* has suggested that an aryl cation, postulated as an intermediate in an aromatic S_N1 mechanism for reactions of aryldiazonium ions, has some extra stability as a triplet ion-radical, which can obtain further stability by conjugation. Some canonical forms of the phenyl cation, as an example of such an ion-radical, are shown in Fig. 11A–D, while stabilisation resulting from conjugation by an electron-releasing substituent *meta* to the point of original attachment of the expelled group is shown by Fig. 11E, F. In the triplet ion-radical a π-electron of the benzene ring has entered with concerted uncoupling into the vacant sp^2 σ-bond orbital formerly occupied by the C–N_α-bond electrons. The presence of radicals has in fact been indicated by ESR measurements, in the decomposition of benzenediazonium tetrafluoroborate in toluene and in nitrobenzene *[2]*, but

References p. 39

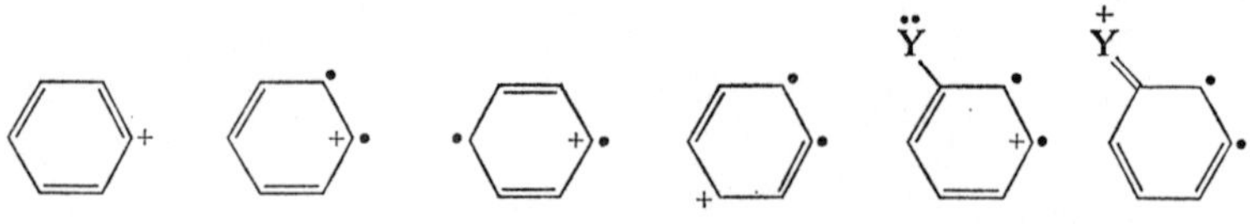

(A) (B) (C) (D) (E) (F)

Fig. 11. Canonical forms of aryl cations showing triplet states.

this in itself establishes neither the existence of the triplet ion-radical nor its occurrence in hydrolysis and similar substitution.

The meagreness of evidence in favour of aromatic S_N1 reactions despite the very substantial body of mechanistic investigations of aromatic nucleophilic substitutions shows that there must be important factors, such as have been suggested above, adverse to a heterolytic S_N1 mechanism, and that even if a triplet-state ion-radical intermediate exists, it does not, in the reactions of almost all classes of compounds, confer sufficient additional stability on the ion to allow such a path to compete with the S_N2 mechanism.

For a considerable time it was believed *[3,5]* that nucleophilic displacement of the diazonium group of aryldiazonium ions proceeds by an S_N1 mechanism analogous to that in aliphatic compounds *[6]*; and that an S_N1 mechanism might be one of several paths for the hydrolysis of some simple halogenobenzenes *[7,8]*; whereas there was no good evidence for other aromatic substitutions proceeding by this path. Some modification of these views has taken place recently: there is evidence that nucleophilic displacement of the diazonium group occurs by a special S_N1 mechanism which does not involve heterolysis to form an aryl cation (Ar^+); and that the only other similar reactions are the S_N1-like copper-catalysed reactions of halogenobenzenes (see Chapter 8, p. 341).

There is no reliable evidence for the occurrence of any other S_N1 mechanism in other uncatalysed aromatic nucleophilic substitutions, though it has been suggested by Hodgson and Leigh for some anomalous reactions of chloronitronaphthalenes with a naphthalenethiol *[9]*, and by Lucas *et al.* for the decomposition of di-*o*-tolyliodonium iodide *[10]*. The results reported by Badger *et al.* *[11]* suggest the possibility of a shift from an S_N2 to an S_N1 mechanism in the reaction of piperidine in benzene with some 1-chloro-4-X-benzenes. In particular chlorobenzene and *p*-chloroaniline produced the same percentage of piperidine derivative in similar conditions. However this amounted to only a little over 1 % of the product and the evidence is inconclusive.

2. Nucleophilic Substitution of the Diazonium Group in Aromatic Rings

First order kinetics have been reported for many decompositions of aryldiazonium ions in water *[3]*, and the rates generally show little or no dependence on the nature and concentration of anions present, even though these are often incorporated to a major extent in the product *[3,5,12–16]*. Decomposing aryldiazonium salts arylate various molecules of very low nucleophilic power, as one might expect from an intermediate highly electrophilic aryl cation. Thus benzene diazonium tetrafluoroborate phenylates nitrobenzene and benzotrifluoride in the *meta*-position, converts bromobenzene into the diphenylbromonium ion, and diphenyl ether into the triphenyloxonium ion *[17]*. This is not conclusive evidence in favour of a free aryl cation however, since diphenyliodonium tetrafluoroborate also acts as a phenylating agent *[17]* and kinetic investigations with weak reagents indicate that it reacts by a bimolecular mechanism in solution *[18]*.

Substituent effects provide evidence for an S_N1-like mechanism, although with special features, and against an S_N2 mechanism. Generally electron-withdrawing substituents decelerate these reactions but many electron-releasing substituents do so too. The argument is somewhat strengthened by kinetic indications favouring an S_N2 mechanism in two or three cases where there is powerful electron-withdrawal.

Lewis and his co-workers *[19]* suggested that for the reaction of bromide ion with *p*-nitrobenzenediazonium ion there is a component S_N2 reaction, but they now offer an alternative explanation *[20]* involving a reversibly formed intermediate (see below). In the case of the *p*-phenylenebis(diazonium) ion, in which the activating *para*-diazonium group is one of the most powerful known for addition–elimination (activated) aromatic S_N2 reactions *[21,22]*, they showed *[23]* that a range of reactions occurs depending on the reagent, *viz.* reagents which do not form stable covalent diazo compounds (*e.g.*

References p. 39

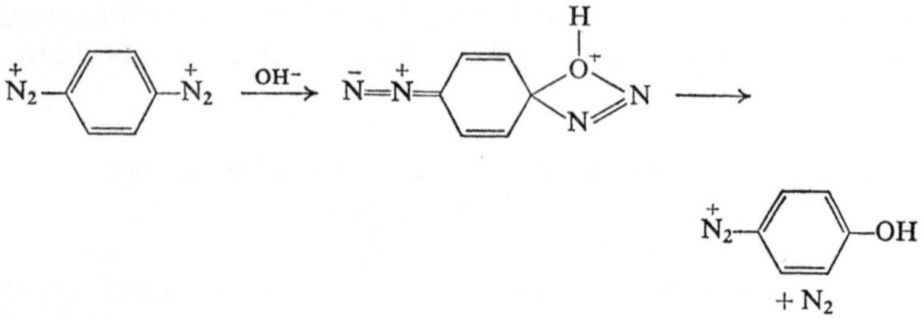

Fig. 12. Bimolecular nucleophilic substitution by side-chain attack followed by an internal nucleophilic substitution.

chloride, bromide, and thiocyanate ions) appear to react by the activated S_N2 mechanism, whereas reagents which form stable covalent diazo compounds appear to react by an alternative but still bimolecular path involving attack on terminal nitrogen, followed by an S_Ni (internal nucleophilic substitution) reaction closely resembling an activated S_N2 reaction, and illustrated in Fig. 12.

They suggest also that diazotised 2-aminopyridine, which is converted especially readily into 2-chloropyridine *[24,25]*, reacts by an S_N2 mechanism, though kinetic studies have yet to be made. In this case activation would be by $\geqslant N^+–H$, and this would activate similarly to the very powerful $\geqslant N^+–Me$ group *[26]*.

In most cases substituent effects, though having some peculiar features, are inconsistent with either the benzyne or normal activated S_N2 mechanisms. These effects are summarised in Table 2, which is based on data of Crossley *et al.* *[15]*, quoted by Bunnett and Zahler *[5]*.

Experimental evidence discussed below has led to the postulate that the nucleophilic displacement of the diazonium group proceeds via two intermediates, one resembling a simple aryl cation (Ar^+), and the other a cyclohexadiene type cation. While this must introduce complexity into the nature of substituent effects in these reactions, they are broadly explicable in terms of reactions proceeding via cationic intermediate species in which at least part of the positive charge is on ring carbon.

In such species the axis of the vacant or partly-filled orbital is at right angles to those of the ring pπ-orbitals. Thus conjugative stabilisation of intermediates by electron-releasing substituents in the *ortho-*

TABLE 2

RATES OF DECOMPOSITION OF ARYLDIAZONIUM SALTS IN AQUEOUS SOLUTION AT 28.8°

	$10^7 k_1$ (sec^{-1}) for position shown[a]		
Substituent	*Ortho*	*Meta*	*Para*
OH	6.8 (0.0092)	9100 (12)	0.93 (0.0013)
OMe	—	3400 (4.6)	0.11 (0.00015)
Ph	1100 (1.5)	1700 (2.3)	37 (0.050)
Me	3700 (5.0)	3400 (4.6)	91 (0.12)
H	740 (1)	740 (1)	740 (1)
CO_2H	140 (0.19)	410 (0.55)	91 (0.12)
SO_3^{2-}	91 (0.12)	150 (0.21)	41 (0.057)
Cl	0.14 (0.00019)	31 (0.042)	1.4 (0.0019)
NO_2	0.37 (0.00050)	0.69 (0.00093)	3.1 (0.0042)

[a] Values in parentheses are relative to H = 1.

and *para*-positions does not occur. On the contrary, conjugative stabilisation of the initial states by such substituents *[27,28]*, illustrated in Fig. 13, raises the activation energy of the reactions, further hindered by increase in double-bond character of the C–N_α bond which is thereby strengthened. Similar effects due to such initial state stabilisation have been reported in other aromatic systems *[28,29]*.

This effect normally swamps any activation by stabilisation of the intermediate by electron-release. Where increased electron density at the point of attachment of an electron-releasing group may be transmitted inductively to the reaction centre, but conjugative stabilisation of the initial state does not occur or is sufficiently reduced, then activation results. This is the norm for electron-releasing *meta*-substituents, and is suggested as a possible explanation of activation by the *ortho*-methyl and -phenyl group (Table 2), in which the conjugation of these

Fig. 13. Canonical forms representing *para*-Y-benzenediazonium salts.

References p. 39

weakly electron-releasing groups in the initial state may be to some extent sterically hindered.

Electron-withdrawing groups in all positions cause deactivation by destabilising the cationic intermediates. The approximately ten-fold greater deactivation by *ortho-* as compared with *para*-chloro and -nitro groups presumably reflects powerful σ-inductive destabilisation more effective from the *ortho*-position. The reactivity order *o-* and *p*-Cl < *o-* and *p*-NO_2 is obscure, but may involve reinforcing factors of stabilising initial states and destabilising the intermediates in the case of the chloro compounds. With carboxyl and sulphonyloxy groups as substituents the presence of anionic forms is assumed as the cause of the opposite order of activity.

The data for *meta*-substituents are more simply explicable. As already indicated electron-releasing substituents activate, while electron-withdrawing substituents deactivate. Degrees of activation are small and do not support the concept of a triplet-state ion-radical intermediate stabilised by conjugative electron-release by substituents *meta* to the point of attachment of the diazonium group (see above, p. 29). The extent of deactivation by *meta*-substituents is also small, except for the *meta*-nitro group: the data correspond, however, to relay by a σ-inductive effect of electron-density changes at the point of attachment of the substituent.

There are clearly uncertainties and speculation in interpretation of the data, which are in any case very meagre. A good deal more work is certainly necessary, but it seems reasonable to assert that substituent effects are explicable in terms of known electronic effects of substituent groups and of intermediates having cationic character.

Despite evidence already mentioned, the concept that these reactions proceed by an aliphatic type S_N1 mechanism has had to be modified. It was questioned by Field and Franklin *[30]* on the grounds that the ion produced by removal of a group from tolyl derivatives to give $C_7H_7^+$ is apparently identical with the ion derived from benzyl derivatives, when studied in the mass spectrometer. The hydrolysis of toluenediazonium ions by an S_N1 mechanism involving intermediacy of a tolyl cation should give rearrangement products, whereas they give unrearranged cresols.

Lewis and Cooper *[20]* investigated the kinetics of reactions of benzenediazonium salts with aqueous thiocyanate ion to give phenols, thiocyanates and isothiocyanates, and derived products. They demonstrated a small increase of rate with increasing SCN^- concentration, in the opposite direction to any expected salt effect, and concluded that there is a highly reactive (unselective) intermediate which reacts with either SCN^- or H_2O at similar rates, and can also revert to diazonium ion. They regard the results as inconsistent with a first-order irreversible formation of an intermediate which subsequently reacts with nucleophiles. Lewis and his co-workers *[31]* elaborated this further. By using ^{15}N they showed that hydrolysis of diazonium salts is accompanied by a slower rearrangement: $Ar{-}^{15}N^+{\equiv}N \rightarrow Ar{-}\overset{+}{N}{\equiv}{}^{15}N$. They eliminated an alternative nitrogen-fixation path *e.g.* $Ar^+ + N_2 \rightarrow Ar{-}N_2^+$ for this change by showing that no $ArCO_2H$ was formed when using the more nucleophilic carbon monoxide ($\overset{-}{C}{\equiv}\overset{+}{O}$), which is isoelectronic with nitrogen ($N{\equiv}N$), and pointed out that carbon monoxide is known to react with some quite stable and relatively unreactive carbonium ions. They also analysed rates in detail, including the acceleration due to added thiocyanate ion, finding an initial rate increase saturated by about 0.6*M* SCN^-, followed by a linear increase to about 3*M*; while the yield of cresol decreased sharply at high thiocyanate concentrations.

All these facts led them to postulate a reaction scheme in terms of *two* intermediates, both of which, however, still contain the two nitrogen atoms of the diazonium ion. This is outlined in Fig. 14. In this scheme, X is an intermediate with the following characteristics:

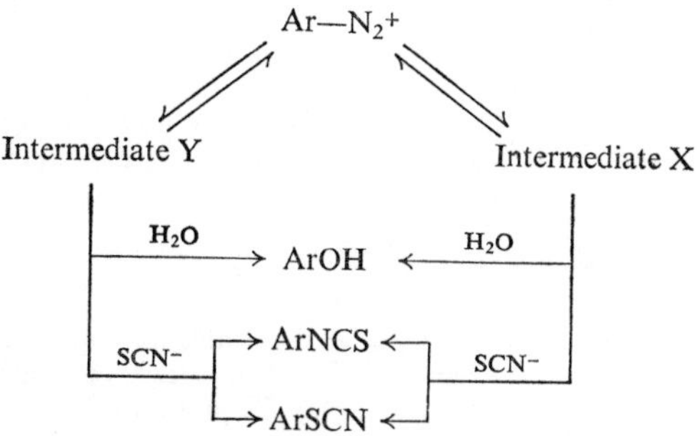

Fig. 14. Reaction paths for nucleophilic substitution of the diazonium group in aromatic diazonium salts (primary products only shown).

References p. 39

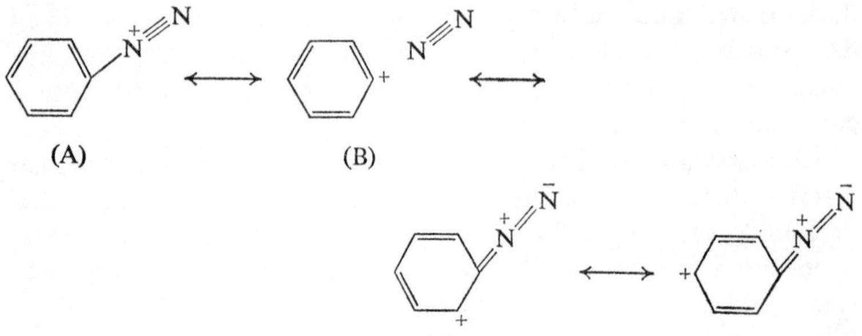

Fig. 15. Canonical forms of an excited state of the benzenediazonium ion.

(*i*) it returns predominantly to ArN_2^+; (*ii*) the two nitrogen atoms are not equivalent (all rearrangement being ascribed to Y); (*iii*) it is reactive and unselective, having a k_{SCN^-}/k_{H_2O} ratio = 2.8, and in this resembles an aryl cation; (*iv*) the C–N_α bond must be weakened since there is an ^{14}N–^{15}N kinetic isotope effect of 1.019. Intermediate Y is less reactive and more selective than intermediate X with a k_{SCN^-}/k_{H_2O} ratio = 470; and the specific rate of return from Y to diazonium ion is twice that of isotopic rearrangement, suggesting the equivalence of the two nitrogens, but not the formation of molecular nitrogen.

They suggest that X is an excited state of the diazonium ion, shown in Fig. 15 in valence bond formulation, with B as the main structure so that the C–N_α bond is weak. This is probably an excited vibrational state with nearly all the activation energy already in it, which reacts

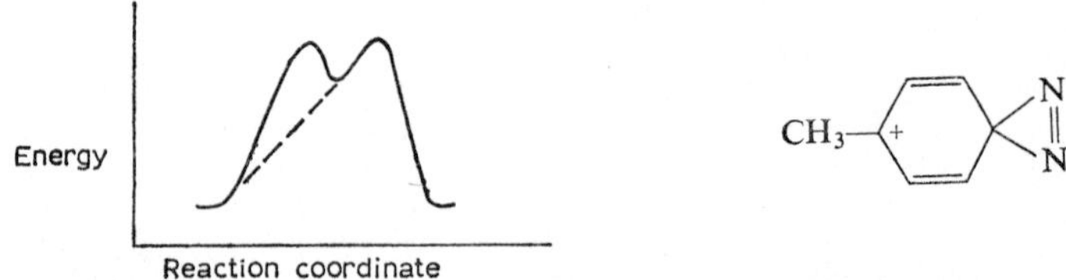

Fig. 16. Plot for reaction through a perceptibly stable intermediate X (solid line) and for a one-step reaction of starting material via a transition state resembling X but containing the nucleophile (dotted line).

Fig. 17. Possible spirocyclic structure for an intermediate in the decomposition of *p*-tolyldiazonium salts.

by a properly oriented collision at the nitrogen-bearing carbon atom, but is deactivated by collisions elsewhere. This species would resemble an aryl cation in having a high reactivity and low selectivity. They indicate that the two step reaction via X is equivalent to a one-step bimolecular reaction in the transition state of which there is little carbon-to-nucleophile bond formation, but also little residual C—N$_\alpha$ bond. This relationship is shown in Fig. 16 (reprinted from ref. *31a*).

The authors suggest that Y may have the spirocyclic structure shown in Fig. 17, for the reaction of the *p*-toluenediazonium salt. The

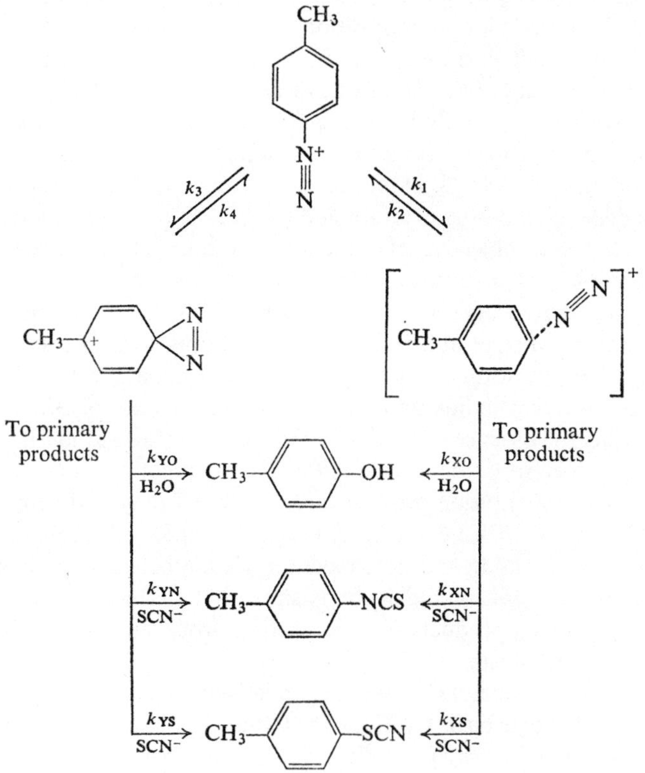

CH_3 · N^+ ≡ N · k_3 · k_4 · k_1 · k_2

To primary products · To primary products

k_{YO}, H_2O → CH_3—C$_6$H$_4$—OH ← k_{XO}, H_2O

k_{YN}, SCN^- → CH_3—C$_6$H$_4$—NCS ← k_{XN}, SCN^-

k_{YS}, SCN^- → CH_3—C$_6$H$_4$—SCN ← k_{XS}, SCN^-

$k_2 \gg k_1, k_{XS}, k_{XN}$; $k_3 > k_4 = 2k$ (rearrangement)
$k_{XO}, k_{XN}, k_{XS} > k_{YO}, k_{YN}, k_{YS}$.

Fig. 18. Detailed scheme for nucleophilic substitution of the diazonium group in *p*-tolyldiazonium salts.

References p. 39

possibility that Y might be formed via X was not excluded, but their experimental results did not settle this point. The scheme in full detail is shown in Fig. 18.

3. Hydrolysis of Halogenobenzenes

Under the conditions utilised by Hale and Britton *[7]* (up to 350°), the rate of hydrolysis of halogenobenzenes was reported to be insensitive to hydroxide ion concentration thus suggesting a unimolecular mechanism, though the extent of by-product formation is sensitive to hydroxide ion concentration. More recently Bottini and Roberts *[32]*, who investigated mainly the halogenotoluenes, which are of similar low reactivity, concluded that these reactions proceed by concurrent normal and benzyne (S_N2) mechanisms (see Chapter 3), the proportions depending on temperature and the basicity of the medium. For example *p*-iodotoluene reacts with aqueous sodium hydroxide at 340° to give 49 % of *m*-cresol and 51 % of *p*-cresol, the rearrangement implying that reaction proceeds largely via the benzyne (S_N2) mechanism, whereas at 250° only *p*-cresol is obtained, corresponding to the direct substitution by the normal S_N2 mechanism. At 340° with aqueous sodium iodide and chloride, no cresols are obtained thus indicating the absence of a reaction proceeding by an S_N1 mechanism. While hydrolysis does occur in mildly basic aqueous sodium acetate, only direct substitution products are found, as compared with those for the strongly basic conditions, which permit initial elimination of the hydrogen halide and thus reaction via the benzyne mechanism. By using ^{14}C, Bottini and Roberts showed that chlorobenzene also gave rearranged products from a reaction proceeding mainly by the benzyne mechanism.

They also suggested that the relatively facile hydrolysis of *o*-chlorophenol in basic solutions *[8]* is due to an internal nucleophilic substitution via a benzyne oxide.

Overall, there is therefore no reliable evidence for a true S_N1 mechanism in these reactions. Hydrolysis proceeds in milder conditions when catalysed by copper, and the S_N1-like mechanism then applicable is discussed in Chapter 8.

4. Summary

There are clearly a few reactions where the evidence is against the occurrence of the common bimolecular mechanisms and favours some form of S_N1 mechanism. In the best investigated of these reactions, the hydrolysis of diazonium salts, the evidence is for an S_N1 mechanism via two intermediates both retaining the nitrogen atoms of the diazonium group. The evidence from uncatalysed hydrolysis of simple halogenobenzenes does not suggest an aromatic S_N1 reaction following the pattern seen with saturated aliphatic compounds, though copper-catalysed reactions which are known (see Chapter 8) are S_N1-like in character.

REFERENCES

1 R. W. Taft, *J. Am. Chem. Soc.*, 83 (1961) 3350.
2 R. A. Abramovitch and G. Tertzakian, *Tetrahedron letters*, (1963) 1551.
3 E. A. Moelwyn-Hughes and P. Johnson, *Trans. Faraday Soc.*, 36 (1940) 948.
4 W. A. Waters, *J. Chem. Soc.*, (1940) 948.
5 J. F. Bunnett and R. E. Zahler, *Chem. Rev.*, 49 (1951) 273.
6 C. A. Bunton, *Nucleophilic Substitution at a Saturated Carbon Atom*, Elsevier, Amsterdam, (1963), p. 3, 8.
7 W. J. Hale and E. C. Britton, *Ind. Eng. Chem.*, 20 (1928) 114.
8 C. F. Boehringer und Söhne, *German pat.*, 269544 (1914); 284533, 286266 (1915), quoted in ref. 7.
9 H. H. Hodgson and E. Leigh, *J. Chem. Soc.*, (1938) 1031.
10 H. J. Lucas, E. R. Kennedy and C. A. Wilmot, *J. Am. Chem. Soc.*, 58 (1936) 157.
11 G. M. Badger, J. W. Cook and W. P. Vidal, *J. Chem. Soc.*, (1947) 1109.
12 A. R. Hantzsch, *Ber.*, 33 (1900) 2517.
13*a* J. C. Cain and F. Nicoll, *J. Chem. Soc.*, 81 (1902) 1412; 83 (1903) 206, 470; *b* J. C. Cain, *Ber.*, 38 (1905) 2511.
14 H. A. H. Pray, *J. Phys. Chem.*, 30 (1926) 1417.
15 M. L. Crossley, R. H. Kienle and C. H. Benbrook, *J. Am. Chem. Soc.*, 62 (1940) 1400.
16 E. S. Lewis, *J. Am. Chem. Soc.*, 83 (1961) 4601.
17 A. N. Nesmeyanov, L. G. Makarova, and T. P. Tolstaya, *Tetrahedron*, 1 (1957) 145.
18 F. M. Beringer and E. M. Gindler, *J. Am. Chem. Soc.*, 77 (1955) 3203 and subsequent papers.
19 E. S. Lewis and W. H. Hinds, *J. Am. Chem. Soc.*, 74 (1952) 304.
20 E. S. Lewis and J. E. Cooper, *J. Am. Chem. Soc.*, 84 (1962) 3847.
21*a* B. A. Bolto and J. Miller, *Chem. and Ind.*, (1953) 640; *b* B. A. Bolto, M. Liveris and J. Miller, *J. Chem. Soc.*, (1956) 750; *c* J. Miller, *Austral. J. Chem.*, 9 (1956) 61.

22 E. S. LEWIS AND H. SUHR, *J. Am. Chem. Soc.*, 82 (1960) 962.
23 E. S. LEWIS AND M. D. JOHNSON, *J. Am. Chem. Soc.*, 82 (1960) 5399, 5408.
24 E. KOENIGS, *Angew. Chem.*, 50 (1937) 911.
25 S. J. ANGYAL AND C. L. ANGYAL, *J. Chem. Soc.*, (1952) 1461.
26 M. LIVERIS AND J. MILLER, *J. Chem. Soc.*, (1963) 3486.
27 E. D. HUGHES, quoted in ref. 5.
28 J. D. DICKINSON AND C. EABORN, *J. Chem. Soc.*, (1959) 3036.
29 K. D. WARREN, *J. Chem. Soc.*, (1963) 598.
30 F. H. FIELD AND J. L. FRANKLIN, *Electron Impact Phenomena*, Academic Press, New York, 1957, p. 232.
31*a* E. S. LEWIS AND J. M. INSOLE, *J. Am. Chem. Soc.*, 86 (1964) 32, 34; *b* E. S. LEWIS AND R. E. HOLLIDAY, *J. Am. Chem. Soc.*, 88 (1966) 5043.
32 A. T. BOTTINI AND J. D. ROBERTS, *J. Am. Chem. Soc.*, 79 (1957) 1458.

Chapter 3

THE BENZYNE OR ELIMINATION–ADDITION MECHANISM

1. Introduction

Both the elimination of HX from vinyl compounds to form alkynes, and the ability of the latter to react with nucleophilic as well as electrophilic reagents are familiar. The idea that aromatic compounds might in certain circumstances undergo overall nucleophilic substitution by initial elimination of HX from neighbouring carbons to form an aryne, followed by a nucleophilic addition to regenerate an aromatic compound, is of fairly recent origin, though an isolated reference appeared in 1902 *[1]* to benzofuran-2,3-yne, the triple bond being in the less aromatic furan ring. In an important series of papers *[2–4]* Roberts and his co-workers have considered aryne formation and also drawn the analogy between aryne and open-chain reactions. Simple cycloalkynes have also been reported recently *[5–7]*.

2. Evidence for Aryne Intermediates

In a series of papers in the 1940's *[8–10]* Wittig and his co-workers brought forward the idea of a di-dehydrobenzene intermediate, with supporting evidence from reactions of fluorobenzene with lithium phenyl. They represented the intermediate in the unsymmetrical dipolar form shown in Fig. 19.

In their review *[11a]*, Bunnett and Zahler referred to a number of nucleophilic substitutions, differing from common activated nucleophilic substitutions, which involve strongly basic reagents at low temperatures, and which result commonly in what they called *cine*-substitution (substitution with rearrangement), more particularly of a vicinal type, *i.e.* the entering group comes in *ortho* to the leaving group. These include a number of reactions of amide and substituted amide ions at low temperatures, as well as some alkali fusion reactions, *e.g.*, the formation of resorcinol (1,3-dihydroxybenzene) from benzene-*p*-

References p. 58

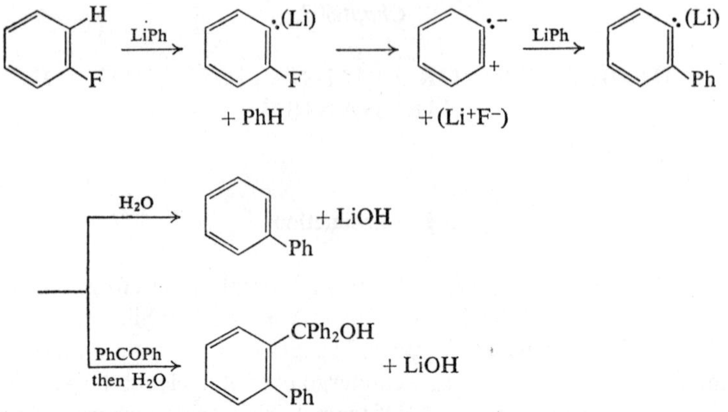

Fig. 19. Representation of an aromatic nucleophilic substitution via an unsymmetrical dipolar di-dehydrobenzene intermediate.

disulphonic acid, in which the absence of solvent greatly enhances the basic strength of the hydroxide ion (see also Chapter 8, p. 317). In some conditions, high temperature hydrolyses of unreactive halogen compounds also fall into this category. Many of these reactions have a long history. Bunnett has also written a short review on benzynes *[11b]*.

Later, Roberts and his co-workers *[2–4]* showed that chlorobenzene-1-^{14}C reacts with amide ion in liquid ammonia to give almost equal amounts of aniline-1-^{14}C and aniline-2-^{14}C. The slight divergence from equality was ascribed to the small ^{12}C–^{14}C isotope effect on product proportions, stemming from the symmetrical intermediate benzyne (Fig. 20). Further, the same ratio was obtained from iodo-

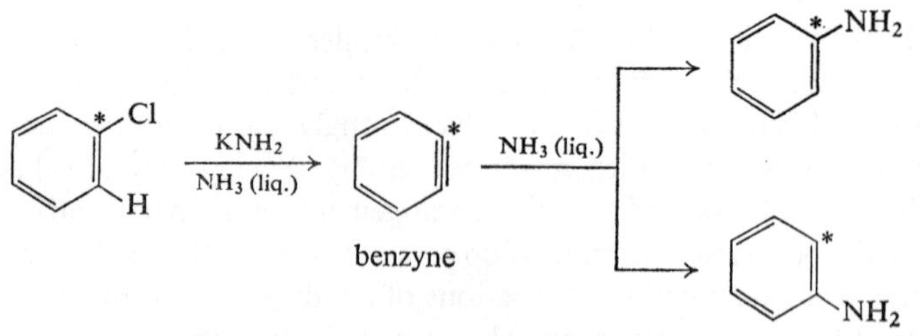

Fig. 20. Formation of aniline-1-^{14}C and aniline-2-^{14}C from chlorobenzene-1-^{14}C.

benzene-1-^{14}C. They showed also that the reactions described by Wittig and co-workers *[8–10]* gave the same statistical results in relation to biphenyl formation. Further support for an elimination–addition or benzyne mechanism comes from the unreactivity of compounds lacking hydrogen *ortho* to the halogen *[3,12a]*.

Especially important is the clear distinction between facility of proton removal and overall hydrogen halide elimination in the halogenobenzene series *[3]*. In reactions of amide ion in liquid ammonia with fluorobenzene, rapid exchange of hydrogen occurs at the *ortho* position, but hydrogen fluoride is not eliminated; bromobenzene exchanges hydrogen more slowly but hydrogen bromide is eliminated; and chlorobenzene is intermediate in behaviour. This is illustrated in Fig. 21.

An interesting result *[11c,12b]* is that for *o*-chlorophenyl anions in liquid ammonia, the specific rate of proton capture (k_{-1}) relative to that for chloride loss (k_2) is increased by all substituents. It is suggested *[11c,12b]* that the increase caused by electron-withdrawing substituents is due to electrostatic stabilisation of the negative charge on ring carbon retarding proton addition less than aryne formation, whereas electron-releasing substituents increase k_{-1}/k_2 by increasing the energy of the transition state for aryne formation.

The elimination stage of the reaction with bromobenzene is thus a concerted E2 reaction. Comparison of bromobenzene and bromobenzene-2-D reveals a large deuterium isotope effect of 5.5, whereas

H (or D)
Hal
+ KNH_2
k_1
k_{-1}
:(K)
Hal
+ NH_3
k_2
NH_2
NH_2^-
NH_3 (liq.)
+ KHal

For PhF, $k_{-1} \gg k_2$; for PhCl, $k_{-1} \sim k_2$; for PhBr, $k_{-1} \ll k_2$.

Fig. 21. Ammonolysis of halogenobenzenes via benzyne.

the two stage E1cB reaction of chlorobenzene has a deuterium isotope effect of 2.7. With diethylamide ion both bromo- and chloro-benzene undergo a concerted E2 elimination and for both the deuterium isotope effect is 5.7.

Huisgen and his co-workers *[13–15]* and Bunnett and Brotherton *[16]* pointed out that whereas piperidine reacts bimolecularly with α- or β-bromonaphthalene by direct substitution to form the corresponding *N*-α or -β-naphthylpiperidine, a mixture of sodium amide and piperidine reacts at a much lower temperature to give roughly the same mixture of 1 part α- to 2 parts β-derivative from either the α- or β-compound. The chloro and iodo compounds behave in the same way. These results suggest that reaction goes through 1,2-naphthalyne (Fig. 22A), which, as an unsymmetrical compound, adds the nucleophile unsymmetrically. The small difference between the proportions of α- and β-piperidino product from the α- and β-halogeno compound is assumed to derive from some 2,3-naphthalyne (Fig. 22B) also being formed in reaction of the β-compound.

1-Fluoronaphthalene gives quite different porportions of α- and β-product corresponding to concurrent aryne and normal direct substitution, whereas 2-fluoronaphthalene reacts via naphthalyne. With potassium amide as reagent, Urner and Bergstrom *[17]* obtained similar results, except in so far as *both* fluoro compounds then reacted by normal direct substitution.

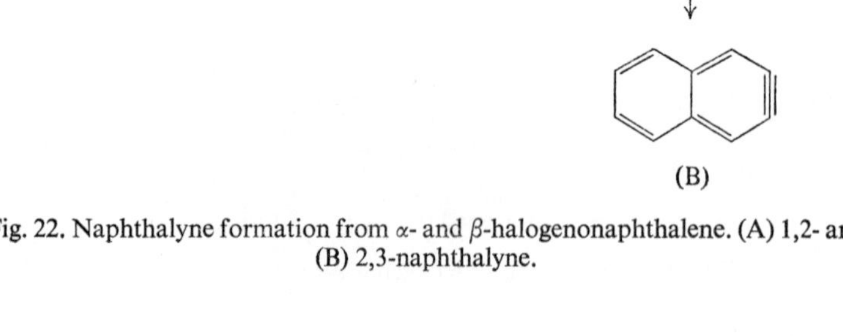

Fig. 22. Naphthalyne formation from α- and β-halogenonaphthalene. (A) 1,2- and (B) 2,3-naphthalyne.

TABLE 3A

RATE CONSTANTS ($l \cdot mole^{-1} \cdot sec^{-1}$) FOR BENZYNE FORMATION WITH LITHIUM PIPERIDIDE IN ETHER AT 20° *a*, WITHOUT PIPERIDINE; *b*, WITH 1 EQUIV. OF PIPERIDINE

		Substrate			
Series	Rate data	Ph–F	Ph–Cl	Ph–Br	Ph–I
(*a*)	$10^4 k_2$	11.0	2.74	4.46	1.72
	Ratio	4.02	1	1.63	0.628
(*b*)	$10^4 k_2$	2.19	2.29	11.6	3.92
	Ratio	0.956	1	5.06	1.71

TABLE 3B

RATE CONSTANTS ($l \cdot mole^{-1} \cdot sec^{-1}$) FOR BENZYNE FORMATION FROM BROMOBENZENE WITH $LiNR_2$ IN THE PRESENCE OF 1 EQUIVALENT OF R_2NH IN ETHER AT 20°

Lithium compound	$10^5 k_2$
Lithium pyrrolidide	144
Lithium piperidide	119
Lithium diethylamide	11
Lithium di-isobutylamide	9.0
Lithium dicyclohexylamide	3.2

Huisgen has summarised his work in a review article written in 1958 *[18]*. This includes valuable kinetic data, some of which are given in Tables 3A and 3B. The results in Table 3A demonstrate that in the absence of free amine the rate is controlled more by the acidity of the *ortho*-hydrogen than the strength of the Ar–Hal bond. Similar relative reactivities are observed in reaction with lithium aryls.

Free amine affects the several Ar–Hal compounds differently, particularly in markedly depressing the rate of elimination from the fluoro compound, and thus favouring the normal activated type of substitution. This is ascribed to reversal by free amine of the original metallation (Fig. 23). It is significant that in liquid ammonia at –33° fluorobenzene undergoes only rapid hydrogen exchange without any

References p. 58

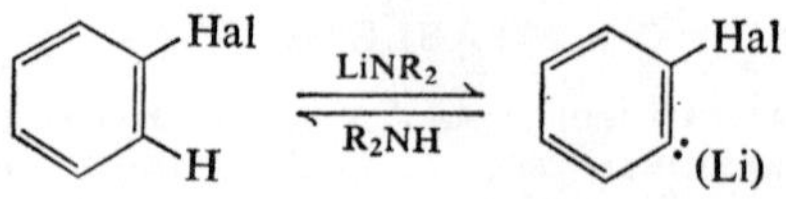

Fig. 23. Metallation of halogenobenzenes and its reversal.

elimination. The relative reactivity of the several amide ions shown in Table 3B suggests modest steric factors as the main cause of the differences.

The formation of tetrafluorobenzyne has been indicated by the work of Tatlow and his co-workers *[19]*. When bromopentafluorobenzene was treated with lithium amalgam, bromine was first eliminated with formation of pentafluorophenyllithium. Above 0° some fluoride ion was also eliminated, and the tetrafluorobenzyne formed was trapped by reaction with furan to give 5,8-epoxy-1,2,3,4-tetrafluoro-5,8-dihydronaphthalene (Fig. 24). Wittig and Pöhmer *[20]* demonstrated the corresponding reaction of *o*-bromofluorobenzene with lithium in furan.

In a number of papers *[21–24]* the formation of pyridynes has been similarly demonstrated during *cine*-substitution reactions of nucleophiles; differences in the proportions of *cine*- and direct substitution between halogenopyridines differing in the position and the identity of the halogen displaced have also been demonstrated. In summary,

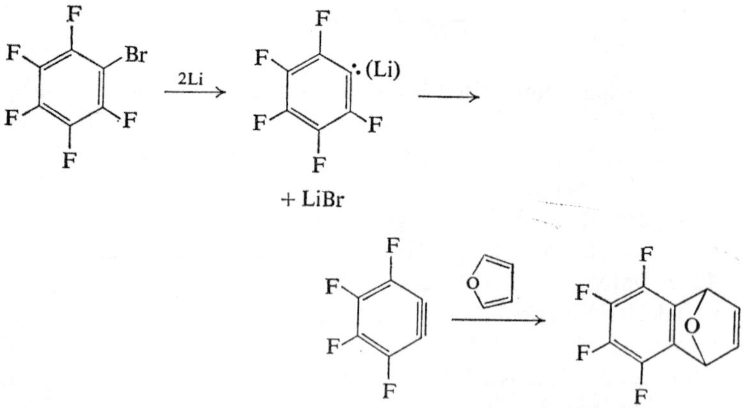

Fig. 24. Trapping of a benzyne intermediate by reaction with furan.

the formation of heterocyclic arynes (hetarynes) by the elimination–addition S_N2 mechanism is readily demonstrable only for the substrates least reactive by the normal addition–elimination S_N2 mechanism. For example 3-chloropyridine reacts by the former, and 3-fluoro- and 4-chloro-pyridine by the latter mechanism. There is evidence for the formation of other hetarynes *[e.g. 25,26]*. The subject has recently been reviewed *[27,28]* and is discussed in Chapter 7, Section 6.

Further support for the existence of aryne intermediates comes from the ability of nucleophiles to compete with aryne-generating bases which are themselves nucleophiles for the addition stage, and in relative reactivities of reagents. Thus the data of Bergstrom and his co-workers *[29]* show that rate and product determination in the reaction of chlorobenzene with amide ion in the presence of triphenylmethide ion occur in separate steps, indicating the formation of an intermediate *[cf. 30]*. They showed that the triphenylmethide ion has almost no reaction with chlorobenzene in liquid ammonia unless some amide ion is added, but the major product is tetraphenylmethane (Fig. 25). The triphenylmethide ion is insufficiently basic to deprotonate the chlorobenzene, although amide ion can do so, but in excess it can compete successfully with the latter as a nucleophile towards benzyne *[3]*.

Huisgen *[18]* has measured rate constants for aryne formation from aryl bromides, and also from a few fluorides. His results, given in Table 4A, indicate the effects of substituents on the acidities of the hydrogen which is eliminated in forming the aryne. The high values for *m*-bromoanisole, where the acidity of the hydrogen is enhanced by *ortho*-bromine and *ortho*-methoxyl are especially noteworthy. The data of Huisgen, shown in Table 4B, fit satisfactorily the expected

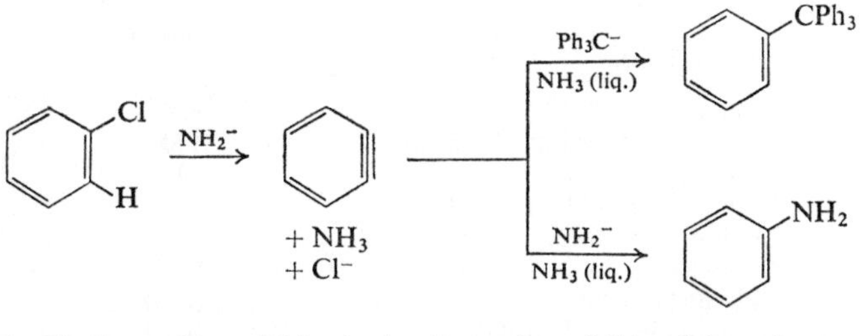

Fig. 25. Competitive addition in the elimination–addition S_N2 mechanism.

TABLE 4A

RATE CONSTANTS ($l \cdot mole^{-1} \cdot sec^{-1}$) FOR ARYNE FORMATION AT 20°. *a*, REACTION OF ARYL BROMIDES WITH LITHIUM PIPERIDIDE AND PIPERIDINE (2 EQUIVALENTS EACH); *b*, REACTION OF ARYL FLUORIDES WITH 4 EQUIVALENTS OF PHENYL LITHIUM

	Series (*a*)		Series (*b*)	
Substituent	Rate constant $10^4\ k_2$	Rate ratio (H = 1)	Rate constant $10^5\ k_2$	Rate ratio (H = 1)
H	11.7	1	9.4	1
o-Me	2.9	0.25	—	
m-Me	4.25	0.36	—	
p-Me	5.3	0.45	—	
p-Ph	29.7	2.5	—	
o-OMe	8.2	0.7		
m-OMe	3300	280		
p-OMe	13.8	1.2		
2,3-fused C_6H_4[a]	39.5	3.4	11.7	1.2
3,4-fused C_6H_4[b]	67.2	5.8	50	5.3
2-Me and 3,4-fused C_6H_4[c]	7.7	0.66	—	—
2,3-fused C_6H_4 and 3,4-fused C_6H_4[d]	228	20	77	8.2

[a] 1-Halogenonaphthalene. [b] 2-Halogenonaphthalene. [c] 2-Halogeno-1-methylnaphthalene. [d] 9-Halogenophenanthrene.

TABLE 4B

RELATIVE RATE CONSTANTS FOR ADDITION TO 9,10-PHENANTHRYNE IN ETHER AT 35°

Reagent	Relative rate constant
Lithium n-butyl (LiC_4H_9)	6300
Lithium thiophenoxide (LiSPh)	1700
Lithium phenyl (LiPh)	1280
Lithium piperidide ($LiNC_5H_{11}$)	100
Lithium phenylacetylide (LiC_2Ph)	13
Lithium dicyclohexylamide [$LiN(C_6H_{11})_2$]	5
Lithium ethoxide (LiOEt)	(0–3)
Lithium phenoxide (LiOPh)	(0–3)

pattern for nucleophilic addition to carbon, though the reactivity of thiophenoxide is somewhat high.

There appears to be no very definite information for comparing reactivities of reagents in different solvents. Whether it be in the elimination or addition stage, the reactivity will be greatly enhanced in dipolar aprotic as compared with protic solvents. This could be of preparative as well as theoretical interest. At the same time, the effect of such a solvent change might also be shown in a switch in path of reaction for substrates which have comparable reactivities by the elimination–addition and addition–elimination paths. Data for relative reactivities of reagents are in any case still relatively scarce. For substrates in which the rate of elimination is known to control the overall rate of reaction, it would be of interest to compare reactivities of reagents in which the nucleophilic atom differs, particularly as between light and heavy nucleophiles, to see whether there is evidence for elimination which involves also nucleophilic attack on carbon, as suggested in a reappraisal of some elimination reactions by Parker, Winstein *et al.* *[31, cf. 32]*, though it would not be expected in these reactions.

There has been much recent work concerned with the detection and isolation of benzyne. It has been produced for example by heating rapidly decomposable substances, and by irradiation procedures. A very suitable substance is the internal salt, benzenediazonium-2-carboxylate, which gives benzyne in mild conditions by eliminating nitrogen and carbon dioxide. It has been detected by its absorption spectrum, and in the mass spectrometer *[33–36]*. By using time of flight mass spectrometry, a species of mass 76 and ionisation potential 9.75 V, regarded as benzyne, was shown to have a life of 250–300 microsec, and its disappearance was matched by the appearance of a peak at mass 152 due to biphenylene (Fig. 26). It was confirmed by substituent labelling that carbon atoms 1 and 2 become equivalent. The numerical value of the ionisation potential is regarded as too

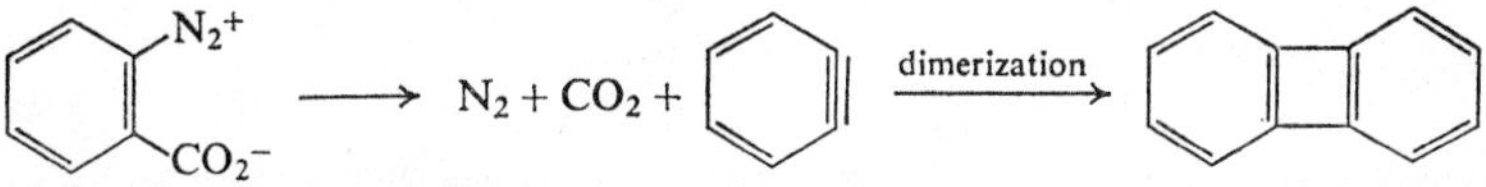

N_2^+, CO_2^- ⟶ $N_2 + CO_2 +$ (benzyne) —dimerization→ (biphenylene)

Fig. 26. Formation of benzyne by pyrolysis, and its dimerisation to biphenylene.

References p. 58

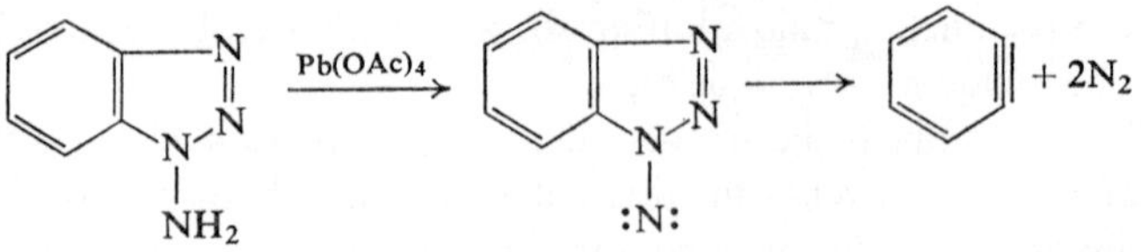

Fig. 27. A synthesis of benzyne by oxidation.

high for a possible acyclic isomer such as CH≡C–CH≡CH–C≡CH, which however appears to be formed in the pyrolysis of *m*- and *p*-di-iodobenzene *[37a]*. A transient absorption spectrum in the 2430 Å region noted in flash photolysis of benzenediazonium-2-carboxylate and of *o*-iodophenylmercuric iodide has been assigned as that of benzyne *[36]*.

A new synthesis of benzyne in mild conditions, which gives a quantitative yield (as shown by formation of its addition product with 2,3,4,5-tetraphenylcyclopentadienone) is the oxidation of 1-aminobenzotriazole with lead tetra-acetate under nitrogen *[38]* (see Fig. 27). The photolysis of *o*-di-iodoarenes *[34]* is also a useful preparative procedure.

An important general method of detecting aryne intermediates is by their participation in such reactions as dimerisation, cycloaddition, and competitive reactions of which examples have been given above.

A stable complex, containing benzyne bonded to nickel has recently been isolated by Gowling *et al.* *[37b]*.

3. Structural Representation of Arynes

There are two current views *[3,11,39,40]* of the nature of the aryne bond. One is that it is a triple bond as in acetylene with the two carbon atoms concerned forming sp σ-bonds, but there is steric strain due to bending of the linear acetylenic σ-bond structure into the hexagonal geometry of the benzene ring.

The other and probably more commonly held view is that sp^2 σ-bonds are used throughout the ring, but that two sp^2 electrons on neighbouring carbons with antiparallel spins occupy an orbital formed by overlap of sp^2 orbitals to give an sp^2 π-bond which is weaker than a pπ-bond. Simmons *[40]* discusses the benzyne bond in terms of an sp^2 mixed π-bond with exchange interaction about one fourth as strong as in a pure pπ-bond; he classes benzyne as a truly aromatic hydrocarbon with one multiple bond of high energy content, resembling a very strained olefin. Coulson *[39]* has suggested bond distances corresponding to some degree of multiple bond fixation.

Some support for the sp^2 π-bond with weak overlap comes from some reactivity ratios for benzyne, 1,2-naphthalyne and 9,10-phenanthryne. Because less reactive species are more selective there should be a correlation between aryne stability and relative reactivities of reagents. Bond distances are in the order benzyne > 1,2 naphthalyne > 9,10-phenanthryne. The shorter bonds should be stronger and thus reagent selectivity should be in the reverse order. Huisgen and his co-workers *[41]* have confirmed this with reactivity ratios $k_{LiPh}/k_{LiNC_5H_{10}} = 4.4$, 5.4 and 12.8 respectively.

Campbell and Rees *[38b]* have recently suggested that benzyne may also be formed in a triplet state.

4. Substituent Effects in the Benzyne Mechanism

Vicinal *cine*-substitution, as described earlier, is common when reaction occurs via aryne intermediates in the elimination–addition mechanism. Roberts and his co-workers *[3]* investigated substituent effects extensively in relation to this, and some of their results and others are given as Table 5, which is from the review by Heaney *[42]*. Yields in many cases are low because of side-reactions, which weaken the arguments somewhat, but the results remain highly significant. Roberts and his co-workers pointed out that an *ortho*-substituted compound can form only a 2,3-benzyne, and *para*-substituted compounds only a 3,4-benzyne, whereas a *meta*-substituted compound can form either or both (Fig. 28). For an E2 elimination stage as in amination of bromobenzene (see above, p. 43) the direction of the elimination from a *meta*-compound depends on which of the hydrogens *ortho* to the halogen is more acidic. For an E1cB elimination stage, as in amination of chlorobenzene, the relative rates of ejection of halogen may affect the orientation, but this is unlikely to be important generally, except where the acidities of the two flanking hydrogens are similar; some information about these acidities is given by data *[3]* on deuterium–protium exchange in Table 6.

TABLE 5

ORIENTATION OF REACTION PRODUCTS OF SUBSTITUTED HALOGENOBENZENES WITH NH_2^-/NH_3 (liq.)

Substituent	Halogen	Total yield	Position yield as % of total		
			ortho	*meta*	*para*
o-OCH_3	Br	33	—	100	—
m-OCH_3	Br	59	—	100	—
p-OCH_3	Br	31	—	49 ± 1	51 ± 1
o-CF_3	Cl	28	—	100	—
m-CF_3	Cl	16	—	100	—
p-CF_3	Cl	25	—	50 ± 5	50 ± 5
p-F	Br	30	—	20 ± 1	80 ± 1
o-CH_3	Cl	66	45 ± 4	55 ± 4	—
o-CH_3	Br	64	48.5 ± 2	51.5 ± 2	—
m-CH_3	Cl	66	40 ± 4	52 ± 4	8 ± 4
m-CH_3	Br	61	22 ± 4	56 ± 4	22 ± 4
p-CH_3	Cl	35	—	62 ± 4	38 ± 4

Fig. 28. Formation of benzynes from monosubstituted *ortho*-, *meta*-, and *para*-halogenobenzenes.

The orientation of the benzyne formed, depending as it does on the hydrogen acidities, appears to be essentially controlled by the inductive effect of the ring substituents (*cf.* Table 4A). A similar control appears to apply also to the subsequent nucleophilic addition, because, it is suggested, the reactant attacks the benzyne bond in the plane of the ring and thus at right angles to the aromatic pπ-system, and so normal conjugative transmission of substituent effects is ineffective. Nevertheless, mesomeric electron release from inductive groups with unshared electrons in appropriate positions may add an adverse component to the changes in electron density resulting from the inductive effect.

Bromobenzenes containing electron-withdrawing *meta*-substituents should yield the 2,3-benzyne, as should chlorobenzenes, provided the electron withdrawing power of the substituent is sufficient to ensure that the hydrogen acidity, rather than the rate of halide elimination, controls the orientation.

Bromobenzenes substituted with a weakly electron releasing *meta*-substituent should give a mixture of 2,3- and 3,4-benzynes with the latter predominating. The corresponding chlorides may give less 3,4-benzyne since the conjugate base form with negative charge *ortho* to both R and Cl will probably lose chlorine more readily and thus favour the formation of 2,3-benzyne. Since only one benzyne may be formed from *ortho* and *para* compounds, this leaving-group effect does not operate.

TABLE 6

FIRST ORDER RATE CONSTANTS (sec^{-1}) FOR DEUTERIUM–PROTIUM EXCHANGE OF SUBSTITUTED DEUTEROBENZENES (C_6H_4DR) WITH 0.6*M* POTASSIUM AMIDE IN LIQUID AMMONIA (AT THE b.p.)

Substituent R	k_1	Substitutent R	k_1
2-F[a]	$> 4 \cdot 10^{-1}$	2-OMe[e]	$1 \cdot 10^{-3}$
3-F[a,b]	$4 \cdot 10^{-4}$	3-OMe[e,f]	$\sim 10^{-7}$
4-F[b]	$2 \cdot 10^{-5}$	4-OMe[f]	$\sim 10^{-8}$
2-CF_3[c]	$6 \cdot 10^{-2}$		
3-CF_3[c,d]	$1 \cdot 10^{-3}$	H	$\sim 10^{-7}$
4-CF_3[d]	$1 \cdot 10^{-3}$		

Rate ratios: [a]2-F/3-F $> 10^3$; [b]3-F/4-F = 20; [c]2-CF_3/3-CF_3 = 60; [d]3-CF_3/4-CF_3 = 1; [e]2-OMe/3-OMe = 10^4; [f]3-OMe/4-OMe = *ca.* 10.

References p. 58

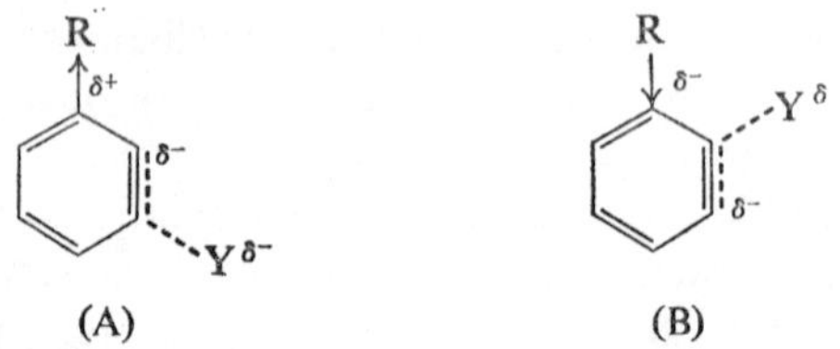

Fig. 29. Transition states for addition to substituted benzynes.

The nucleophilic attack in the addition stage is assumed to take place at the point of least electron density. In terms of transition state theory, electron-withdrawing (activating) substituents favour reaction at the more distant position since then the partially unshared aryne bond electrons are closer to the electron-deficient carbon at the point of attachment (Fig. 29A). Correspondingly, with electron-releasing (deactivating) substituents, reaction is preferred at the nearer position since then the partially unshared electrons are further from the electron, rich carbon at the point of attachment (Fig. 29B). The situation discussed above is summarised in Fig. 30, which is taken, with modification, from a paper by Roberts *[3]*.

It will be seen that all the halogen compounds with electron-withdrawing *ortho*- and *meta*-substituents give 100% *meta*-substituted anilines. The *ortho*-methyl compounds however give 50% *o*-anisidine and 50% *m*-anisidine; rather more *ortho*-compound might have been expected, and the reason for the observed result is not clear, since steric effects with this substituent and reagent are likely to be minimal. Steric hindrance can occur however as is shown when the alkyl group is isopropyl *[18,43]* and the amide is piperidide, in which case 96% of the product is *meta*. Nevertheless these steric factors are not very large, and various reagents adding to 1,2-naphthalyne all give product ratios of about 1/3 α- to 2/3 β-product. Only very large amides give greater percentages of β-product *[16,18]*.

The *meta*-methyl compounds are expected to give both benzynes and rather more than a statistical proportion of *meta*-anisidine compound and less of *para*-anisidine, as is found. The small differences in results for *m*-bromo- and *m*-chloro-toluene seem to reflect the greater proportion of 2,3-benzyne from the chloro-compound as suggested earlier.

The proportion of *meta*- to *para*-substituted aniline from *para*-substituted halogen compounds seems to be correlated well with relative electron-withdrawal illustrated by k_m/k_p values for deuterium–protium exchange. These are approximately equal for the CF_3 substituent (Table 6), and correspondingly, equal amounts of *meta*- and *para*-substituted anilines are formed. For *para*-F this ratio is about 20, and though the *meta*- to *para*-ratio of anilines does not exactly parallel this, there is a preponderance of the *para*-compound.

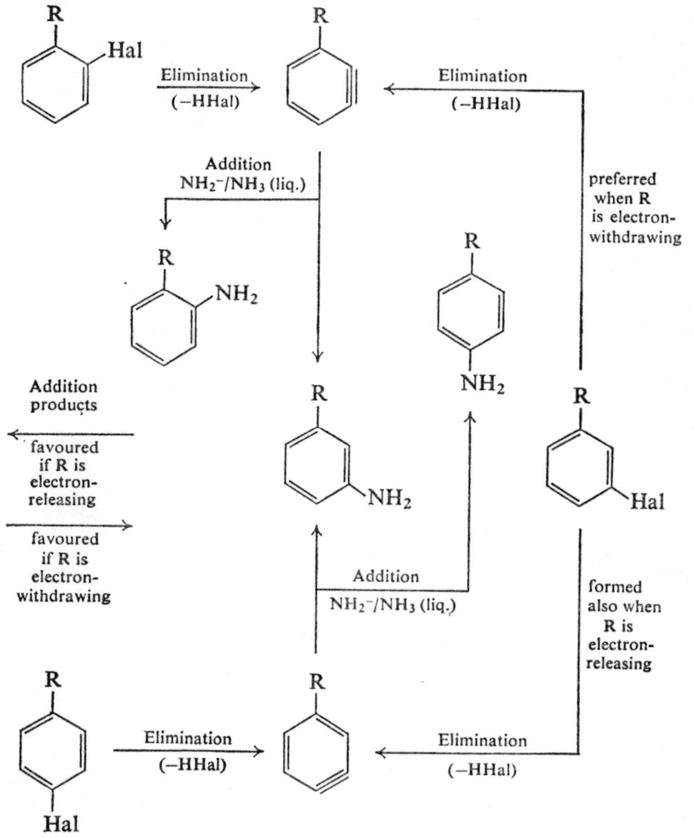

Fig. 30. Orientation of products from *ortho*-, *meta*-, and *para*-substituted halogenobenzenes in reaction via the elimination–addition S_N2 mechanism.

References p. 58

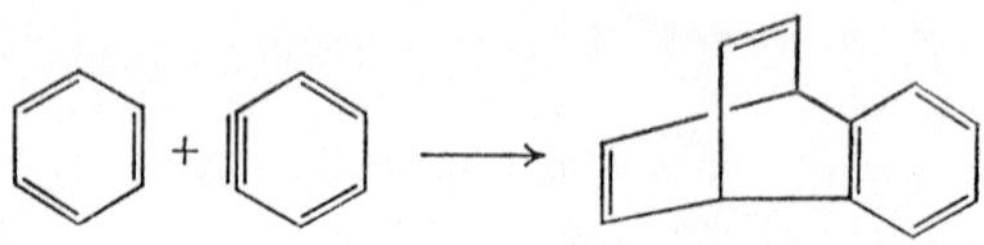

Fig. 31. Addition of benzyne to benzene to form benzobicyclo[2,2,2]octatriene.

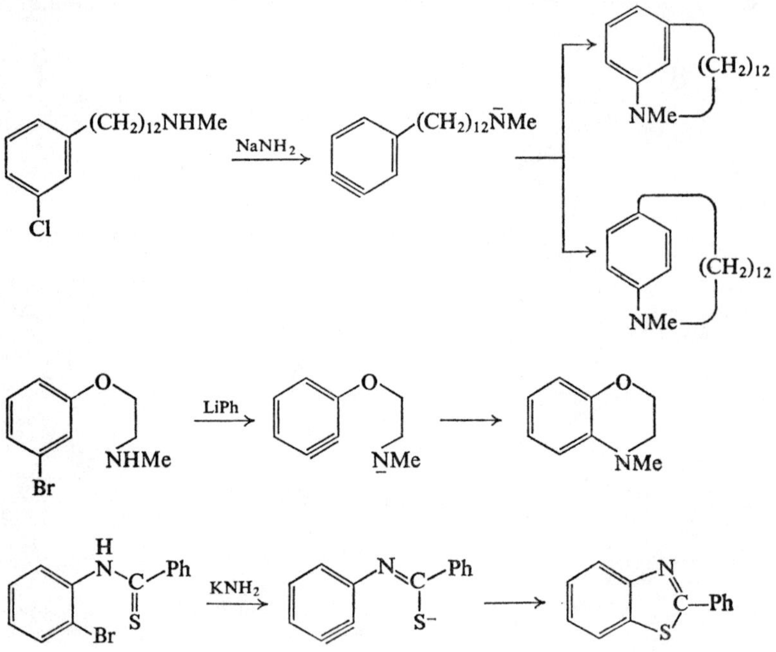

Fig. 32. Some ring closure reactions via aryne intermediates.

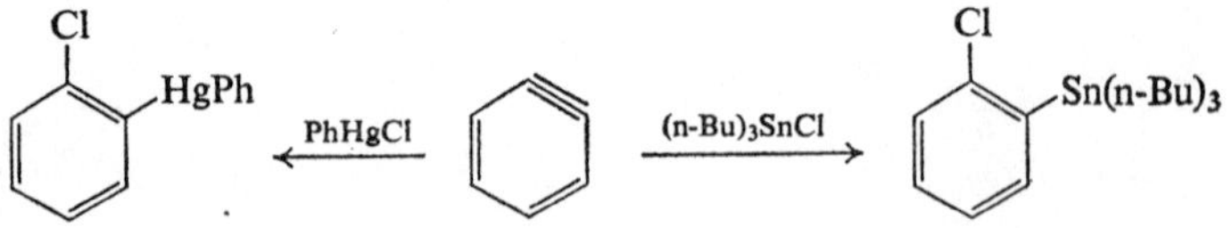

Fig. 33. Some organometallic syntheses via benzyne.

5. Preparative Aspects of Aryne Chemistry

The occurrence of vicinal *cine*-substitution, dimerisation (and polymerisation), cycloaddition, and arylation have already been mentioned. The synthetic applications of these reactions are obvious and many examples are given in recent reviews *[11b,18,42]*. An interesting reaction illustrating the striking dienophilic properties of benzyne is 1,4-addition to benzene of benzyne formed from benzenediazonium-2-carboxylate, to give benzobicyclo[2,2,2]-octatriene as a major product *[44]* (Fig. 31).

A particularly valuable development stems from the work of Huisgen and his co-workers *[45]*, and Bunnett and his co-workers *[46]* in which ring closures result from the formation of an intermediate with an aryne bond and a suitably located nucleophilic centre in a side chain. Some examples are shown in Fig. 32.

Another interesting development is the synthesis of organic compounds of boron and other elements via benzyne. For example Tseng *et al.* *[47]* treated benzyne formed from fluorobenzene and phenyllithium with boron, mercury or tin compounds at low temperatures to give corresponding aryl-metal derivatives. Two examples are given in Fig. 33.

Alkaloids have also been synthesised via aryne intermediates *[48]*.

6. Limits of the Benzyne Mechanism

Factors favouring the benzyne or elimination–addition mechanism are: (*i*) the presence of a hydrogen atom *ortho* to a group X, which is sufficiently electronegative for the 1,2-elimination of HX to take place (hydrogen being eliminated as a proton and X with its bonding electrons); (*ii*) the use of reagents which are strongly basic rather than nucleophilic, though it should be recalled that in the overall reaction, the reagent is involved in both the elimination of HX, and addition to the benzyne; (*iii*) low reactivity of the substrate via the normal, *i.e.* addition–elimination S_N2 mechanism. In relation to these factors an interesting point is that fluoro compounds are expected to be the first among the halogens to switch from the benzyne to the normal mechanism as conditions are changed, when the nucleophilic atom in

References p. 58

the reagent is an element in the first row of the Periodic Table (Li to Ne), but not generally otherwise. This is not a definitive rule but reflects approximately the influence of the reagent on the reactivity of fluorine relative to that of other halogens in the normal mechanism (see Chapter 5, Section 3*b*). An alternative to (*i*) and (*ii*) is the presence of two *ortho*-X (or X and X′) groups which can be removed by a reactive metal or other suitable reagent to give the benzyne, which then reacts with a nucleophile as in the usual form of the substitution reaction via a benzyne intermediate, or may react with other reagents, *e.g.* in a Diels–Alder type reaction.

A normal substitution (without rearrangement) of a substrate of low reactivity by a nucleophile which is also a strong base, followed by base-catalysed isomerisation as discussed in Chapter 9, p. 383, is difficult to distinguish from an aryne substitution. A detailed study of such a reaction, in the thiophene series, has been made by Adickes and Reinecke *[49]*. They referred specifically to this difficulty.

REFERENCES

1 R. Stoemer and B. Kahlert, *Ber.*, 35 (1902) 1633.
2 J. D. Roberts, H. E. Simmons, L. A. Carlsmith and C. W. Vaughan, *J. Am. Chem. Soc.*, 75 (1953) 3290.
3 J. D. Roberts, D. A. Semenov, H. E. Simmons and L. S. Carlsmith, *J. Am. Chem. Soc.*, 78 (1956) 601, 611.
4 E. F. Jenny and J. D. Roberts, *Helv. Chim. Acta*, 38 (1955) 1248.
5*a* L. K. Montgomery and J. D. Roberts, *J. Am. Chem. Soc.*, 82 (1960) 4750; *b* L. K. Montgomery, F. Scardiglia and J. D. Roberts, *J. Am. Chem. Soc.*, 87 (1965) 1917.
6 J. H. Ridd, *Ann. Rep. Progr. Chem., Chem. Soc. (London)*, 57 (1960) 192.
7*a* G. Wittig and E. R. Wilson, *Chem. Ber.*, 98 (1965) 451; *b* G. Wittig, J. Weinlich and E. R. Wilson, *Chem. Ber.*, 98 (1965) 458.
8 G. Wittig, G. Pieper and G. Fuhrmann, *Ber.*, 73 (1940) 1193.
9 G. Wittig and H. Witt, *Ber.*, 74 (1941) 1474.
10 G. Wittig, (a) *Naturwiss.*, 30 (1942) 696; (b) *Angew. Chem. Intern. Ed.*, 4 (1965) 731.
11*a* J. F. Bunnett and R. E. Zahler, *Chem. Rev.*, 49 (1951) 273; *b* J. F. Bunnett, *J. Chem. Educ.*, 38 (1961) 278; *c* J. A. Zoltewicz and J. F. Bunnett, *J. Am. Chem. Soc.*, 87 (1965) 2640.
12*a* R. A. Benkeser and W. E. Buting, *J. Am. Chem. Soc.*, 64 (1952) 3011; *b* B. C. Challis, *Ann. Rep. Progr. Chem., Chem. Soc. (London)*, 62 (1965) 263.
13 R. Huisgen and H. Rist, *Naturwiss.*, 41 (1954) 358.
14 R. Huisgen and H. Zirngibl, *Chem. Ber.*, 91 (1958) 1438.

15 J. Sauer, R. Huisgen and A. Hauser, *Chem. Ber.*, 91 (1958) 1461.
16 J. F. Bunnett and T. K. Brotherton, *J. Am. Chem. Soc.*, 78 (1956) 155, 6265.
17 R. S. Urner and F. W. Bergstrom, *J. Am. Chem. Soc.*, 67 (1945) 2108.
18 R. Huisgen, *Theoretical Organic Chemistry*, Kekulé Symposium, Chem. Soc., London, 1958, p. 158; *b* R. Huisgen and J. Sauer, *Angew. Chem.*, 72 (1960) 91; *c* R. Huisgen and H. Zeiss, *Organometallic Chemistry*, Reinhold, 1960, Chapter 2.
19 P. L. Coe, R. Stephens and J. C. Tatlow, *J. Chem. Soc.*, (1962) 3227.
20 G. Wittig and L. Pöhmer, *Angew. Chem.*, 67 (1955) 348; *Chem. Ber.*, 89 (1956) 1334.
21 R. Levine and N. W. Leake, *Science*, 121 (1955) 780.
22 T. H. Kauffman and F. P. Boettcher, *Angew. Chem.*, 73 (1961) 65; *Chem. Ber.*, 95 (1962) 949, 1528.
23 M. J. Pieterse and H. J. den Hertog, *Rec. Trav. Chim.*, 80 (1961) 1377.
24*a* R. J. Martens and H. J. den Hertog, *Tetrahedron Letters*, (1962) 643; *b* R. J. Martens and H. J. den Hertog, *Rec. Trav. Chim.*, 83 (1964) 621.
25 T. H. Kauffman, F. P. Boettcher and J. Hansen, *Ann.*, 659 (1962) 102; *Chem. Abstr.*, 59 (1963) 13974[b].
26 H. C. van der Plas and G. Geurtsen, *Tetrahedron Letters*, (1964) 2093.
27 H. J. den Hertog and H. C. van der Plas in A. R. Katritzky (Ed.), *Advances in Heterocyclic Chemistry*, Vol. 4, Academic Press, 1965, p. 121–144.
28 T. Kauffman, *Angew. Chem.*, 77 (1965) 557; *Angew. Chem., Intern. Ed.*, 4 (1965) 543.
29*a* R. E. Wright and F. W. Bergstrom, *J. Org. Chem.*, 1 (1936) 179; *b* R. A. Seibert and F. W. Bergstrom, *J. Org. Chem.*, 10 (1945) 544.
30. C. K. Ingold, *Structure and Mechanism in Organic Compounds.* Cornell Univ. Press, Ithaca, N.Y., 1953, p. 337.
31*a* A. J. Parker, S. Winstein *et al.*, unpublished work. *b* A. J. Parker and S. Winstein, *ANZAAS 39th Congress*, Jan. 1967, Section B, Abstracts, p. 15.
32*a* B. D. England, *ANZAAS 39th Congress*, Jan. 1967, Section B Abstracts, p. 17; *b* G. M. Frazer and M. R. Hoffmann, *J. Chem. Soc., B,* (1967) 425.
33 E. Le Goff, *J. Am. Chem. Soc.*, 84 (1962) 3786.
34*a* J. A. Kampmeier and E. Hoffmeister, *J. Am. Chem. Soc.*, 84 (1962) 3787; *b* N. S. Kharasch and R. K. Sharma, *Chem. Commun.*, (1967) 492.
35 G. Wittig and R. W. Hoffmann, *Chem. Ber.*, 95 (1962) 2728.
36*a* R. S. Berry, G. N. Spokes and M. Stiles, *J. Am. Chem. Soc.*, 82 (1960) 5240; 84 (1962) 3570; *b* R. S. Berry, J. Clardy and M. E. Schafer, *J. Am. Chem. Soc.*, 86 (1964) 2738.
37*a* I. P. Fisher and F. P. Lossing, *J. Am. Chem. Soc.*, 85 (1963) 1018; *b* E. W. Gowling, S. F. A. Kettle and G. M. Sharples, *Chem. Commun.*, (1968) 21.
38*a* C. D. Campbell and C. W. Rees, *Proc. Chem. Soc.*, (1964) 296; *b Chem. Commun.*, (1965) 192.
39 C. A. Coulson, *Chem. Soc. Special Publ.* (*London*), 12 (1958) 100.
40 H. E. Simmons, *J. Am. Chem. Soc.*, 83 (1961) 1657.
41 R. Huisgen, W. Mack and L. Möbius, *Tetrahedron*, 9 (1960) 29.
42 H. Heaney, *Chem. Rev.*, 62 (1962) 81.
43 R. Huisgen, W. Mack, K. Herbig, N. Ott and E. Anneser, *Chem. Ber.*, 93 (1960) 412.
44*a* R. G. Miller and M. Stiles, *J. Am. Chem. Soc.*, 85 (1963) 1798; *b* M. Stiles, U. Burckhardt and G. Freund, *J. Org. Chem.*, 32 (1967) 3718.

45 R. Huisgen and H. König, *Angew. Chem.*, 69 (1952) 268 and subsequent papers.
46 B. F. Hrutfiord and J. F. Bunnett, *J. Am. Chem. Soc.*, 80 (1958) 2021, 4749 and subsequent papers.
47*a* C. L. Tseng, S. H. Tung and K. M. Chang, *Wuhan Daxue Xuebao* (*Natural Sci. Ed.*), (1963) 112; *b* C. L. Tseng, S. H. Tung and K. M. Chang, *Scientia Sinica*, 13 (1963) 1170; *c* C. L. Tseng, S. H. Tung and K. M. Chang, *Chem. Abstr.*, 61 (1964) 7035h, 16084h.
48 T. Kametani and K. Ogasawara, *J. Chem. Soc.*, *C*, (1967) 2208.
49 H. W. Adickes and M. G. Reinecke, *J. Am. Chem. Soc.*, 90 (1968) 511.

Chapter 4

SUBSTITUENT EFFECTS

1. Introduction

As in other aromatic substitutions the effects of substituents have been a major subject of investigation. In electrophilic aromatic substitution there are commonly several points of substitution and much of the quantitative work has been devoted to the directional effect of substituents and proportions of products, with corresponding rates estimated only qualitatively, although many rate measurements have also been made.

Except in polyhalogenobenzenes (Section 8, p. 124) and similar poly-substituted compounds, the very much greater ease of replacement of halogen and other common leaving groups than of hydrogen means that in aromatic nucleophilic substitution there is often only one point at which replacement occurs and effects of substituents can be simply and precisely related to their effect at that point. Thus instead of the partial rate factor commonly used in S_E reactions, there is the substituent rate factor (S.R.F.) *[1]* of aromatic S_N reactions, defined as the relative rates of reaction of X–Ar–H and X–Ar–R under the same conditions, where X is the displaced group. For convenience, standard abbreviations, *e.g.* $f_p^{NO_2}$ for the S.R.F. of the nitro group in the *para*-position, are used.

The effects of substituents in addition–elimination aromatic S_N2 reactions are concerned with polar reactions proceeding via transition states in which the bond to the reagent is largely formed and the bond to the expelled group is unbroken, or in which the bond to the reagent is fully formed and the bond to the expelled group is broken to a small extent. For such reactions substituent effects are large.

Ignoring the few but very interesting pan-activating substituents*, one can say qualitatively that a substituent which activates an aromatic

* A term now suggested by the author to denote a substituent which is able to activate electrophilic, nucleophilic and radical substitution in an aromatic ring to which it is attached.

References p. 132

ring for nucleophilic substitution deactivates it for electrophilic substitution and *vice versa*, but the magnitude of activation is much greater than the magnitude of the corresponding deactivation since only the former involves stabilisation by the further conjugation possible when reaction occurs (electromeric or E effect) *[2a]*. This is strictly true only when systems with similar susceptibility to substituent effects (similar ρ-values) are concerned, so that, for example, for comparison with mono-substituted benzenes in S_E reactions, one could legitimately consider mono-substituted halogenobenzenes in S_N reactions.

There are several approaches to the problem of activating power. The most attractive in principle is perhaps by *a priori* quantum mechanical calculations, but as Dewar *[3]* has pointed out this is not yet possible. Approximations are necessary in order to make calculations practicable and the parameters required have usually to be evaluated by experiment or estimated. Many of the authors have dealt only with π-electron polarisations and ignored the σ-bonds, based on Hückel's hypothesis *[4]* that, to a first approximation, only the π-electrons need be considered for such molecules as these. This is inadequate, since Dewar has pointed out at least five distinct processes by which a substituent (R) can affect a distant reaction centre. Three are initiated by the polarity of the Ar–R bond (the inductive or I effect); and two by resonance interactions (the conjugative or T effect). The first consists of a direct field effect (D), acting independently of intervening bonds; the second by successive polarisation of σ-bonds (σ-inductive effect); the third, the polarisation by the substituent of a neighbouring conjugated atom, relayed by the π-electron system (inducto-electromeric or π-inductive effect). The other two are the familiar permanent polarisation of the π-electron system by resonance interaction (mesomeric or M effect); and additional conjugation between the substituent and the reaction centre due to bonding with the reagent (electromeric or E effect).

By using molecular orbital calculations based on the reaction proceeding through a fully-bonded activated complex (Chapter 1, p. 9) Wheland *[5]* predicted qualitatively orders of substituent effects in electrophilic, radical, and nucleophilic aromatic substitution. For example, he predicted that in the S_N reactions the nitro and nitroso groups are activating especially at the *ortho-* and *para-*positions, and that the methyl group is deactivating more particularly at the *ortho*- and *para*-positions. The review by Brown in 1952 *[6]* is still a reasonable summary of the situation. He shows that by one approximation, dealing with π-electrons only, there can be obtained quantities such as charge densities, bond orders and free valences, which can be correlated with reactivity. Another method, used by Wheland *[5]*, although it too considers only the π-electron system, deals with the calculation of localisation energies with which activation energies can be correlated. Such energy calculations imply either that the entropies of activation in a series of reactions are constant or vary simply with the activation energy. This is often the case in the absence of major steric factors but is by no means a general rule.

The account of a recent symposium *[7]* summarises the major recent developments. These include a paper by Nagakura *[8]* in which he considers a complex formed by initial electron transfer (charge-transfer or π-complex) followed by electron localisation to form a fully bonded addition complex (σ-complex). On this model he discusses activation energies in terms of (*i*) the ionisation potential and electron affinity of reagent and substrate, as appropriate to the type of substitution, (*ii*) the localisation energy of benzenide (for S_N) or benzenium ions (for S_E), and (*iii*) the rehybridisation energies of the ring atom which changes from sp^2 to sp^3 in forming the addition complex. This procedure has very close similarities to the semi-empirical thermochemical calculations of Miller *[9]* which, however, also take into account solvation energies. Brown [ref. 7, p. 376–377], in commenting on Nakagura's paper, specifically mentioned the inadequacy of any theoretical treatment which did not incorporate quantitative recognition of the role of solvent. He also pointed out that the transition states are unsymmetrical even though the intermediate complex is symmetrical.

Abe *[10]* made quantum mechanical calculations of the energy of transition states and intermediate complex relative to the aromatic ether and methoxide ion, in reactions of *p*-nitro-, 2,4-dinitro- and 2,4,6-trinitro-anisole. His results agreed with those of Miller *[9]* as regards the relationship of the intermediate complex to transition states but not to the initial (and final) state, and he ascribed this to his ignoring the effects of solvent.

By and large these methods are most useful in explaining in a

References p. 132

qualitative way (*i*) the comparative effect of substituents in different kinds of substitution; (*ii*) the positional order of substituent effects; and (*iii*) relative reactivities, in different kinds of substitution, of mono- and poly-cyclic benzenoid and heteroaromatic systems.

2. Transition State Theory

In terms of transition state theory the effects of substitution on rate depend on the value of $\Delta\Delta G^{\ddagger}$ resulting from a differential effect on the initial state (I.St.) and the rate-limiting transition state (T.St.) caused by the substitution. For a nucleophilic reagent, stabilisation of the transition state involves stabilising a pair of electrons withdrawn from the reaction centre, within or preferably beyond the ring. Thus in Fig. 34 (using the reaction of a nucleophilic anion with a neutral substrate as an example) both (B) and (C) involve activation: the former by a change from carbon to a more electronegative ring atom, and the latter by an attached substituent.

The transition states are benzenide (cyclohexadienide) in structure and, in the case illustrated, the formation of the first transition-state is rate-limiting. Structurally the situation is similar if the second transition-state is rate-limiting, the relative bonding of Y and X being reversed (see Fig. 5, p. 9). The magnitude of substituent effects indicates that the extent of Y–Ar bonding is substantial. Calculations by Miller *[9]* of the relative energy-levels of initial and final states,

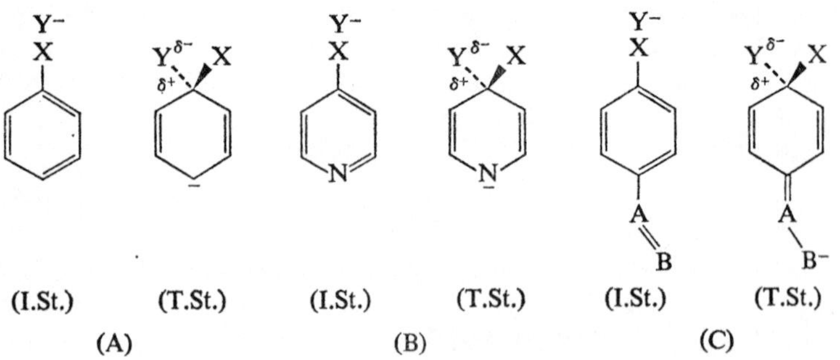

Fig. 34. Comparison of initial and transition states in some addition–elimination S_N2 reactions (*cf.* Fig. 5, Chapter 1, p. 9).

of the intermediate complex and the two transition states, which agree with experiment, lead to the same conclusion; while results of Miller and his co-workers *[11]* on the reactions of phenoxides with 1-chloro-2,4-dinitrobenzene and with picryl chloride show that in this system the Hammett reaction constant (ρ) is very similar to that in the aqueous ionisation of phenols. They concluded that the extent of bond formation was similar in both. Norman and his co-workers *[12]* came to the same conclusions from similar experiments. At the present time the most satisfactory and comprehensive way of considering the effects of substituents is to assess qualitatively their ability to withdraw or release electrons and thus to stabilise or destabilise the rate-limiting transition-state relative to the initial state.

Before commencing a detailed discussion of substituent effects from different positions in the ring it is necessary to consider ways in which one can compare substituent effects measured in reactions which may differ in reagent, substrate, solvent and temperature, etc. Furthermore in comparing *ortho*- or *meta*- with *para*-substituent effects as standard, it is necessary to consider factors other than differences in the polar effects in the three positions, which may affect the comparison. These are considered in the next two sections.

3. Linear Free-Energy Relationships

Reference has already been made to inequalities in magnitude of substituent effects in aromatic S_N and S_E reactions, though they are in general opposite in direction. There are a few pan-activating substituents which violate this general rule. Free-radical substitution is much less sensitive to substituent effects than the polar reactions, and the pattern of substituent effects is different. Aromatic side-chain reactions are also less sensitive to substituent effects as regards polar character than corresponding substitutions at a ring atom, and differ also in relation to the contribution of conjugative effects. The magnitude of substituent effects may vary even in closely related reaction series, *e.g.* substituent effects are substantially greater in S_N reactions of 1-halogeno-4-X-benzenes than in those of 1-halogeno-2-nitro-4-X-benzenes *[13]*.

Since one of the primary aims of chemistry is to discover, explain

References p. 132

and predict the effects of reagent and substrate structure, both for theoretical understanding and practical applications, many workers have sought and discussed such relationships.

The most famous and still most widely applicable of these relationships is that now known as the Hammett equation *[14]*, *viz.* $\log k/k_0 = \rho\sigma$, applied originally to side-chain reactions of *meta*- and *para*-substituted aromatic substrates. In this equation k values are rate or equilibrium constants for reactions of *meta*- or *para*-substituted benzene derivatives, and k_0 those for the unsubstituted reference compound. The ρ (or reaction constant) value is a measure of the sensitivity of the substrate system to substituent effects in the chosen conditions, while the σ (or substituent constant) values depend only on the substituent and its position. By definition the ρ-value for ionization of benzoic acids in water at 25° is unity and the σ-value for hydrogen is zero.

The inapplicability of the original equation to substitutions taking place on ring atoms, to *ortho*-substituted systems, and to aliphatic systems led to several modifications. One of these was the use of an alternative larger σ-value, then known as σ^* but now as σ^-, based on phenol and anilinium-ion acidities. This made allowances for a substantial degree of conjugative interaction of some substituents with the ring; essentially by those which *withdraw* electrons by the conjugative effect (–T groups). In 1953, Jaffé *[15]* listed more than 200 reactions, under a wide variety of conditions, which followed the Hammett equation, using σ- or σ^--values, with an acceptable degree of correlation.

The equation is a relationship between variations of the values of log rate constant ($\log k$) or log equilibrium constant ($\log K$) with structure and conditions of reaction. Such values are linearly related to the standard free energy of activation or reaction, ($\Delta G^{\ddagger}$ or ΔG^0) as shown in equations (1) to (4) below; and thus the Hammett and similar equations are examples of linear free energy relationships:

$$\mathrm{R}T\ln K = -\Delta G^{\circ} \qquad (1)$$

$$\mathrm{R}T\ln k = -\Delta G^{\ddagger} + \mathrm{R}T\ln \kappa\frac{\mathrm{R}T}{Nh} \qquad (2)$$

$$\Delta G^{\circ} = \Delta H^{\circ} - T\Delta S^{\circ} \qquad (3)$$

$$\Delta G^{\ddagger} = \Delta H^{\ddagger} - T\Delta S^{\ddagger} = \Delta E^{\ddagger} - \mathrm{R}T + p\Delta V^{\ddagger} - T\Delta S^{\ddagger} \qquad (4)$$

In equation (2) R is the molar gas constant, N, Avogadro's number, h is Planck's constant, and κ the transmission coefficient of transition state theory, usually taken as unity. In solution reactions $\Delta H^{\ddagger}$ and $\Delta E^{\ddagger}$ (E_{act}) have very similar values.

For a series of compounds for which the entropy of activation is constant, as is often the case in aromatic S_N reactions, log$k$ is correlated with E_{act} ($\Delta E^{\ddagger}$) and reactivity can profitably be discussed in terms of potential energy differences between initial and transition states, correlated with inductive or conjugative stabilisation or destabilisation. Leffler *[16]* has also found many series where there is a linear relationship between enthalpies ($\Delta H^{\ddagger}$) and entropies of activation ($\Delta S^{\ddagger}$), and the Hammett equation can then still apply. Taft has also considered the Hammett equation at length *[17a]* and regards entropy of activation variations in such series as partly polar in origin.

Attempts have been made, with considerable success, to produce useful, if less general, modifications of the Hammett equation. Separation of σ-values into inductive and resonance contributions has been considered by Taft and Lewis *[18]*. A similar effort has been made by van Bekkum, Verkade and Wepster *[19]*. Both groups of workers obtained inductive terms by study of selected series of reactions in which only inductive effects are possible.

Taft *[17b]* developed a Hammett-like equation for reactions of *ortho*-substituted aromatic and aliphatic compounds and used values from the latter in estimating inductive components of σ-values in aromatic reactions *[17]*. Farthing and Nam *[20]* have also discussed a Hammett-type equation for *ortho*-substituents, separating values into electronic and steric components. More recently, Solomon and Filler *[21]* have determined some *ortho* σ-values from measurements on *ortho*-substituted propiolic acids. Pearson and his co-workers *[22]* considered the application of the Hammett equation to substitution at ring atoms, and showed that when using ordinary σ-values, the equation had validity for *meta*- but not *para*-substituents, particularly electron-releasing ones. Similar comments were made by De la Mare *[23]*.

Miller *[13]* showed that σ-values, and $\sigma^{\star}$ (now known as σ^{-}) values where these differ from σ-values, are satisfactory for electron-withdrawing and weakly electron-releasing (inductive) substituents

References p. 132

in S_N reactions at ring atoms, but not for the majority of electron-releasing substituents. Berliner and Monack *[24]*, considering S_E reactions, suggested that a new set of σ (now known as σ^+) values might be necessary for ring-atom substitutions. Elaborating this, Miller *[13]* suggested that for ring substitution σ- and $\sigma^\star$ (σ^-)-values are satisfactory for all except conjugatively electron-releasing (+T) substituents, and that $\sigma^{\star\star}$-values, such as those suggested by Berliner, are required for +T substituents, but they require subdivision for nucleophilic and electrophilic reactions, *i.e.*, in more modern terminology into σ_N^- and σ_E^+ values, the latter being larger.

In a series of papers *[e.g. 25–28]* Brown and his co-workers discussed aromatic S_E reactions and introduced a set of σ^+ values for electrophilic substitution at ring atoms, based on substituent effects in S_N1 reactions of a series of compounds, $ArCMe_2Cl$, on the grounds that these reactions proceed via a carbonium ion intermediate in which conjugative electron-release plays a major part in stabilising the intermediates, as it does the benzenium ion intermediates of aromatic S_E reactions.

Several valuable attempts have also been made to include solvent effects and nucleophilicity in linear free energy relationships *[29–31]*, and the general subject is covered very comprehensively in a recent review by Wells *[32]*.

In the following sections correlation of substituent effects is based on the simple Hammett equation by using σ-, σ^--, and where necessary σ_N^--values which fit it.

4. Multiple Substituents

In considering the effects of two or more substituents in an aromatic system it is tacitly assumed that the substituents act independently and additively. Nevertheless there are a number of limitations to this and the more important are discussed below.

(a) *Susceptibility factor*

There is ample evidence in aromatic S_E as well as S_N reactions that substituents *meta* to each other do act essentially independently, but not necessarily completely additively.

The most obvious exception to additivity is for multiple substituents which are powerful activating groups acting by conjugative stabilisation. With these there is a substantially greater effect from the first of such groups. This seems reasonable when it is recalled that such a substituent both extends conjugation beyond the ring and places an electron-deficiency or surplus, according to whether it can activate S_E or S_N reactions, on an atom other than carbon. A second substituent of the same type simply provides an additional and alternative location for the same electron deficiency or surplus, and for additional conjugation outside the ring. This is a less radical change and this will show up in the value of ρ, the reaction constant.

An alternative approach is that of the Hammond postulate *[33]*. In terms of this postulate the more reactive the aromatic substrate, the less the transition state is removed from the initial state and thus the smaller the substituent effect, for this is a differential effect in stabilising or destabilising the transition state relative to the initial state. Miller's calculations *[9]* based on the addition–elimination mechanism, and utilising Hammond's postulate, illustrate this more precisely.

S_N reactions of aromatic halogenonitro compounds with methoxide ion in methanol provide examples of changes in susceptibility as discussed above. In reactions of 1-fluoro-4-X-benzenes (quoted in ref. *13*) the reaction constant (ρ) is 7.55 at 50° (8.68 at 0°), and rates for *p*-fluoronitrobenzene correspond to an $f_p^{NO_2}$ value of $1.33 \cdot 10^9$ at 50°. Miller and Wan *[34]* showed that for 1-chloro-4-X-benzenes the reaction constant (ρ) has the similar but slightly larger value of 8.47 at 50°, from which the $f_p^{NO_2}$ value is $7.06 \cdot 10^{10}$ (this is an estimate since the rates of reaction of chlorobenzene were obtained indirectly). The corresponding reactions with *ortho*-compounds have been studied *[35]* from which the $f_o^{NO_2}$ value at 50° in the 1-fluoro-2-X-benzene series is $8.54 \cdot 10^8$, very close to the value for the *para*-nitro group. The ratio $f_p^{NO_2}/f_o^{NO_2}$ in the chlorobenzene series is similarly only 3.37 *[1,34]*, so that these systems have similar sensitivity to *ortho*- and *para*-substituent effects. If the reactivity of chlorine in 1-chloro-2,4-dinitrobenzene is compared with that in both 1-chloro-2-nitro- and 1-chloro-4-nitro-benzene as parent compounds, one obtains *f* values

References p. 132

for the *ortho*- and *para*-nitro group in systems where another activating nitro group is already present. The 1-Cl-2-NO_2-4-X-benzene series is seen to have the substantially lower reaction constant (ρ) value of only 3.90 at 50° (4.59 at 0°) *[13]*. Correspondingly the $f_p^{NO_2}$ value in this case is only $1.14 \cdot 10^5$ at 50°, compared to $1.33 \cdot 10^9$ in the more susceptible 1-fluoro-2-X-benzene series. The more substituted system too has similar susceptibility to *ortho*- and *para*-substituents, and the $f_o^{NO_2}$ value is only $3.39 \cdot 10^4$ compared to $8.54 \cdot 10^8$ in the 1-fluoro-2-X-benzene series. There is a further but still smaller reduction of the effect of a third nitro group when two others are present. Full data are not available, and with methoxide as reagent, reactions are complicated by the formation of a *stable* intermediate complex between product and reagent. However, the ρ value for 1-chloro-2,6-$(NO_2)_2$-4-X-benzenes with this reagent at 0° is 3.80, as compared with 4.59 at 0° for the reactions with 1-chloro-2-NO_2-4-X-benzenes *[13]*.

(*b*) *Cross-conjugation*

When substituents are *para* to each other, particularly when one is electron-releasing and the other electron-attracting, marked interactions result, and these affect dipole moments, for example *[2b]*. With one of these substituents activating in a particular reaction, the other substituent will then compete with the reaction centre to supply or withdraw electrons (according to whether an S_N or S_E reaction is involved) and thus the effect of the activating group is reduced. Some examples are given on pp. 117 and 124. The same situation exists for substituents *ortho* to each other, but then there are also complicating steric factors.

(*c*) *Primary and secondary steric effects*

Ingold *[2c]* has defined primary steric effects as short-range repulsive forces due to electron exchange. As a kinetic phenomenon this involves differential consideration of the non-bonding compressional energy in initial and transition states. This can accelerate as well as hinder reaction. When the non-bonding compression exerts its influence indirectly, by interfering with some internally transmitted polar effect, such effects are called secondary steric effects.

It should be noted that in addition to kinetic steric effects one may observe significant differences in non-bonding energy between the initial and final states of a reaction and this will affect the equilibrium. This is a thermodynamic steric effect.

These two factors have been discussed at considerable length for many reactions *[e.g. 36–38]*, and in aromatic S_N reactions are considered below as they arise, in assessment of substituent effects at different positions in the ring. Later discussion also compares polar substituent effects from *ortho*- and *meta*-positions with those from the *para*-position.

Substituent effects in aromatic S_N reactions have been considered at length in several reviews *[e.g. 39–42]*. In this Chapter, substituents are classified in eight groups, largely following an earlier classification by Miller in a review *[40]*, with initial consideration given to *para*-substituents. In relation to these, the differences observable when the substituent is in the *meta*- or *ortho*-position can be discussed more simply (see also Chapter 8, Section 4, p. 348).

5. *Para*-Substituent Effects

Class 1. Substituents attached by an atom bearing a positive charge

Such groups withdraw electrons powerfully and so stabilise a benzenide (cyclohexadienide) transition state. However, if the substituent acts entirely via the inductive effect (Fig. 35A) stability results only from the increase in the electro-negativity of the ring carbon atom, on which the benzenide electron pair is located, and to which the substituent is attached, with electrostatic stabilisation resulting from neighbouring positive and negative centres. Where however the atom bearing the positive charge can increase its outer electronic shell beyond an octet (Fig. 35B), conjugative stabilisation results and such a group is more effective in activation. This is clearly shown by the data for the $-NMe_3^+$ and $-SMe_2^+$ group (Table 7).

TABLE 7

REACTION OF OMe^- IN MeOH WITH 1-Cl-2-NO_2-4-X-BENZENES FOR WHICH $\rho = 3.90$ AT 50° *[1,13,43]*

4-Substituent	Rate constant k_2 (°C) ($1 \cdot mole^{-1} \cdot sec^{-1}$)	Substituent rate factor (S.R.F. or f)	Substituent constant (σ^-)	$\Delta E^{\ddagger}$ (kcal·mole^{-1})	$\Delta S^{\ddagger}$ (e.u.)
H	$2.5 \cdot 10^{-6}$ (50°)[a]	1	0	23.6_5	−13.1
	$1.5 \cdot 10^{-7}$ (25°)[b]	1	—	—	—
NMe_3^+	$5.36 \cdot 10^{-2}$ (50°)[a]	$2.13 \cdot 10^4$	1.110	22.2	+2.2
	$8.23 \cdot 10^{-4}$ (25°)[b]	$5.49 \cdot 10^3$	—	20.3	−6.3
SMe_2^+	$7.96 \cdot 10^{-1}$ (50°)[a]	$3.16 \cdot 10^5$	1.410	17.7	−6.4

[a] Refs. *1* and *43*; corrected to zero ionic strength.
[b] Ref. *44*; not corrected to zero ionic strength.

The evidence from measurements of the acidities of $Me_3\overset{+}{N}$– and $Me_2\overset{+}{S}$–CH_2CO_2H *[45a]*, the *meta*-directing power of –$CH_2\overset{+}{N}Me_3$ and –$CH_2SMe_2^+$ in aromatic electrophilic substitution *[46,47]*, and the values of Pauling's electronegativity indices *[48]* all suggest that the –*I* effect of N^+ is greater than that of S^+. The substantially greater activation by the sulphonium group in the nucleophilic substitutions is clearly therefore conjugative in origin, and there is other evidence for this effect *[45]*. Values of the Arrhenius parameters are also highly

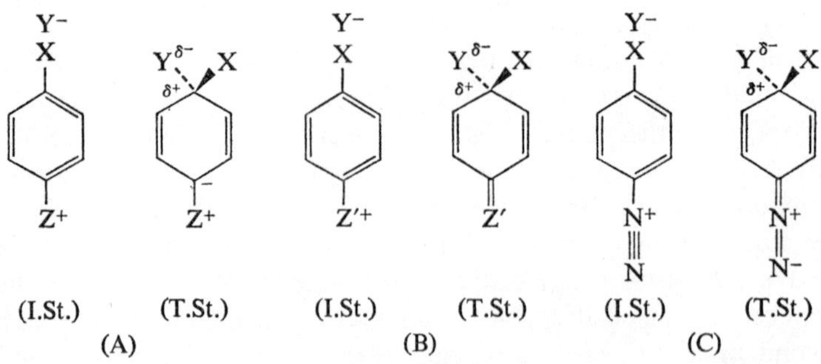

Fig. 35. Initial and transition states illustrating activation by substituents attached to the ring by an atom bearing a positive charge.

significant: the $\Delta\Delta E^{\ddagger}$ value is only -1.45 kcal·mole^{-1} for the $-NMe_3^+$ but -5.95 kcal·mole^{-1} for the $-SMe_2^+$ group. In fact much of the activation by the $-NMe_3^+$ group in the reactions with methoxide ion as nucleophile is due to the high positive value of $\Delta S^{\ddagger}$, consequent on the reaction being between an anion and a cation. This factor also contributes substantially to the high reactivity of the sulphonium compound, but is proportionately much less because a large part of the increase in reactivity is due to a large negative $\Delta\Delta E^{\ddagger}$ value. The components of the σ^--values due to $\Delta\Delta E^{\ddagger}$ alone are 0.251 (from 1.110) for the $-NMe_3^+$ and 1.025 (from 1.410) for the $-SMe_2^+$ group. A still more direct demonstration of the activating effect of conjugative stabilisation, possible in a saturated group when it can accommodate more electrons in its outer-shell, is obtained in the comparison of the $\overset{+}{N}Me_3$ and $\overset{+}{P}Me_3$ group (both from Group V of the Periodic Table) made by Miller and Wan *[34b]*. They showed that the latter group is more activating, even though its inductive effect must be less; and also demonstrated that it fits into the expected activating order, $\overset{+}{N}Me_3 < \overset{+}{P}Me_3 < \overset{+}{S}Me_2$, with σ^--values at 50° of 1.110, 1.377 and 1.410 respectively. As is characteristic of groups activating by virtue of a conjugative effect, the activating power of the trimethylphosphonium group is due in the main to a large decrease in $\Delta E^{\ddagger}$. Ridd *[45b]* has shown also in aromatic S_E reactions that non-first row "onium" groups, such as $\overset{+}{P}Me_3$ attached to the ring, can withdraw electrons conjugatively.

Another indication of the relative importance of *I* and *T* effects comes from a comparison of the $-\overset{+}{N}Me_3$ and $-\overset{+}{N}{\equiv}N$ group. The comparison is not as direct as it would seem, because relative rates for OH^- in H_2O and OMe^- in MeOH are involved, and the ionic strength corrections are values for other, though similar, compounds. It also suffers because of the ease of decomposition of diazonium compounds, with elimination of nitrogen, and because the equilibrium $ArN_2^+ + OH^- \rightleftharpoons ArN_2OH$ lies well to the right *[49]*. The reactivity noted for the $-\overset{+}{N}{\equiv}N$ group is thus a *minimum* value and substantially less than its true value. Table 8 gives some data for this group.

Since the reaction constant (ρ) is large, the high activating power of $\nless 1.68 \cdot 10^{16}$ is equivalent to a minimum σ^--value of only 1.87, of

TABLE 8

REACTION OF OMe^- IN MeOH AND OH^- IN H_2O WITH SOME *p*-SUBSTITUTED FLUOROBENZENES AT 0°, FOR WHICH $\rho = 8.68$ AT 0° *[13,50,51a]*

p-Substituent	Rate constant k_2 (0°) ($l \cdot mole^{-1} \cdot sec^{-1}$)	Substituent rate factor (S.R.F. or *f*)	Substituent constant (σ^-)
H[a]	$1.01 \cdot 10^{-16}$	1	0
H[b]	$7.94 \cdot 10^{-17}$	1	0
N_2^+[a]	$1.70 \cdot 10^{0}$[c]	$\not< 1.68 \cdot 10^{16}$ ($\not< 2.14 \cdot 10^{16}$)	$\not< 1{\cdot}87$ ($\not< 1.88$)

[a] Ref. *50*; [b] Ref. *51a*; [c] Estimated for OMe^- in MeOH from the reaction with OH^- in H_2O, and using the ρ-value = 8.68 at 0° *[13]*.

which 1.45 is due to the large negative value of $\Delta\Delta E^{\ddagger}$, typical of conjugative activating groups. This may be compared with 1.11_0 for the *p*-NMe_3^+ group*, of which only 0.25 is due to a negative value of $\Delta\Delta E^{\ddagger}$. Even allowing for the electronegativity difference of sp and sp^3 nitrogen, the major part of the difference between the two groups must be due to conjugation in the diazonium group (Fig. 35C). The linear relationship between the Hammett substituent constant (σ) and electronegativity shown by Sager and Ritchie *[52]*, and applied by Lewis and his co-workers to sp and sp^2 nitrogen *[53]*, may similarly be estimated to give a part-σ^--value of about 0.4 for the sp and sp^3 difference in nitrogen. The component of the difference in values of *p*-N_2^+ and *p*-$\overset{+}{N}Me_3$ due to conjugation in the former, as shown by $\Delta\Delta E^{\ddagger}$ values, may then be estimated as $\not<(1.87 - 0.4 - 0.25)$, *i.e.* $\not<1.22$.

A recent example of very powerful activation by the diazonium group is the occurrence of Br–Cl exchange *[54]* in 2-chloro-5-nitro- and 4-chloro-3-nitro-benzenediazonium ions at 25° in water. In 2,4-

* Most of this is due to a substantial positive $\Delta\Delta S^{\ddagger}$ value, which seems to be about the same in the two series (Tables 7 and 8), though one has a ρ-value of 3.90 at 50° and the other 8.68 at 0°. The σ^--value for *p*-N_2^+ was calculated in the 1-fluoro-4-X-benzene series, with $\rho = 8.68$ at 0° *[13]* and the σ^--value for *p*-NMe_3^+ can therefore be recomputed from 1.110 (=0.251 + 0.859) to 0.637 (=0.251 + 0.386).

dinitrohalogeno compounds, commonly used as examples of reactive compounds, this halide exchange only takes place at elevated temperatures even in facilitating dipolar aprotic solvents *[55]*.

Activation by positive poles is greater still when they are in the ring, as in comparisons of the pyridinium with the benzene ring system. This is referred to later and discussed in Chapter 7.

Activation by Class 1 type substituents occurs also in acid-catalysed reactions, which are often facile even though strong nucleophiles are absent in such conditions. It is most common in heteroaromatic S_N reactions (see Chapter 7) as when weak nucleophiles react with protonated forms of azines formed by added acid or acid produced during reaction *[e.g. 56,57]*.

An interesting example in the benzene system involves the protonated azo group *[58]*, which activates replacement of an alkoxy group by water—a very weak nucleophile. The unprotonated azo group is itself moderately activating (see pp. 77, 79), but protonation clearly results in an enormous increase of activating power (see also references *59,60*).

Class 2. Substituents attached by an atom bearing a negative charge

Groups in this class markedly destabilise a benzenide (cyclohexadienide) transition state. The paucity of experimental data is itself an indication of this. Berliner and Monack *[24]*, reporting on reactions of 1-Br-2-NO_2-4-X-benzenes with piperidine as reagent and solvent, showed that deactivation by the hydroxyl group is substantially greater than by methoxyl and ethoxyl, and lies between values for the dimethylamino and amino group (see Table 15,a). This suggests that the reaction observed is due only to that proportion of the substrate present as the free phenol and that the phenoxide in equilibrium with it does not react at all in the conditions used.

Clark and Ball *[61]* showed that in reaction with methoxide ion in methanol, in conditions where chlorobenzene undergoes 4% replacement of chlorine, the *p*-chlorophenoxide ion does not undergo detectable substitution.

De Crauw *[62]* also studied a series of chlorobenzenes containing activating and deactivating groups and showed that the O^-, S^-, NH_2 and OMe derivatives are inert to this reagent at 180° for 8 hours. Powerful deactivation by the *p*-O^- group may also be inferred from the fact that even *m*-O^- has been shown to be strongly deactivating (see p. 117).

Class 3. Substituents attached by the positive end of a dipole

This class, which is not confined to substituents with a formal dipole, contains the most studied substituents in aromatic nucleophilic substitution, headed by the nitro group. Examples are:

$-\overset{+}{N}(=O)O^-$ $-\overset{+}{S}(=O)(O^-)R$ $-C(=O)R$ $-C{\equiv}N$ $-N{=}O.$

Apart from their inductive effects all are able to stabilise a benzenide transition state by conjugative withdrawal of electrons into the group. A fairly comprehensive list of such substituents and their influence on reactivity is given as Tables 9–11.

Of these, Table 9, based on results obtained by Miller and his co-workers *[1,13,63–67]*, lists the main substituents of Class 3 for which rate data are available. Some similar measurements, which are in good agreement, are available from the work of Bunnett and his co-workers *[44,68]*, and of Eliel and Nelson *[69]*. Table 10, from the work of Brieux and his co-workers *[70]*, lists the effects of the substituents in the same series on reactions with piperidine in benzene. Despite this major change in reagent and solvent, substituent effects remain much the same except for the carboxylate group. This is because the reaction with methoxide is one between two anions and the rate is consequently reduced. Table 11, from results of Miller and his co-workers *[13,64,71,72]*, lists results for a reaction series similar to that in Table 9 but involving activation by an extra nitro group. Substituent effects are all slightly weaker but otherwise closely similar.

The Arrhenius parameters show a striking pattern in which there is generally little variation in $\log_{10} B$ (and thus $\Delta S^{\ddagger}$) for reactions of the same charge type, and a marked reduction of $\Delta E^{\ddagger}$ of about 3 to 7.5

TABLE 9

REACTIONS OF 1-Cl-2-NO_2-4-X-BENZENES WITH OMe^- IN MeOH AT 50° FOR WHICH $\rho = 3.90$ *[1,13,63–67]*.

4-Substituent	Rate constant k_2 (50°) ($l \cdot mole^{-1} \cdot sec^{-1}$)	Substituent rate factor (S.R.F. or f)	Substituent constant (σ^-)	$\Delta E^{\ddagger}$ (kcal·mole^{-1})	$\log_{10} B$
H	$2.52 \cdot 10^{-6}$	1	0	23.6_5	10.4
CO_2^- [a]	$8.50 \cdot 10^{-6}$	3.37	0.135	21.1	9.2
$CONH_2$	$6.58 \cdot 10^{-4}$	$2.62 \cdot 10^2$	0.627	21.3	11.2
$CONMe_2$ [b]	—	—	0.477	—	—
CO_2Me	$3.93 \cdot 10^{-3}$	$1.56 \cdot 10^3$	0.819	18.2_5	9.9_5
COMe	$5.00 \cdot 10^{-3}$	$1.99 \cdot 10^3$	0.874	18.7_5	10.3_5
COPh	$6.68 \cdot 10^{-3}$	$2.66 \cdot 10^3$	0.879	18.1_5	10.1
CHO [c]	—	—	0.939	—	—
CN [c]	—	—	0.997	—	—
SO_3^- [a,d]	—	—	0.186	—	—
SO_2NH^-	$6.56 \cdot 10^{-5}$	$2.60 \cdot 10^1$	0.363	21.5	10.4
SO_2NMe_2	$1.90 \cdot 10^{-2}$	$7.57 \cdot 10^3$	0.994	19.2_5	11.3
$SO_2NC_5H_{10}$	$1.53_5 \cdot 10^{-2}$	$6.09 \cdot 10^3$	0.970	18.8	10.9
SO_2NMePh	$2.05 \cdot 10^{-2}$	$8.14 \cdot 10^3$	1.003	18.6_5	10.9
SO_2Me	$3.22 \cdot 10^{-2}$	$1.28 \cdot 10^4$	1.049	18.6_5	11.1_5
SO_2Ph [e]	$4.64 \cdot 10^{-2}$	$1.84 \cdot 10^4$	1.117	—	—
$N{=}\overset{+}{N}{=}\overset{-}{N}$ [f]	$7.15 \cdot 10^{-6}$	2.84	0.116	24.2	11.2_5
N=N-Ph	$1.04_5 \cdot 10^{-3}$	$4.16 \cdot 10^2$	0.672	20.3_5	10.8
$N{=}\overset{+}{N}(O^-)$-Ph [f,g]	$6.36 \cdot 10^{-4}$	$2.52 \cdot 10^2$	0.616	—	—
$\overset{+}{N}(O^-){=}N$-Ph [f]	$2.58 \cdot 10^{-3}$	$1.02_5 \cdot 10^3$	0.772	19.5	10.6
$\overset{+}{N}$<(CH=C-O^-)(N—O) (ring) [e,h]	—	—	0.717	—	—
NO_2	$2.88 \cdot 10^{-1}$	$1.14 \cdot 10^5$	1.270	17.4_5	11.2_5
NO [e]	$1.58 \cdot 10^0$	$6.27 \cdot 10^5$	1.486	16.2	11.1_5

[a] Corrected to zero ionic strength. The component of the σ^--value for CO_2^- due to $\Delta\Delta E^{\ddagger}$ alone is 0.443. [b] σ^--value after allowing for side reaction, *viz.* displacement of NMe_2. [c] Computed from measurements with SCN^- on corresponding iodo compounds. There may be some minor interaction of reagent with the carbon of CHO and CN, so that σ^--values may be a little low. [d] Estimated from measurements with OH^- in H_2O. [e] Estimated from measurements in MeOH–C_6H_6. [f] σ^--values slightly different from values previously computed for 0°. [g] Estimated from measurements with OH^- in dioxane–water. [h] Computed from measurements with SMe^- in MeOH.

References p. 132

TABLE 10

REACTIONS OF 1-Cl-2-NO_2-4-X-BENZENES WITH PIPERIDINE IN BENZENE AT 45° FOR WHICH $\rho = 4.08$ *[70]*

4-Substituent	Rate constant k_2 (45°) ($l \cdot mole^{-1} \cdot sec^{-1}$)	Substituent rate factor (S.R.F. or f)	Substituent constant (σ^-)	$\Delta E^{\ddagger}$ ($kcal \cdot mole^{-1}$)	$\Delta S^{\ddagger}$ (e.u.)
H	$3.63 \cdot 10^{-6}$	1	0	13.9	−40
CO_2^-	$5.20 \cdot 10^{-5}$	$1.43 \cdot 10^{1}$	0.283	10.4	−46
CO_2Et	$3.35 \cdot 10^{-3}$	$9.22 \cdot 10^{2}$	0.729	9.4	−41
CN	$2.14 \cdot 10^{-2}$	$5.89 \cdot 10^{3}$	0.925	10.0	−35
N_2Ph	$1.84 \cdot 10^{-3}$	$5.07 \cdot 10^{2}$	0.664	8.6	−44
NO_2	$5.49 \cdot 10^{-1}$	$1.51 \cdot 10^{5}$	1.270	6.5	−42

kcal·mole^{-1} compared with $\Delta E^{\ddagger}$ for the parent compound, depending mainly on the overall ability of the groups to withdraw electrons conjugatively. The $\Delta S^{\ddagger}$ values vary over a range of only 4–5 e.u. (up to 1 in terms of $\log_{10} B$). Values of $\Delta S^{\ddagger}$ are typical for the anion–dipole and dipole–dipole reactions respectively. The charged carboxylate compound is naturally an exception, and in the reaction with methoxide ion has an expectedly more negative value of $\Delta S^{\ddagger}$.

TABLE 11

REACTIONS OF 1-Cl-2,6$(NO_2)_2$-4-X-BENZENES WITH OMe^- IN MeOH AT 0° FOR WHICH $\rho = 3.80$ *[13,64,71,72]*

4-Substituent	Rate constant k_2 (0°) ($l \cdot mole^{-1} \cdot sec^{-1}$)	Substituent rate factor (S.R.F. or f)	Substituent constant (σ^-)	$\Delta E^{\ddagger}$ ($kcal \cdot mole^{-1}$)	$\log_{10} B$
H	$4.98 \cdot 10^{-5}$	1	0	17.5_5	9.7_5
CO_2^{-} [a]	$2.03 \cdot 10^{-4}$	4.08	0.160	14.7_5	8.1
$CONH_2$	$1.19 \cdot 10^{-2}$	$2.39 \cdot 10^{2}$	0.627	15.3	10.3
$CONMe_2$	$3.22 \cdot 10^{-3}$	$6.47 \cdot 10^{1}$	0.477	15.8_5	10.2
CO_2Me [b]	$3.61 \cdot 10^{-2}$	$7.25 \cdot 10^{2}$	0.753	16.5_5	11.8
COPh	$4.07 \cdot 10^{-2}$	$8.17 \cdot 10^{2}$	0.767	13.9	9.7
SO_3^{-} [c]	—	—	0.336	—	—

[a] Corrected to zero ionic strength. [b] Possibly affected by some concurrent demethylation of the CO_2Me group. [c] Estimated from the reaction with OH^- in H_2O.

TABLE 12

REACTION OF 1-Cl-4-X-BENZENES WITH OMe^- IN MeOH AT 50° FOR WHICH $\rho = 8.47$ *[34]*

4-Substituent	Rate constant k_2 (100°) ($l \cdot mole^{-1} \cdot sec^{-1}$)	Substituent constant (σ^-)
H	($1.20 \cdot 10^{-16}$)	(0)
$COCH_3$	$3.22 \cdot 10^{-9}$	0.874
NO_2	$8.47 \cdot 10^{-6}$	1.270

The nitro group is a very important and convenient reference group, $\Delta\Delta E^{\ddagger}$ for this group in methanolysis in the series of Table 9 is –6.2 $kcal \cdot mole^{-1}$ and $\Delta\Delta S^{\ddagger}$ is only +4 e.u. ($\Delta \log_{10} B = 0.8_5$). In the methanolysis of the less reactive 1-fluoro-4-X-benzene series, with its much larger value of ρ (8.68 at 0°), $\Delta\Delta E^{\ddagger}$ for the nitro group is correspondingly larger, *viz.* -13.7_5 $kcal \cdot mole^{-1}$, and $\Delta\Delta S^{\ddagger}$ is –1 e.u. ($\Delta \log_{10} B = -0.2$).

Two Class 3 substituents were included by Miller and Wan in an investigation *[34]* of some *para*-substituents in chlorobenzenes. Data are given in Table 12 and were used to estimate values for chlorobenzene itself.

Miller *[9]* has suggested that the effect of an additional *ortho*- or *para*-nitro group in comparing 1-X-2- or -4-NO_2-benzenes with 1-X-2,4-dinitrobenzenes, or the latter with 1-X-2,4,6-trinitrobenzenes, is to increase stabilisation by about 9 $kcal \cdot mole^{-1}$. Only about 6 $kcal \cdot mole^{-1}$ of this appears in $\Delta E^{\ddagger}$ because the extent of bond formation, in forming the transition-states of the more reactive compounds with extra nitro groups, is less.

On examining the substituent effects in more detail some interesting patterns emerge. Among multiple-bonded nitrogen-containing groups the greater ability of nitroso than azo groups to stabilise the transition state *[2d]* is shown by comparison of azophenyl and nitroso groups for which σ^- values are 0.672 and 1.486; and β-azoxyphenyl and nitro groups for which σ^--values are 0.772 and 1.270. The difference is less in the second pair and this is no doubt due to differing magnitudes of

References p. 132

the internal conjugation which is greater in the nitro than in the azoxy group, and which reduces the overall ability to stabilise the

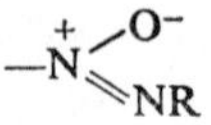

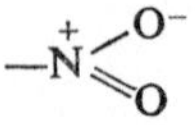

$$-\overset{+}{N}\overset{\nearrow O^-}{\searrow NR} \qquad\qquad -\overset{+}{N}\overset{\nearrow O^-}{\searrow O}$$

transition state. This also explains the smaller activating power of the nitro than of the nitroso group *[2e]* (σ^--values 1.270 and 1.480), which results entirely from a difference in $\Delta E^{\ddagger}$. The $-I$ effect of the nitro group must be greater than that of the nitroso group and thus the greater activating effect of the latter must be due to its superior conjugative power.

The same effect of internal conjugation also shows up in comparing the azo with the azoxy group, but because the internal conjugation in the latter is less marked, the reactivity of the azophenyl group is not greater than but lies between those of the α- and β-azoxyphenyl group (σ^- values are 0.616 0.672, and 0.772). The close resemblance of the

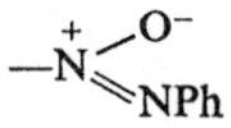

$$-\overset{+}{N}\overset{\nearrow O^-}{\searrow NPh} \qquad -N{=}\underset{O^-}{\underset{|}{\overset{+}{N}Ph}} \qquad -\overset{+}{N}\overset{\nearrow CH=C-O^-}{\searrow N-O}$$

β-azoxyphenyl 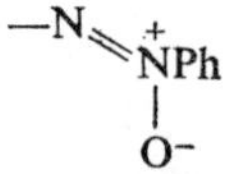 $\qquad$ α-azoxyphenyl 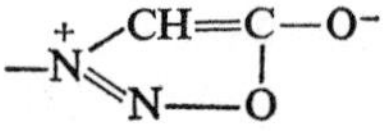 $\qquad$ α-sydnone

sydnone ring to the β-azoxy group and the possibility that conjugation in that ring might be little different from that in the β-azoxyphenyl group was shown by Miller and his co-workers *[64]*. This suggestion is strongly supported by the relative reactivity of aromatic halogen activated by a sydnone ring attached by its 1-nitrogen atom to the 4-position in the benzene ring, and that of halogen activated by 4-α- and 4-β-azoxyphenyl groups. The σ^--values are 0.616, 0.717 and 0.772 respectively. Close similarity in the effects on the Arrhenius parameters *[67]* also confirm the analogy.

Internal conjugation in the azido group ($-N{=}\overset{+}{N}{=}\overset{-}{N}$) places a fractional negative charge on the α-nitrogen, and it is noteworthy in relation to the discussion of Class 2 substituents that this almost removes activating power and the accompanying reduction of $\Delta E^{\ddagger}$, so that the σ^--value is only 0.116. The $\Delta E^{\ddagger}$ value is in fact slightly higher than for the unsubstituted compound, but the difference is close to the experimental error.

The effect of internal conjugation in reducing the external effect of a group has been discussed by Ingold *[2d]*, particularly with regard to $-COX$ groups. The same situation applies also to the $-SO_2X$ series. Both of these series have been investigated as activating groups in aromatic S_N reactions (Tables 9 and 11) and the pattern predicted from the expected electron-releasing power of X is very clearly followed. The order is: $CO_2^- < CONR_2 < COR < CHO$; SO_3^-, $SO_2NH^- < SO_2NR_2 < SO_2R$. Abnormally small activation by the 4-$CONMe_2$ but not by the 4-SO_2NMe_2 group is regarded *[64]* as a consequence of steric hindrance to coplanarity in the bulky groups, even when the neighbouring (3- and 5-)positions are occupied by hydrogen. The twisting out of the plane of the ring inhibits the conjugation of the $p\pi$–$p\pi$ carbon–oxygen double bond with the ring; but not that of the $p\pi$–$d\pi$ sulphur–oxygen double bond. Therefore only the activating power of the $CONMe_2$ group is affected. The effect described is a secondary steric effect.

The order of activating power $NO_2 > SO_2Me > COMe$ is of interest in relation to the relative importance of inductive and conjugative effects in stabilising the transition state. It may be compared *[64]* with relative electron-withdrawing effects revealed by (*a*) the acidity of XCH_2CO_2H and $XCH_2CH_2CO_2H$: *viz.*, $NMe_3^+ > SO_2Me > COMe$, and $NO_2 > COMe$; (*b*) the acidity of XOH: *viz.*, $SO_2Me > NO_2 > COMe$; (*c*) the acidity of XNH_2: *viz.*, $NO_2 > SO_2Me > COMe$; (*d*) the acidity of XCH_3: *viz.*, $NO > NO_2 > COMe > SO_2Me$. These are in increasing relative importance of the conjugative effect rising from zero in series (*a*). Activated aromatic S_N reactions follow pattern (*c*). Nucleophilic condensation reactions follow pattern (*d*), see p. 381.

Data for confirming the expected activating order, $C{=}O > C{=}NR$ are lacking. In contrast, the similar activating power of the $C{=}O$ and the $C{\equiv}N$ group is seen in Table 9, and similarities in the general chemistry of carbonyl compounds and of the nitriles are well-known. Miller and Parker *[66]* suggested that the conjugative power of $C{=}O$ and of $C{\equiv}N$ are approximately equivalent, as a consequence of the virtual independence of movement of the two pairs of π-electrons in the triple bond. Charge displacement resulting from the movement of the conjugated electrons, tending to a state of higher potential energy, is compensated in the triple bond by the movement of the second pair of π-electrons in the other direction. The total polarisation in forming the transition state can thus involve a greater displacement of the conjugated π-electrons than is possible in a double bond. Some part is played, however, by the difference in electronegativity of sp^2 and sp carbon, and from Sager and Ritchie's paper *[52]* this may be estimated as equivalent to about 0.2 in terms of σ^-. Thus conjugative electron-withdrawal by C═O is in fact somewhat greater than that by C≡N, in accord with more detailed comparisons of the chemistry of carbonyl and of cyano compounds.

In recent papers Sheppard and his co-workers *[73]* have reported on a number of polyfluorinated substituents, including the interesting SO_2CF_3 group for which σ^--values, based on phenol and anilinium acidities, of 1.36 and 1.65 respectively are reported. This would clearly class it as among the most powerful electrically neutral activating groups, comparable with the NO group ($\sigma^- = 1.486$).

While no direct quantitative measurements on the influence of the C═S group in aromatic S_N reactions seem to have been made, the order of conjugative effects, C═S > C═O *[2d]*, associated with the weak tendency of the larger atoms to form double bonds with carbon, suggests the greater activating power of –CS·X than –CO·X groups in aromatic S_N reactions.

The effect of the –S═O group (using –SOMe as example) has been considered by Bordwell and Boutan *[75]* and they conclude that it withdraws electrons effectively *[cf. 76]*. In this it resembles SO_2X and SR_2^+ groups *[43,45,63]*. The *p*-SOMe group has a σ^--value, from phenol acidities, of 0.73 very similar to that of the COMe group (Table 9, p. 77), which it may therefore be assumed to resemble as an activating group in aromatic S_N reactions. It is of interest that the SOMe group appears to be less effective than the SO_2Me group (Table 9) in electron-withdrawal, whereas the NO group is more effective than the NO_2 group. This is perhaps because the proportion of the inductive contribution to the polar effect is larger in the sulphur-containing groups and thus the internal conjugation is consequently less important in the sulphonyl than in the nitro group. The methylsulphinyl group has a pair of unshared electrons on sulphur and is thus, like the nitroso group, expected to be a pan-activating substituent (p. 61).

Miller and Wan *[34b]* have recently confirmed this assessment of the effect of the methylsulphinyl group in aromatic S_N reactions. They obtained a σ^--value of 0.827 at 50°, a value similar to that from phenol acidities (see above). As expected, the value is substantially less than that of the methylsulphonyl group in the same series (and that in Table 9) *viz.* 1.049; while it is similar to that of the methoxycarbonyl group, *viz.* 0.874 (Table 9).

The σ-value of 0.617 at 0° for the p-$\overset{+}{N}Me_2\overset{-}{O}$ group may be estimated in an alkaline methanolysis reaction, from the results of Bevan *et al.* *[51b]*. It may be compared with the expected larger value of 0.924 at 0° for p-$\overset{+}{N}Me_3$ from the same work [*cf.* Table 7]. Neither group can exert a conjugative effect and in each case activation is by a small reduction in $\Delta E^{\ddagger}$. For the cation $\log_{10} B$ is higher.

Class 4. Substituents attached by an electrically neutral atom bearing neither unshared electrons nor forming part of a conjugated side chain

The most common substituents of this class are alkyl and substituted alkyl groups. An aliphatic sp^3 carbon is less electronegative than an aromatic sp^2 carbon. Alkyl groups are thus weakly electron releasing and slightly destabilise transition states in aromatic S_N reactions. The presence of electron-attracting groups on the aliphatic carbon will increase its electronegativity. This is marked in the CF_3 group, and Lagowski *[77]* has estimated the electronegativity of the group as 3.3 on Pauling's scale *[48]*, *i.e.* it is more electronegative than Cl. Such a group can stabilise aromatic S_N transition states quite substantially. There is ample evidence to confirm these conclusions and some relevant data are shown in Table 13. In it, the ρ-values have been obtained by using literature σ- (or σ^--)values of several electron-withdrawing substituents and the standard zero value for hydrogen, which give good linear Hammett plots. Values for alkyl and substituted alkyl groups are obtained from the plots.

Alkyl groups are slightly deactivating, with the CH_3 compounds less reactive than the CMe_3 compounds. This difference could be ascribed to hyperconjugative electron release, or as a more recent alternative, to solvation effects *[82–85]*. The moderate activating power of the CF_3 group with similar σ^--values of 0.746 and 0.668

References p. 132

TABLE 13

REACTIONS OF 1-Cl- AND 1-Br-2-NO_2-4-X- AND 1-Cl-2,6-$(NO_2)_2$-4-X-BENZENES: (*a*) 1-Cl-2-NO_2-4-X-BENZENES WITH OMe^- IN MeOH AT 50° FOR WHICH $\rho = 3.90$ *[13,78]*; (*b*) 1-Cl-2,6-$(NO_2)_2$-4-X-BENZENES WITH OMe^- IN MeOH AT 50° FOR WHICH $\rho = 3.36$ *[64b,79,80]*; (*c*) 1-Cl-2-NO_2-4-X-BENZENES, WITH PIPERIDINE IN C_6H_6 AT 45° FOR WHICH $\rho = 4.08$ *[70]*; (*d*) 1-Br-2-NO_2-4-X-BENZENES WITH AN EXCESS OF PIPERIDINE AT 25° FOR WHICH $\rho = 4.95$ *[24]*; (*e*) 1-Br-2,6-$(NO_2)_2$-4-X-BENZENES WITH I^- IN Me_2CO AT 80° *[81]*

Series	Substituent 4-X	Rate constant k_2 ($l \cdot mole^{-1} \cdot sec^{-1}$)	Substituent rate factor (S.R.F. or *f*)	Substituent constant (σ^-)
(*a*)	H	$2.52 \cdot 10^{-6}$	1	0
at 50°	CH_3	$2.99 \cdot 10^{-7}$	0.119	– 0.237
	CH_2OMe	$3.32 \cdot 10^{-6}$	1.32	0.031
	CF_3	$2.05 \cdot 10^{-3}$	813	0.746
(*b*)	H	$7.40 \cdot 10^{-3}$	1	0
at 50°	CH_3	$1.34 \cdot 10^{-3}$	0.181	–0.221
	CMe_3	$2.39 \cdot 10^{-3}$	0.322	–0.147
(*c*)	H	$3.63 \cdot 10^{-6}$	1	0
at 45°	CH_3	$5.3 \cdot 10^{-7}$	0.15	– 0.20
	CMe_3	$6.1 \cdot 10^{-7}$	0.17	– 0.19
	CF_3	$1.91 \cdot 10^{-3}$	526	0.668
(*d*)	H	$4.83 \cdot 10^{-5}$	1	0
at 25°	Me	$7.05 \cdot 10^{-6}$	0.146	– 0.169
	CMe_3	$8.23 \cdot 10^{-6}$	0.171	– 0.155
(*e*)	H	$2.40 \cdot 10^{-4}$	1	—
at 80°	Me	$2.95 \cdot 10^{-5}$	0.123	—
	CMe_3	$3.80 \cdot 10^{-5}$	0.158	—

(mean, 0.707) in two series, confirms the marked effect of replacing α-hydrogen by fluorine discussed above. Miller and his co-workers *[78]* suggest that hyperconjugation of the $-E$ type occurs, and such an

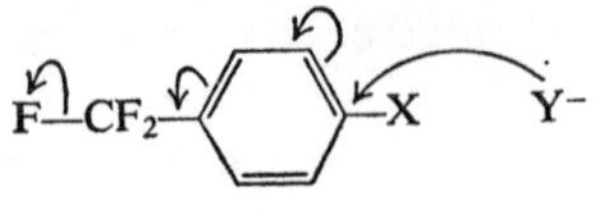

F—CF_2—C_6H_4—X Y⁻

effect in other work was previously suggested by Roberts and his co-workers *[86]*.

It is interesting to note (Table 14) that the pattern of effects of CH_3, H and CF_3 as illustrated in series (*a*) of Table 13, is similar to that when substituents are attached to a carboxyl group, so that the inductive effect is mainly responsible for the substituent effects.

Because of the large ρ-values of aromatic S_N reactions, even minor differences in substituent effects are readily measured. Thus Heppolette and Miller *[64]* have studied a series of $-CH_2X$ substituents. The change from $-CH_2-H$ to $-CH_2-OMe$ results in an 11-fold increase in rate and a change from weak deactivation to slight activation: the S.R.F. or *f* values in series (*a*) of Table 13 are 0.119 and 1.32 respectively.

TABLE 14

SUBSTITUENT EFFECT ON LEAVING GROUP MOBILITY AND ACIDITY

Substituent	S.R.F. or *f* [a] (0°)	S.R.F. or *f* [a] (50°)	Acidity ratio [b] (25°)
CF_3	2510	813	2777
H	1	1	1
CH_3	0.040	0.119	0.100

[a] Series (*a*) of Table 13 at 0° and 50°. [b] Series $X-CO_2H$.

Little experimental work has been carried out on the many other groups classifiable in Class 4. Aryl sulphurpentafluorides have been prepared by Sheppard *[73]* and he has shown that the SF_5 group is *meta*-directing in aromatic electrophilic substitution. From various physical measurements the σ- and σ^--values are 0.68 and 0.86 respectively. This may be compared with the σ^--values of the CF_3 group in Table 13, *viz.*, 0.746 and 0.668, (mean 0.707). One can predict therefore that the SF_5 group will be an activating group in aromatic S_N reactions, probably somewhat more effective than the CF_3 group.

Silicon-containing groups corresponding to alkyl groups are of special interest. The element is less electronegative than carbon, but on the other hand, like other elements in the second and higher horizontal rows of the Periodic Table, it can have an outer shell containing

References p. 132

more than an octet of electrons *[87]*. It might then offset the electronegativity factor by conjugative stabilisation of the transition state. Since results of Miller and his co-workers *[43]* suggest that such an effect does not occur with the SMe group, though it does for the SMe_2^+ group, there is some uncertainty as to whether it occurs in the SiR_3 group. However, Eaborn and Parker *[88]* have shown that the $SiMe_3$ group is a weakly activating group in alkaline desilylation of substituted benzyltrimethylsilanes, a series in which substituent effects should resemble those in aromatic S_N reactions. On this basis $SiAlk_3$ groups may be weakly activating in aromatic S_N reactions. The trifluorosilyl group should certainly be comparable in activating power with the trifluoromethyl group, with the likelihood of an additional activating conjugative component.

The situation in boron compounds is rather speculative. In $ArBX_2$ compounds, an electron-withdrawing stabilisation should be possible since there is only a sextet in the outer shell of the boron, while the element is the least electropositive in its group. In contrast the charge on boron in $ArBX_3$ is negative, and such substituents are classifiable in Class 2.

Class 5. Substituents attached by an electrically neutral atom bearing unshared electrons, but not forming part of a conjugated side chain

As in aromatic electrophilic substitution there are conflicting factors to consider in predicting the effect of groups in this class.

The position is fairly simple for amino and substituted amino groups. Saturated (sp^3) 3-covalent nitrogen has but moderate electronegativity and will conjugate readily with an aromatic system. The consequent very powerful activation by such groups in electrophilic substitution is well known *[e.g. 89]*, and while most of this can only become evident in stabilising the electron-deficient benzenium (cyclohexadienium) transition state, there must be a quite substantial permanent conjugation (+*M* effect) which will correspondingly *de*stabilise the benzenide aromatic S_N transition state. Results for these and other Class 5 substituents are given in Table 15.

The deactivating power of the amino group in aromatic S_N reactions is known *[13,24,70,90]* and kinetic results clearly show it to be quite strongly deactivating. All workers stress difficulties and inaccuracy in measurements, but in the series of Table 15, with ρ-values of about 4, *para*-amino groups deactivate between 10^3 and 10^4 times. The substituent-constant values corresponding to this, which can be called σ_N^- values*, are all in the range –0.6 to –0.9.

Similar arguments apply to alkoxy groups but the electronegativity is higher and electron-release smaller. In the series of Table 15, *para*-alkoxy groups deactivate by a factor of about 10^2, and σ_N^- values in the range –0.4 to –0.6 correspond to this.

The results of Miller and his co-workers *[43]* permit the important comparison of methoxyl and thiomethoxyl groups. The markedly inferior ability of second and higher row elements to provide an unshared pair of electrons to conjugate with aromatic systems is well known *[2f]*. For example, despite the fact that sulphur is less electronegative than oxygen, the thiomethoxyl group is substantially less activating than methoxyl in aromatic S_E reactions *[91]*.

Recalling that sulphur is more electronegative than carbon and that there is little permanent (mesomeric) electron release, one would predict that in aromatic S_N reactions a thioalkoxyl group would be weakly activating. This is confirmed by the results of Miller and his co-workers *[13]*, shown in series (*c*) of Table 15, which demonstrate activation of the same order of magnitude as for the heavier halogens.

Sheppard's interesting work on some polyfluoro substituents *[73]* also included a study of trifluoromethoxyl and trifluorothiomethoxyl groups. His results suggest that replacement of β-hydrogen by fluorine converts the weakly deactivating methoxyl group (σ_N^- about –0.5) into the weakly activating trifluoromethoxyl group, resembling the halogens with σ^- about +0.4. This may be compared with the data of Table 13 indicating that replacement of α-hydrogen by fluorine converts the slightly deactivating methyl group (both σ^- and σ_N^- about –0.2) into the moderately activating trifluoromethyl

* Electron-releasing groups are given special σ^+-values in aromatic S_E reactions. Miller *[13]* had suggested differentiating such substituent-constant values in S_N and S_E reactions as σ_N^{**} and σ_E^{**} values. In view of modern nomenclature it might be convenient to call these σ_N^- and σ_E^+ values. The subscript N then refers to special values used for electron-releasing groups in aromatic S_N reactions. The experimental values of substituent constants from Table 15 for these electron-releasing substituents are thus called σ_N^- values. The less negative σ- and σ^--values from the literature do not fit a Hammett plot *[13]*.

TABLE 15

REACTIONS OF 1-Cl- AND 1-Br-$(NO_2)_n$-4-X-BENZENES ($n = 1,2$): (*a*) 1-Br-2-NO_2-4-X-BENZENES WITH AN EXCESS OF PIPERIDINE AT 25° FOR WHICH $\rho = 4.95$ *[24]*; (*b*) 1-Cl-2-NO_2-4-X-BENZENES WITH PIPERIDINE IN BENZENE AT 45° FOR WHICH $\rho = 4.08$ *[70]*; (*c*) 1-Cl-2-NO_2-4-X-BENZENES WITH OMe^- IN MeOH AT 50° FOR WHICH $\rho = 3.90$ *[13]*; (*d*) 1-Cl-2,6-$(NO_2)_2$-4-X-BENZENES WITH OMe^- IN MeOH AT 50° FOR WHICH $\rho = 3.36$ *[13]*

Series	Substituent 4-X	Rate constant k_2 ($l \cdot mole^{-1} \cdot sec^{-1}$)	Substituent rate factor (S.R.F. or f)	Substituent[a] constant (σ^- or σ_N^-)
(*a*)	H	$4.83 \cdot 10^{-5}$	1	0
at 25°	NH_2	$6.00 \cdot 10^{-9}$	$1.24 \cdot 10^{-4}$	−0.789
	NMe_2	$5.87 \cdot 10^{-8}$	$1.22 \cdot 10^{-3}$	−0.589
	OH[b]	$2.82 \cdot 10^{-8}$	$5.84 \cdot 10^{-4}$	−0.654
	OMe	$8.70 \cdot 10^{-7}$	$1.80 \cdot 10^{-2}$	−0.353
	OEt	$7.30 \cdot 10^{-7}$	$1.51 \cdot 10^{-2}$	−0.368
	F	$1.26 \cdot 10^{-5}$	$2.61 \cdot 10^{-1}$	−0.118
	Cl	$2.70 \cdot 10^{-4}$	5.59	0.151
	Br	$3.79 \cdot 10^{-4}$	7.85	0.181
	I	$2.62 \cdot 10^{-4}$	5.43	0.148
(*b*)	H	$3.63 \cdot 10^{-6}$	1	0
at 45°	NH_2	*ca.* $3 \cdot 10^{-9}$	*ca.* $8 \cdot 10^{-4}$	*ca.* −0.8
	OMe	$9.1 \cdot 10^{-8}$	$2.5 \cdot 10^{-2}$	−0.39
	OEt	$7.8 \cdot 10^{-8}$	$2.1 \cdot 10^{-2}$	−0.41
	Cl	$2.29 \cdot 10^{-5}$	6.31	0.196
	Br	$3.49 \cdot 10^{-5}$	9.61	0.241
	I	$4.41 \cdot 10^{-5}$	12.15	0.266
	H	$2.52 \cdot 10^{-6}$	1	0
(*c*)	F	$2.25 \cdot 10^{-6}$	0.895	−0.015
at 50°	Cl	$3.51 \cdot 10^{-5}$	14.0	0.265
	Br	$3.87 \cdot 10^{-5}$	15.4	0.289
	I	$4.39 \cdot 10^{-5}$	17.4	0.318
	SMe	$5.47 \cdot 10^{-5}$	21.7	0.343
(*d*)	H	$7.40 \cdot 10^{-3}$	1	0
at 50°	NH_2	$8.7 \cdot 10^{-6}$	$1.2 \cdot 10^{-3}$	−0.87
	OMe	$1.48 \cdot 10^{-4}$	$2.00 \cdot 10^{-2}$	−0.506
	Cl	$5.86 \cdot 10^{-2}$	7.12	0.267
	Br	$1.01 \cdot 10^{-1}$	13.6_5	0.338
	I	$7.50 \cdot 10^{-2}$	10.1	0.299

[a] For groups attached by nitrogen and oxygen these are σ_N^- values (see footnote to page 87 and ref. *13*). For the halogens these are σ^- values. It is reasonable to include fluorine with the other halogens, although it is very mildly deactivating.

[b] Because of a proportion of ionised phenol, the hydroxyl group appears to be more deactivating than it really is.

group (σ^- about +0.7). In each case $\Delta\sigma$ is about 0.9, but this is coincidental since in the latter case the substitution is on the atom attached to the ring. The thiomethoxyl group is itself activating and the fluorine substitution simply strengthens this—the σ^--value of thiomethoxyl is 0.343, and Sheppard's value of σ^- for trifluorothiomethoxyl is 0.64. The $\Delta\sigma$-value between O and S is only 0.3. Such a difference is perhaps to be expected, since in the former the structural change affects not only an inductive, but also a substantial conjugative effect. In the latter however, it is mainly a change in inductive effect. The difference of $\Delta\sigma$ (both essentially inductive in origin) of about 0.3 for SCH_3 and SCF_3, and about 0.9 for CH_3 and CF_3 attached directly to the ring is reasonable, and indicates a transmission coefficient (see p. 90) across –S– of about one third.

The halogens form a complete series of special interest and importance. Fluorine is the most electronegative (and thus has the largest activating $-I$ effect) but is best able to conjugate its unshared pair of electrons (in a deactivating $+M$ effect), and there is a smooth gradation through the series. According to the relative weighting of these two factors, any reactivity order for the halogens could result. The evidence from the reactivity of the methoxyl and thiomethoxyl groups in S_N reactions, and what is known of the halogens in S_E reactions *[2f]*, clearly suggests that the factor controlling the relative activating power is conjugative electron-release. In the S_N reactions, the two factors approximately cancel for fluorine, but the I effect slightly outweighs the M effect with the other three. The balance is so close for these that the order in relation to Cl, Br, I is temperature-variant in conditions of moderate reactivity. The experimental pattern of activation is, however, usually $F \leqslant H < Cl < Br \sim I$. Values of σ^- for the halogens seem to be a little lower with piperidine than with methoxide ion as reagent.

The work of Berliner, of Brieux, and of Miller and their co-workers *[13,24,70]* shows that compounds with deactivating NR_2 and OR groups have enhanced $\Delta E^{\ddagger}$ values as might be expected. $\Delta S^{\ddagger}$ values are low for NR_2 groups.

The weakly activating substituents SMe, Cl, Br and I all follow a pattern of small $\Delta E^{\ddagger}$ changes and somewhat raised $\Delta S^{\ddagger}$ values. The accuracy of these results may be insufficient for these small changes to have much validity, and it seems pointless to speculate on possible reasons.

Class 6. Substituents containing carbon–carbon multiple bonds conjugated with the ring

Heteropolar multiple bonds, and homopolar multiple bonds between hetero atoms, have been classified as Class 3 substituents. Class 6 consists of conjugated ethylenic and acetylenic groups.

There are two factors to consider, *viz.* the intrinsic electronic effect of the homopolar system and the extent to which it will transmit the effect of a terminal (or ω-) substituent.

In the conjugated alkenes the double bond sp^2 carbons are of the same electronegativity as aromatic sp^2 carbons, but the π-bond electrons can conjugate to some extent with the ring to stabilise transition states in both electrophilic and nucleophilic substitution (*i.e.* the substituent is a $\pm T$ group). In aromatic electrophilic substitution –CH=CH–X groups, where X is electron-withdrawing, are known to be *ortho–para* directing, and weakly deactivating *[92]*, but the –CH=CH_2 group itself may be weakly activating. In aromatic S_N reactions one would expect the –CH=CH_2 group to be weakly activating ($-T$ stabilisation of the transition state). Electron-withdrawing ω-substituents (X) in –CH=CH–X, by increasing the electronegativity of the β-C, would increase this activation somewhat. Some data of Heppolette and Miller *[64b,93]*, given in Table 16, show this clearly.

Values for three ω-substituents are shown and it is clear from these that ω-substituent effects are small. Although there are results for only three compounds, the estimated value for the –CH=CH_2 group is regarded as reliable, and must be close to that for the –CH=CH–CO_2^- group, since the activating power even of the directly attached CO_2^- group in series with large ρ values is small, and close to that for hydrogen (*i.e.* the parent compound). Because ionic strength corrections were made only for the cinnamate ion at 110.6°, this temperature has been used for the comparison. The ρ value for 4-ω-substitution is 1.30, and for 4-substitution is 3.26. The ratio of these is a measure of the transmission of polar effects across the system and has been called the transmission coefficient *[94]*. Its value here at 110.6° is 0.40.

Bowden *[94]* has discussed transmission of polar effects at some length and quotes other recent references. His conclusions in essence

TABLE 16

COMPARATIVE REACTIVITY OF (*a*) 1-Cl-2-NO_2-4-X- AND (*b*) 1-Cl-2-NO_2-4-CH=CH—X-BENZENES WITH OMe^- IN MeOH AT 110.6° FOR WHICH THE VALUE OF ρ FOR 4-SUBSTITUENTS IS 3.26 *[13,64b,93]*. THE VALUE OF ρ DETERMINED FOR ω-SUBSTITUTION IS 1.30 (see text)

Series	Substituent *4-X*	Rate constant k_2 ($l \cdot mole^{-1} \cdot sec^{-1}$)	Substituent rate factor (S.R.F. or f)	Substituent constant (σ^-)	Terminal substituent rate factor (ω-S.R.F. or ω-f)
(*a*)	H	$8.45 \cdot 10^{-4}$	1	0	
	CO_2^-[a]	$1.55 \cdot 10^{-3}$	1.83	0.135	
	$CONH_2$	$1.20 \cdot 10^{-1}$	142	0.627	
	CO_2Me	$3.48 \cdot 10^{-1}$	412	0.819	
	4-ω-X				
(*b*)	(H)[b]	$(2.74 \cdot 10^{-3})$	3.29	0.143	(1)
	CO_2^-[a]	$3.98 \cdot 10^{-3}$	4.76	0.188	1.45
	$CONH_2$	$1.95 \cdot 10^{-2}$	13.7	0.392	7.20
	CO_2Me	$2.96 \cdot 10^{-2}$	35.5	0.445	10.8

[a] At zero ionic strength. [b] Value of rate constant obtained by Hammett plot of the rest of the series.

are that transmission of substituent effects across the two-carbon -CH=CH- system occurs to an extent approximately equal to that across a -CH_2- group. He quotes the average value of the transmission coefficient for *trans*-CH=CH- as 0.449, quite close to the value 0.40 referred to above.

Acetylenic groups

The acetylenic sp carbon is more electronegative than aromatic sp^2 carbon, as is indicated for example by the dipole moment of phenylacetylene, *viz.* 0.78 *D [2g]*. In aromatic S_N reactions therefore -C≡C-X groups are expected to resemble Class 3 substituents and cause activation, though probably substantially weaker than most Class 3 substituents. As yet there is no quantitative information to support this, though it may be inferred from the ready formation of 1,4-dialkynyltetrafluorobenzenes by reaction of lithium alkynes and hexafluorobenzene (Section 8(b), p. 130).

Bowden [94] quotes a lower transmission coefficient for –C≡C– than –CH=CH–. Fuchs [95] has discussed this and reinvestigated a series of reactions (alkaline hydrolysis of ethyl propiolates), results for which had suggested a higher value of the transmission coefficient. His results confirm the lower value. The reasons for the less effective transmission of substituent effects across the acetylenic system are not clear: Fuchs suggests that π-bond interactions involving sp carbon are weaker than those involving sp^2 carbon.

Class 7. Fused and attached aromatic rings as substituents

A fused carbocyclic aromatic ring should be able to stabilise transition states of both electrophilic and nucleophilic aromatic substitution by extending the delocalisation of either electron deficiency or surplus. In this section, to avoid consideration of steric effects, fused rings *meta* and *para* to the replaced group are considered.

Berliner and his co-workers [96] investigated the reactions of 2-chloro-, 2-bromo- and 2-iodo-naphthalene with piperidine as reagent and solvent, *i.e.* compounds with a fused benzene ring in the 3,4-positions to the halogen and found that it activates only 2- to 3-fold. Amstutz and his co-workers [97] in similar experiments with chloro- and bromo-compounds showed that a fused benzene ring in the 3,4-positions to the halogen, activates 4.68 times in chlorine replacement, 2.40 times in bromine replacement. A 3,4-fused naphthalene system increases the rate of chlorine replacement by 6.46, a larger value, but indicating nevertheless that the first ring substituent to some extent insulates the second ring substituent. Wepster and his co-workers [98] showed that 2-Br-1-NO_2-naphthalene is considerably more reactive than 2-Br-3-NO_2-naphthalene, despite the possibility of steric hindrance of a nitro group *ortho* to the fused ring. This is because the transition state for the former compound retains one ring benzenoid whereas that for the latter does not. Wepster and his co-workers did not make measurements on *o*-bromonitrobenzene under the same conditions so that S.R.F. values cannot be calculated.

As regards substitution in the second ring, activation from the 6- and 8-position was shown to be more effective than from the 5- and 7-position. These factors of substitution in bicyclic rings are discussed at length in Chapter 7.

Wepster states that the precision of his results *[98]* is not high, and there is attachment at two points, so that no satisfactory calculation of a transmission coefficient is possible. It appears nevertheless that transmission across fused rings is substantially reduced, compared to that within a single benzene ring.

Systems with an attached benzene ring should resemble both the fused systems but with less effect, and those with attached ethylenic groups. Table 17, Series (*a*), which gives results for the 4-phenyl and 4-*p*-nitrophenyl substituent confirms this. By comparing ρ-values for 4- and 4′-substituents in replacement of chlorine by methanolic methoxide ion at 50° *[13,50c]*, *viz.* 3.34 and 0.855, one obtains a value of the transmission coefficient of 0.256.

Similarly for reactions with piperidine in dioxane at 25° *[99]* (Table 17, Series (b)) one obtains the ρ-value 0.858 for 4′-substituents. Assuming $\rho = 4.95$ (the value is based on slightly different conditions) for 4-substituents in the series *[24]* a similar but somewhat lower value of 0.173 is obtained for the transmission coefficient.

The values of transmission coefficients are similar to those quoted by

TABLE 17

REACTIONS OF (*a*) 1-Cl-2,6-$(NO_2)_2$-4-X-BENZENES WITH (*i*) OMe^- IN MeOH AND (*ii*) OMe^- IN MeOH-C_6H_6 ($\rho = 3.34$) AT 50° *[13,50c]*; (*b*) 1-Br-2-NO_2-4-X-BENZENES WITH PIPERIDINE IN DIOXANE AT 25° *[99]* ($\rho \cong 4.95$); (*c*) 1-Cl-2-NO_2-4-X-BENZENES WITH PIPERIDINE IN C_6H_6 AT 45° *[70]* ($\rho = 4.08$)

Series	Substituent 4-X	Rate constant k_2 ($l \cdot mole^{-1} \cdot sec^{-1}$)	Substituent (4-S.R.F. or *f*)	Rate factors (4′-S.R.F. or *f*)	Substituent constant (σ^-)
(*a*) (*i*)	H	$7.45 \cdot 10^{-3}$	1	—	0
	Ph	$1.21 \cdot 10^{-2}$	1.62	—	0.062
(*a*) (*ii*)	Ph	$1.12 \cdot 10^{-2}$	(1.62)	1	—
	p-$NO_2C_6H_4$	$1.37 \cdot 10^{-1}$	(19.8)	12.2	0.386
(*b*)	H	$1.30 \cdot 10^{-5}$	1	—	0
	Ph	$2.87 \cdot 10^{-5}$	2.21	1	0.070
	p-$NO_2C_6H_4$	$3.52 \cdot 10^{-4}$	27.1	12.3	0.290
(*c*)	H	$3.63 \cdot 10^{-6}$	1	—	0
	Ph	$7.58 \cdot 10^{-6}$	2.1	—	0.079

Bowden *[94]* from investigations of transmission across the benzene ring in side-chain reactions, although the individual ρ-values are larger for the ring substitution reactions.

The activating power of the attached benzene ring, like that of the fused benzene ring, is unexpectedly small. Similar results in three series (Table 17) show that the *p*-phenyl group has a very small f_p-value. An average value of about 2 was obtained for selective series with ρ-values of approximately 4. It appears possible that the low activating power reflects the requirement for *both* rings to change from benzenoid to benzenide conjugation if the second ring is to activate by conjugation. The low activating power of the ring, and associated low transmission coefficient, then parallels the markedly lower activating power of a 3-substituent than of a 1-substituent in a 2-X-naphthalene (see above, p. 92 and Chapter 7, p. 266). Streitwieser *[100]* has referred to weak conjugation in arylolefins and biaryls, implying a low transmission coefficient.

Class 8. Substituents attached by a metal atom

Attachment of an organic group to a metal is well-known to cause substantial changes in reactivity of the organic moiety, so that, for example, whereas alkyl halides are characteristically susceptible to nucleophilic substitution, Grignard and similar organometallic compounds are characteristically susceptible to electrophilic substitution.

Correspondingly, attachment of an aromatic ring to a metal atom must affect substitution reactions in the ring. This could be discussed at great length, but it is not warranted in the context of this Chapter. Essentially, simple attachment of a ring by a bond to an electropositive metal deactivates the ring for nucleophilic substitution, whereas attachment to a metal by donation of ring π-electrons into vacant orbitals of a metal results in activation of the ring. There is good evidence for the latter, for example, in the results and discussion of Brown, Whiting and their co-workers *[101,102]* on reactions of arene tricarbonylchromium compounds $[XC_6H_4Cr(CO)_3]$. Their results demonstrated the close equivalence of the tricarbonylchromium and *ortho-* or *para-*nitro groups in activating power. Their action depends almost entirely on a large reduction in $\Delta E^{\ddagger}$, with $\Delta S^{\ddagger}$ scarcely affected. The results also show the high reactivity of fluorine compared to chlorine as a leaving group with first-row nucleophiles in protic solvents.

This aspect of aromatic S_N reactions is a very large, virtually unexplored field, which would repay full investigation.

6. *Ortho*-Substituent Effects

With *para*-substituent effects as standard it is possible to consider (*a*) differences in polar effects of *ortho*- and *para*-substituents, and (*b*) special effects caused by a neighbouring substituent.

The model of transition states in aromatic S_N reactions is one in which entering and leaving groups are well out of the plane of the ring but in which activating substituents utilising pπ bonding in conjugation require to be in the plane of the ring in order to exert their maximum effect.

For maximum stability the initial state similarly requires substituents to be in the plane of the ring. Although there is an increase of one in the number of attached groups in forming the transition state, this merely forms an aliphatic type centre with four bonded groups, and further any steric repulsions in the initial state are likely to be relieved by the shift from planar to non-planar geometry at the reaction centre. Steric retardations due to *ortho*-substituents are therefore not normally likely to be of great magnitude. Similarly, steric acceleration is unlikely to be of common occurrence.

Since the inductive effect, and more especially its associated field effect, which Dewar *[3]* considers to be the major component, is more effective at the *ortho*-position, *ortho*-groups having such polar effects will be more effective. When this is an electron-withdrawal a relative *ortho*-acceleration results. The conjugative effect is more effective at the *para*-position. When it is electron-release the *ortho*-position is less deactivated. In each case reactivity at the *ortho*- is greater than at the *para*-position.

It is necessary to consider intrinsic differences in *ortho*- and *para*-transition states (Fig. 36A and B), ignoring electrostatic considerations. This is a matter upon which agreement has not been finally reached. Though his argument was concerned with S_E reactions,

References p. 132

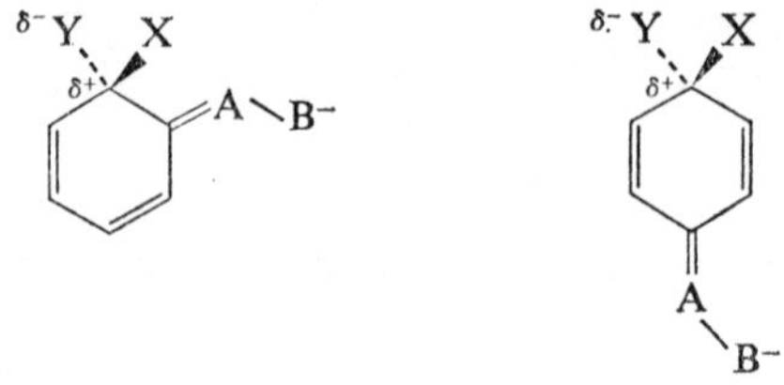

Fig. 36. Comparison of transition states having *ortho*- or *para*-substituents.

Ingold *[2h]* pointed out that the greater effectiveness of *para*-conjugative substituents implies that transition states corresponding to (B) are more stable, quoting in support the greater stability of *para*- than *ortho*-benzoquinones, which have similar structures. Green *[103]* by a valence bond calculation suggests the order $o \geqslant p(> m)$ for S_N and $p > o(> m)$ for S_E reactions. From the very large amount of evidence available in S_E reactions, it is generally agreed that conjugative stabilisation is greatest from the *para*-position.

If some estimate can be made of this difference in effectiveness of conjugative substituents in the *ortho*- and *para*-positions, then some disentangling of the various factors which contribute to *ortho*-substituent effects is possible.

Investigation of a reaction in which the *ortho*-substituent has low steric requirements and activates by conjugative stabilisation of the rate-limiting transition state is required. The linear cyano group with its small atoms is a possible choice. Unfortunately an otherwise reasonably precise and reliable comparison, with methanolic methoxide ion as reagent *[63,104a]*, was vitiated by imido-ester formation with the *para*- but not the *ortho*-cyano compound *[104]*. Although an estimate was made for the *para*-cyano group from another reaction *[65]*, the most that can be said from this work is that the *ortho*- and *para*-groups are approximately equivalent. The estimated σ^--values are *para*-CN, 0.997 and *ortho*-CN, 0.966 (assuming ρ is the same for reactions of 1-Cl-2-NO_2-4X- and 1-Cl-2-X-4-NO_2-benzenes). In view of the approximations involved the difference in σ^--values cannot be regarded as significant. For the *o*-cyano group, $\Delta\Delta E^{\ddagger} = -6.45$ kcal·mole^{-1} and $\Delta \log_{10} B = -0.6$: these are values very similar to those for *para*-groups in this series, activating by conjugation.

The –C≡C–X groups may be used similarly but are complicated by field effects, according to the results of Newman *[105]* and Roberts *[106]* and their co-workers, in investigations of reactivity of phenylpropiolic acid derivatives. Charton *[107]* has investigated a total of 14 reaction series involving –C≡C–X groups, and has concluded that conjugative activation from the *ortho*-position is about 3/4ths of that from the *para*-position.

Comparisons of reactivities of 2- and 4-chloropyridines (Liveris and Miller *[108]*) is a convenient way of comparison of conjugative transmission in an aromatic ring from *ortho*- and *para*-positions. They showed that activation by the nitrogen in the 2-position is about 3/4ths that from the 4-position.

It seems that differences between *ortho*- and *para*-substituent effects in aromatic S_N reactions are in no major way consequential on an intrinsic difference between the magnitude of conjugative effects from *ortho*- and *para*-positions, but what difference there is favours the *para*-position. The relative value of 3:4 seems about correct. Where there is an important *ortho*-effect one should therefore look to inductive effects, field and electrostatic effects, hydrogen bonding, or primary and secondary steric effects for the explanations. These are discussed below where relevant.

A simple quantitative measure of the overall difference in *ortho*- and *para*-substituent effects is given by the ratio of the S.R.F. or f values in corresponding *ortho*- and *para*-series, utilised by Miller and Williams *[79a]**. Validity implies near equivalence of ρ-values in the very closely similar series and this is virtually certain.

The classes previously referred to† are now discussed in turn.

(*a*) Ortho-*substituent effects in Classes 1 and 2*, viz. *Substituents attached by charged atoms*

These groups are expected to show distinct *ortho*-effects. A good example is the diazonium group. Its steric requirements are very low

* These workers have called the ratio f_p/f_o the Steric Index or S.I., though it might without implications have been called the Ortho Index.

† For a full description of the substituent classes see Section 5.

References p. 132

but there should be substantial differences in inductive, and field effects, plus an electrostatic effect when these compounds react with anionic nucleophiles.

Miller and his co-workers *[50,79b]* have measured rates of reaction of both *ortho*- and *para*-fluorophenyldiazonium salts with OH^- in H_2O at 0°, corrected to zero ionic strength. The f_p/f_o ratio is 0.104. The corresponding σ^--value for the *ortho*-diazonium group (estimated for 0°) is then $\nless 1.983$ compared with $\nless 1.870$ for the *para*-diazonium group (see p. 73). The difference is small considering the absence of steric effects, and the higher I and D effects. This supports the view that the conjugative stabilisation, which forms the major part of the *para*-effect (see p. 71) does act more effectively from the *para*- than the *ortho*-position, and thus largely cancels out the inductive and field effect, which are more effective from the *ortho*-position. Analogously, Liveris and Miller *[108]* have compared the reactivity of *N*-methyl 2- and 4-chloropyridinium salts, in which activation by 2- or 4-ring $\overset{+}{N}$Me is measured. With *p*-nitrophenoxide ion in methanol at 50° the f_p/f_o value is rather low, *viz.* 0.0331, as one might expect where steric effects are almost negligible. Corresponding σ^--values are 2.492 and 2.317. However of these, 2.079 and 2.164 respectively are due to reduction in $\Delta E^{\ddagger}$ and the rest to increases in $\Delta S^{\ddagger}$, so that more effective *para*-conjugation may again be inferred.

Another cationic group for which reliable data are available is the trimethylammonio group *[50c]*. In this case one would expect a steric retardation of about the same order as with a tert-butyl group. With the bulky reagent piperidine, the f_p/f_o value for tert-Bu is >100 times larger than for the methyl group *[81]*, (see p. 105). The difference should be substantially less with methoxide ion as reagent. The f_p/f_o value for the trimethylammonio group is somewhat temperature-dependent: it varies from 0.828 at 0° to 0.734 at 50°. It appears that the component due to polar effects alone is probably comparable with actual values for the diazonium and ring $\overset{+}{N}$Me group, but is approximately cancelled by a steric retardation. Values of σ^- for *o*- and *p*-$\overset{+}{N}Me_3$ in activation of replacement of ring chlorine were computed by Liveris and Miller *[108]* in a comparison with reactions of chloropyridines, -pyridine-*N*-oxides and *N*-methyl chloropyridinium salts for which the Hammett ρ-value is 8.47 at 50°. Values of σ^- are:

o-$\overset{+}{N}Me_3$, 0.762 (of which 0.436 is due to a $\Delta E^{\ddagger}$ reduction and the rest to a $\Delta S^{\ddagger}$ increase consequent on reaction occurring between an anion and a cation); p-$\overset{+}{N}Me_3$, 0.647 (of which 0.251 is due to a $\Delta E^{\ddagger}$ reduction and the rest to $\Delta S^{\ddagger}$ increase). The total value of σ^- for p-$\overset{+}{N}Me_3$ is less than that quoted earlier (p. 72). This is because the entropy component of the σ^--value seems to be independent of the magnitude of the Hammett ρ-value, so that it is relatively larger as the value of ρ decreases *[108]*.

One can conclude that inductive and field effects of positively charged substituents are more effective from the *ortho*-position. Anion–cation interactions of reagent and substrate do not seem to have a very marked influence in the comparison of activation from the two positions: the $\log_{10}B$ value is raised by similar amounts in both cases. Nevertheless it would be advantageous to have kinetic data comparing these cationic substituents with an electrically neutral substrate.

Apart from the polar effect of a negatively charged group, the transition state is destabilised by electrostatic interactions when the reagent is also negatively charged. In contrast it is stabilised when the reagent is neutral since it forms a fractional positive charge close to the negatively charged *ortho*-substituent in forming the transition state. A more important factor favouring an *ortho*-substituent is hydrogen-bonding, which is common where the neutral reagent is of the type $\ddot{Y}$–H. This is illustrated in Fig. 37.

Through lack of data there is no evidence for these conclusions for a true Class 2 substituent, but there is supporting evidence for a negatively-charged substituent of Class 3, and it is convenient to quote the evidence here. In methanolysis of 1-chloro-4-nitro-2-X-benzenes the *ortho*-CO_2^- group is deactivating, whereas in methanolysis of

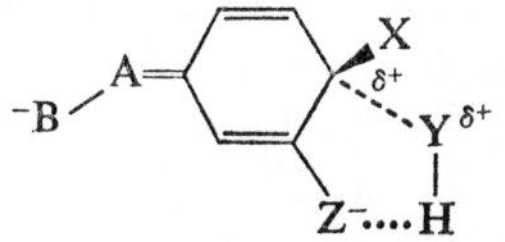

Fig. 37. Transition state for S_N reaction of a neutral nucleophile $\ddot{Y}$–H with an aromatic substrate containing a negatively-charged *ortho*-substituent. (◄, full bond above plane of ring; ----, partial bond below plane of ring; ....., hydrogen bond).

1-chloro-2-nitro-4-X-benzenes the *para*-CO_2^- group is weakly activating. The value of f_p/f_o uncorrected for ionic strength effects is 19.1 at 50° *[109]*, whereas its value for the methyl group, which is of similar size, is about unity *[79]*. A similar but more striking result is shown in a series where there is an additional standard substituent (NO_2) in the other *ortho*-position, and thus steric factors are enhanced. The *para*-CO_2^- group is still weakly activating, but the *ortho*-CO_2^- group is substantially more deactivating than before, and the f_p/f_o value is large, 1640 at 0° *[104a]*; whereas in the same series the f_p/f_o value for the methyl group is still only 1.3 *[79]*.

For the CO_2^- group, data are also available for reaction with a neutral reagent, *viz.* piperidine. Bunnett and his co-workers *[110a]* have studied the reactions of this with 1-chloro-4-nitro-2-X- and 1-chloro-2-nitro-4-X-benzenes, and compared their results with those of Miller and his co-workers for the methoxide ion *[63]*. Their results showed an *increase* in activating power of the *ortho*-CO_2^- group, with an f_p/f_o value of 0.106. They referred to this activating effect (see Fig. 37) as "built in solvation".

The Hammett σ^--values for CO_2^- are included in Table 18.

(*b*) Ortho-*substituent effects in Class 3*, viz. *Substituents attached by the positive end of a dipole*

An interesting pattern of *ortho*-substituent effects is observable with common activating groups of this class such as the nitro group.

Bunnett and Morath *[110b]* commented that in reactions of 2,4-dihalogenonitrobenzenes with a variety of nucleophiles the 2-halogen is preferentially replaced and concluded that, other things being equal, *ortho*-substitution is favoured. It must be pointed out however that chlorine in the 2-position has a secondary steric effect in the transition state of the substitution of the 4-chlorine atom since it prevents the full co-planarity of the nitro group with the ring; whereas there is no such effect of chlorine in the 4-position on the substitution of the 2-chlorine atom. This reduction of conjugative stabilisation by the nitro group reduces the rate of substitution of the 4-chlorine atom. This interpretation is supported by considering the halogen reactivity in 2,4-dichloroquinoline, where in the absence of this effect the 4-

chlorine is preferentially replaced by methoxide ion, the 4-Cl/2-Cl mobility ratio being 1.9 *[111]*.

Miller and his co-workers *[35,112]* have compared the ease of replacement of halogen by methoxide ion in methanol in the *ortho*- and *para*-halogenonitrobenzenes and showed that in all cases the *ortho*-nitro group is slightly less activating than the *para*-nitro group, the f_p/f_o values varying only over the small range of 2.00 to 3.66 at 100°. In the transition state, electrostatic interactions between ---$OMe^{\delta-}$ and the dipolar $\overset{+\,-}{NO_2}$ group probably amount to very little. For the reaction of piperidine with *ortho*- and *para*-chloronitrobenzene, Bunnett and Morath *[110b]* have shown that the *ortho*-nitro group is clearly more activating, and have ascribed this to built-in solvation between the $N^{\delta+}$ of the piperidine and the negative oxygens of the nitro group, possibly also to hydrogen-bonding between the *N*-hydrogen of piperidine and oxygen of the nitro group (*cf.* Fig. 37).

The concept that hydrogen-bonding plays a major part in determining the greater activating power of an *ortho*- than a *para*-nitro group in S_N reactions with an electrically neutral agent of the type $\ddot{Y}$–H, is suggested by the results of Greizerstein and Brieux *[113]*. They studied the reactions of *o*- and *p*-chloronitrobenzene with piperidine in methanol and in benzene, and showed a marked change in reactivity of the *para*-nitro compound in the change from aprotic to protic solvent, but little change in the reactivity of the *ortho*-nitro compound for which "built-in" *i.e.* internal solvation is postulated.

In similar reactions Chapman and his co-workers *[114]* had specifically ascribed the relative *ortho*-activation to hydrogen-bonding. Hawthorne *[115]* argued against this on the ground that *N*-deuteropiperidine ought to react at a different rate if hydrogen bonding is significant. This does not seem a powerful argument and strong support for hydrogen-bonding as the major factor has come again from recent work by Ross and Finkelstein *[116]*. For reaction of piperidine with *o*- and *p*-chloronitrobenzene in benzyl alcohol their f_p/f_o value is 0.179 at 150°.

For reaction with a tertiary amine [1,4-diazabicyclo(2.2.2)octane], which, lacking an N–H group, is unable to partake in hydrogen-bonding, the reactions in benzyl alcohol at 150° were shown to have an f_p/f_o value of 250. This strongly supports (*a*) the importance of hydrogen-bonding in reactions of primary and secondary amines, (*b*) the assumption of an intrinsically greater activating power of *para*- than *ortho*-nitro groups. The product of the tertiary amine reaction undergoes a fast subsequent reaction closely resembling the ring-fission of 2,4-dinitrophenylpyridinium compounds (see Chapter 9, p. 380), but this is not relevant to the present discussion.

There is some contrary evidence suggesting that the total inductive and conjugative effect of an *ortho*-nitro group is greater than that of a *para*-nitro group. In the iodine exchange of *o*- and *p*-iodonitrobenzene in acetonitrile the *ortho*-compound is reported *[117]* to be more reactive, giving an f_p/f_o value of 0.248 at 207°. This, however, may well involve some specific effect of the solvent, while the formation of free iodine is mentioned, indicating the occurrence of side-reactions.

In the absence of special effects such as have been described, it appears that the increase in inductive stabilisation due to an *ortho*-substituent is commonly more than counterbalanced by the smaller conjugative stabilisation from that position .

Heppolette and Miller *[64b]* have compared the reactivities of 1-halogeno-2,6-dinitro-4-X-benzenes and 1-halogeno-4,6-dinitro-2-X-benzenes (not including fluoro compounds). This provides comparison systems in which primary steric effects are enhanced and are thus noticeable even with reagents of low steric requirements. They found that f_p/f_o values are substantially larger than in comparisons made with corresponding mononitro compounds [see above (ref. *35*)], and the magnitude of the increase depends on the size of the halogen—a typical steric pattern. For example, values of $f_p^{NO_2}/f_o^{NO_2}$ at 50° are 38.9, 41.7 and 116 for replacement of chlorine, bromine, and iodine respectively, as compared with values 3.36, 4.05, 3.68 in the corresponding mononitro series at 50° *[35]*. Similar data for reagents of greater steric requirement would be most valuable.

Miller and his co-workers *[104a,109]* have compared the activation from the *ortho*- and *para*-positions of a series of COX groups and most interesting results emerge. The CO_2^- group has been included although it was referred to also when considering charged groups of Classes 1 and 2.

For 4-substituents in both 1-Cl-2-NO_2-benzene and 1-Cl-2,6-$(NO_2)_2$-benzene the theoretical order H < CO_2^- < $CONH_2(CONR_2)$ < CO_2R < COR is found, and both series follow the Hammett equation, using the usual σ^--values. Further, the difference in ρ-values of the two series is quite small and the degrees of activation are similar in both series (see p. 104). When the COX groups are present as 2-substituents in both 1-Cl-4-NO_2- and 1-Cl-4,6-$(NO_2)_2$-benzenes a special pattern emerges, which is the same in both series except that in the latter a marked additional primary steric effect is imposed. Some data are given in Table 18, which also gives corresponding σ^--values.

In the mononitro series (*a*) the *ortho*-substituent order of reactivity is $CO_2^- < H < COPh < CO_2Me < COMe < CONH_2$. In the dinitro series (*b*) the *ortho*-substituent order is $CO_2^- < COPh < CONMe_2 < H < CO_2Me < CONH_2$ *i.e.* the same order except that all are shifted down relative to hydrogen by a primary steric effect. Even in the mononitro series these results indicate primary steric effects lowering the reactivity of compounds containing a COPh or a CO_2^- group ($CONMe_2$ was not measured in this series), with the former most affected; while a specific accelerative effect for $CONH_2$ is indicated.

In the dinitro series an additional primary steric effect equivalent to a further rate reduction of 10- to 15-fold for the electrically neutral groups, and 85.9 for the negatively charged CO_2^- group is observed. The $CONMe_2$ group is seen to be quite unlike the $CONH_2$ group and exhibits marked steric hindrance. The $CONH_2$ group, though no longer exhibiting an overall *ortho*-acceleration, because of the additional primary steric effect, is still the most activating of the *ortho*-COX groups considered. The suggested steric effects are clearly correlated with size, or for the CO_2^- group, electrostatic repulsion, and this is very much magnified in the dinitro series.

The high reactivity for the $CONH_2$ group has been ascribed *[109]* to stabilising Cl · · · · H–N hydrogen bonding in the transition state. This is confirmed also by the high activating power of the CONHPh group. The small enhancement due to CONHPh compared with that for a $CONH_2$ group is a straightforward polar effect; the conjugation of the nitrogen with the phenyl group lessens the substantial internal conjugation of $\dot{N}$–C=O which causes the amides to be the least activating of the neutral COX groups as *para*-substituents (Table 9).

Results are also available for SO_3^- and SO_2Me groups in the mononitro series *[79b]* and are included in Table 18. The pattern closely resembles that of the COX compounds. The *ortho*-SO_3^- group is deactivating and the *para*-SO_3^- activating with an f_p/f_o value of 10.6.

TABLE 18

REACTION OF OMe^- IN MeOH WITH (*a*) 1-Cl-2-NO_2-4-X-BENZENES, FOR WHICH $\rho = 3.90$ AT 50° *[13]* AND 1-Cl-2-X-4-NO_2-BENZENES *[104a,109]* (ASSUMING ρ IS ALSO 3.90); (*b*) 1-Cl-2,6-$(NO_2)_2$-4-X-BENZENES, FOR WHICH $\rho = 3.80$ AT 0° *[13]* AND 1-Cl-2-X-4,6-$(NO_2)_2$-BENZENES *[104a,109]* (ASSUMING ρ IS ALSO 3.80)

Series	Substituent	Rate constant k_2 ($l \cdot mole^{-1} \cdot sec^{-1}$)	Substituent rate factor (S.R.F. or f)	Substituent constant (σ^-)	f_p/f_o [a]	(f_p/f_o) Series (*b*) / (f_p/f_o) Series (*a*)
(*a*)	2-H	$8.47 \cdot 10^{-6}$	1	0	1	—
at 50°	4-H	$2.52 \cdot 10^{-6}$	1	0		
	$2\text{-}CO_2^{-}$ [b]	$3.16 \cdot 10^{-6}$	0.373	−0.110	19.1	—
	$4\text{-}CO_2^{-}$ [b]	$1.79 \cdot 10^{-5}$	7.12	0.218		
	2-$CONH_2$	$3.90 \cdot 10^{-3}$	462	0.683	0.567	—
	4-$CONH_2$	$6.58 \cdot 10^{-4}$	262	0.627		
	2-CONHPh	$1.44 \cdot 10^{-2}$	1700[e]	0.828	—	—
	2-CO_2Me	$1.47 \cdot 10^{-3}$	174	0.574	8.97	—
	4-CO_2Me	$3.93 \cdot 10^{-3}$	1560	0.819		
	2-COMe	$2.08 \cdot 10^{-3}$	246	0.613	8.09	—
	4-COMe	$5.00 \cdot 10^{-3}$	1990	0.847		
	2-COPh	$1.65 \cdot 10^{-4}$	21.6	0.342	123	—
	4-COPh	$6.68 \cdot 10^{-3}$	2660	0.874		
	2-CHO[c]	"$2.41 \cdot 10^{-3}$"	"285"	"0.630"	"7.87"	—
	4-CHO[c]	"$5.64 \cdot 10^{-3}$"	"2240"	"0.860"		
	$2\text{-}SO_3^-$	—	0.512[d]	−0.074	10.6	—
	$4\text{-}SO_3^-$	—	5.44[d]	0.189		
	2-SO_2Me	$1.95 \cdot 10^{-2}$	2310	0.862	5.55	—
	4-SO_2Me	$3.22 \cdot 10^{-2}$	12800	1.050		
(*b*)	2-H	$2.00 \cdot 10^{-3}$	1	0	1	1
at 0°	4-H	$4.98 \cdot 10^{-5}$	1	0		
	$2\text{-}CO_2^{-}$ [b]	$1.40 \cdot 10^{-5}$	0.00702	−0.567	1640	85.9
	$4\text{-}CO_2^{-}$ [b]	$5.70 \cdot 10^{-4}$	11.5	0.280		
	2-$CONH_2$	$5.58 \cdot 10^{-2}$	27.9	0.381	8.60	15.2
	4-$CONH_2$	$1.19 \cdot 10^{-2}$	240	0.627		
	2-$CONMe_2$[f]	$3.75 \cdot 10^{-4}$	0.188	−0.190	344	—
	4-$CONMe_2$	$3.22 \cdot 10^{-3}$	64.7	0.477		
	2-CO_2Me	$1.11_5 \cdot 10^{-2}$	5.57	0.196	131	14.6
	4-CO_2Me	$3.61 \cdot 10^{-2}$	728	0.753		
	2-COPh	$1.22 \cdot 10^{-3}$	0.613	−0.0559	1340	10.9
	4-COPh	$4.07 \cdot 10^{-2}$	821	0.768		

[a] *i.e.* $f_{4subst.}/f_{2subst.}$. [b] At approx. the same ionic strength. Both values are lower at zero ionic strength. [c] Reduced values owing to concurrent reversible hemiacetal formation. Since this is not equal in extent for *o*- and *p*-CHO, the value of the positional f values is not the true value. [d] Estimated values for zero ionic strength for measurements with OH^- in H_2O. [e] S.R.F. or f ratio, CONHPh/$CONH_2$ = 3.67 (due to reduction in internal conjugation by competition). [f] Estimated from measurements in dioxane/MeOH.

Like the NO_2 and COMe group, the *ortho*-SO_2Me group is less activating than the *para*-SO_2Me group, and the f_p/f_o value is 5.55 (between that for the NO_2 and COMe group). As with the NO_2 group the relative *ortho–para* reactivity for SO_2Me is reversed with the neutral reagent piperidine, with an f_p/f_o value of 0.0635 *[118]*, though these workers report a reversal for the SO_2Me group even with ethoxide ion. Kloosterziel and Backer *[119]* have shown that the activating power of an SO_2Me group is not affected by a methyl group *ortho* to it (*i.e.* there is no secondary steric effect), no doubt because sulphur uses d-orbitals in multiple bonding, and conjugation is not seriously affected by any rotation about the aryl–S bond *[cf. 64,120]*.

The σ-values of 0.532 and 0.617, at 0°, may be estimated for *o*- and *p*-$\overset{+}{N}Me_2\overset{-}{O}$ groups in different alkaline methanolysis reactions *[51b]*. The lower value for the *ortho*-group is of interest, for with *o*- and *p*-$\overset{+}{N}Me_3$ groups, also unable to exert a conjugative effect, the value of σ is greater for the *ortho*-group.

(*c*) Ortho-*substituent effects in Class 4*, viz. *Substituents attached by an electrically neutral atom bearing neither unshared electrons nor forming part of a conjugated side chain*

Miller and his co-workers *[78b]* have made some measurements in the same series as (*a*) and (*b*) of Table 18, with compounds containing methyl and trifluoromethyl groups, and these are given in Table 19.

These results show that with methoxide ion as reagent, the methyl group has an almost negligible steric effect. The trifluoromethyl group has an f_p/f_o value of 4.72. This is small, but nevertheless indicates some factor outweighing preferred inductive activation from the *ortho*-position. It seems reasonable to suggest an electron-withdrawing hyperconjugation as the main cause, and such an effect has been suggested for other reactions *[86,121,122]*. A possible contributory cause is repulsion between the electronegative fluorine atoms of the *ortho*-trifluoromethyl group and the reagent.

There is evidence of steric deactivation in the case of the tert-butyl group, with ample supporting evidence from work in other fields *[e.g. 123,124]*. Fierens and Halleux *[81]* made measurements of the substituent effect of *ortho*-Me, -Et, -iso-Pr and -tert-Bu, and compared results from *ortho*-Me and -tert-Bu with *para*-Me and -tert-Bu by

TABLE 19

REACTION OF OMe^- IN MeOH WITH (*a*) 1-Cl-2-NO_2-4-X-BENZENES FOR WHICH $\rho = 3.51$ AT 100° *[13]*, AND 1-Cl-2-X-4-NO_2-BENZENES (ASSUMING $\rho = 3.51$); AND (*b*) 1-Cl-2,6-$(NO_2)_2$-4-X-BENZENES FOR WHICH $\rho = 3.35$ AT 50° *[64b]* AND 1-Cl-2-X-4,6-$(NO_2)_2$-BENZENES (ASSUMING $\rho = 3.35$) *[104a,109]*

Series	Substituent	Rate constant k_2 ($l \cdot mole^{-1} \cdot sec^{-1}$)	Substituent rate factor (S.R.F. or f)	Substituent constant (σ^-)	f_p/f_o[a]	(f_p/f_o) Series (*b*) / (f_p/f_o) Series (*a*)
(*a*)	2-H	$1.28 \cdot 10^{-3}$	1	0	1	—
at 100°	4-H	$3.50 \cdot 10^{-4}$	1	0		
	2-CH_3	$3.35 \cdot 10^{-4}$	0.262	−0.166	1	—
	4-CH_3	$9.18 \cdot 10^{-5}$	0.262	−0.166		
	2-CF_3	$9.69 \cdot 10^{-2}$	75.7	0.535	4.72	—
	4-CF_3	$1.25 \cdot 10^{-1}$	357	0.727		
(*b*)	2-H	$2.88 \cdot 10^{-1}$	1	0	1	1
at 50°	4-H	$7.40 \cdot 10^{-3}$	1	0		
	2-CH_3	$2.81 \cdot 10^{-2}$	0.0976	−0.301	1.86	1.86
	4-CH_3	$1.34 \cdot 10^{-3}$	0.181	−0.221		

[a] *i.e.* $f_{4subst.}/f_{2subst.}$.

using the reactions of iodide ion in acetone with 1-bromo-2,6-dinitro-4-X- and 1-bromo-4,6-dinitro-2-X-benzenes. Results are given in Table 20. The ρ-values are assumed to be the same in both dinitro series.

The value of f_p/f_o for the methyl group is slightly less than unity and the origin of this is uncertain. It might involve some differential solvation effect favouring the *ortho*-substituent transition state, or a polarisability factor such as has been suggested by Bunnett *[125,126]*. In reactions with methoxide ion in methanol of corresponding chloro compounds (Table 19), the value of the f-ratio, though also close to unity, slightly exceeds it.

With bulky reagents steric effects are more noticeable especially when both *ortho*-positions are occupied. This is clearly shown by the results given in Table 18 and is further supported by results of Brieux and Deulofeu *[127]*. They reported data leading to an f_p^{Me}/f_o^{Me} value of 27.4, when compared as a 4- or 6-substituent in 1-chloro-2-nitrobenzenes in reactions with piperidine in benzene at 100°.

TABLE 20

REACTION OF I^- IN Me_2CO WITH 1-Br-2,6-$(NO_2)_2$-4-X- AND 1-Br-2-X-4,6-$(NO_2)_2$-BENZENES AT 80° *[78]*

Substituent	Rate constant k_2 ($l \cdot mole^{-1} \cdot sec^{-1}$)	Substituent rate factor (S.R.F. or f)	f_p/f_o[a]
2-H	$2.8 \cdot 10^{-3}$	1	1
4-H	$2.4 \cdot 10^{-4}$	1	
2-Me	$6.2 \cdot 10^{-4}$	0.22	
4-Me	$2.9_5 \cdot 10^{-5}$	0.123	0.56
2-Et	$3.6 \cdot 10^{-4}$	0.13	—
2-iso-Pr	$4.3 \cdot 10^{-4}$	0.153	—
2-tert-Bu	$4.8 \cdot 10^{-6}$	0.017	
4-tert-Bu	$3.8 \cdot 10^{-5}$	0.158	93

[a] *i.e.*, $f_{4subst.}/f_{2subst.}$.

(*d*) Ortho-*substituent effects in Class 5*, viz. *Substituents attached by an electrically neutral atom bearing unshared electrons, but not forming part of a conjugated side chain*

In contrast with the difficulty in detailed prediction of their intrinsic behaviour as polar substituents, the gross *comparative* effects of Class 5 substituents acting from *ortho*- and *para*-positions are easier to predict. Irrespective of whether conjugative destabilisation outweighs inductive stabilisation, it is generally accepted that the former operates more effectively from the *para*- than the *ortho*-position, and the latter operates more effectively from the *ortho*- than the *para*-position, though this is less definite and applies more strictly to the pure σ-inductive, and direct or field effect. Both conjugative and inductive effects reinforce each other in differentiating substituent effects in aromatic S_N reactions, and lead to higher reactivity of a Class 5 substituent from the *ortho*-position. The marked changes in relative importance of inductive and conjugative effects in substituents, attached by atoms of Groups V, VI or VII of the Periodic Table, would be expected to lead to an increase in the *ortho–para* difference in the group order: VII < VI < V (*e.g.* F < OMe < NH_2), while

References p. 132

relationships such as F > Cl result from the operation of the stereo-electronic effect *[2f]*. However, where conditions are such as to lead to steric hindrance, as when both positions *ortho* to the reaction centre are occupied, the f_p/f_o values will be increased and tend to obscure the polar factors. Some data are given in Table 21 which also includes σ^- and σ_N^- values.

The results for chlorine, methoxyl and amino groups, for example, (Table 21) confirm the conclusions reached above. The similarity of f_p/f_o-values of the methoxyl group, for which data for both OMe^- in MeOH and piperidine in C_6H_6 are available, is noteworthy. The value in series (*c*) is a little higher, *viz.* 0.236 as against 0.111, and this is ascribed to the fact that in series (*c*) the *ortho*-substituent effect is seen in compounds where *both ortho*-positions are occupied and thus the primary steric effect is enhanced (see above). The f_p/f_o value of the amino group is nearer to unity than that of the methoxyl group, but this is almost certainly a consequence of the inaccuracy of measurements with the 4-amino compound and cannot be regarded as significant. It is also clear that steric effects are not large for any of these three groups with these reagents.

Bunnett and Zahler *[39]* analysed some results of Sandin and Liskear *[128]* and showed that one *ortho*-iodine increases rates of replacement of an iodine atom but a second one decreases it, whereas for chlorine and bromine the first one increases the rate and the second has little effect. The relatively minor role of steric effects is thus again confirmed. It generally requires both the *ortho*-substituent and reagent to be bulky in order to obtain a substantial steric effect.

(*e*) Ortho-*substituent effects in Class 6,* viz. *Substituents containing carbon–carbon multiple bonds conjugated with the ring*

Evidence has been given to support the classification of –CH=CH–X groups as weakly activating by an electron-withdrawing conjugative effect (–*E*). Such a mode of action will markedly differentiate *para*- and *ortho*-substituents, favouring the former (*cf.* the diazonium ion, p. 98). There is very little information at hand, but in accord with this view, Heppolette and Miller *[64b]* have obtained an f_p/f_o value of 8.4 at 100° for the $-CH{=}CH{-}CO_2^-$ group in halogeno-

TABLE 21

REACTIONS OF (*a*) 1-Cl-2-NO_2-4-X-BENZENES FOR WHICH $\rho = 3.51$ AT 100° *[13]* AND 1-Cl-2-X-4-NO_2-BENZENES WITH OMe^- IN MeOH *[104a,109]* (ASSUMING $\rho = 3.51$); (*b*) 1-Cl-2,6-$(NO_2)_2$-4-X-BENZENES FOR WHICH ρ-VALUES ARE 3.80; 3.35 AND 3.03 AT 0°, 50° AND 100° RESPECTIVELY, AND 1-Cl-2-X-4,6-$(NO_2)_2$-BENZENES WITH OMe^- IN MeOH *[50c,104a]* (ASSUMING THE SAME ρ-VALUES, 3.80, 3.35, 3.03); (*c*) 1-Cl-2-NO_2-4-X- AND 1-Cl-2-X-4-NO_2-BENZENES WITH PIPERIDINE IN BENZENE *[127]*; (*d*) 1-Cl-2 AND -4-X-6-NO_2-BENZENES WITH PIPERIDINE IN BENZENE *[127]*

Series	Substituent	Rate constant k_2 ($l \cdot mole^{-1} \cdot sec^{-1}$)	Substituent rate factor (S.R.F. or f)	Substituent constant σ^- or σ_N^-	f_p/f_o [a]	(f_p/f_o) Series (*b*) / (f_p/f_o) Series (*a*)
(*a*)	2-H	$1.28 \cdot 10^{-3}$	1	0	1	—
at 100°	4-H	$3.50 \cdot 10^{-4}$	1	0		
	2-Cl	$1.00_5 \cdot 10^{-2}$	7.84	0.255	1.43	—
	4-Cl	$3.92 \cdot 10^{-3}$	11.2	0.299		
(*b*)	2-H[d]	$2.00 \cdot 10^{-3}$	1	0	1	1
	4-H[e]	$4.98 \cdot 10^{-5}$	1	0		
at 0°	2-Cl	$7.44 \cdot 10^{-3}$	3.72	0.150	2.28	1.59[c]
	4-Cl	$4.23 \cdot 10^{-4}$	8.49	0.244		
at 50°	2-OMe	$4.67 \cdot 10^{-2}$	0.159	−0.238	0.126	—
	4-OMe	$1.48 \cdot 10^{-4}$	0.0200	−0.506		
at 100°	2-NH_2	$5.89 \cdot 10^{-2}$	0.00538	−0.745	0.396	—
	4-NH_2[b]	$6 \cdot 10^{-4}$	0.002	-0.8_9		
(*c*)	2-H	$1.68 \cdot 10^{-6}$	1	0	1	—
at 100°	4-H	$7.79 \cdot 10^{-5}$	1	0		
	2-OMe	$7.0 \cdot 10^{-7}$	0.416	−0.126	0.111	—
	4-OMe	$3.59 \cdot 10^{-6}$	0.0461	−0.440		
(*d*)	2- and 4-H	$7.79 \cdot 10^{-5}$	1	—	1	—
at 100°	2-OMe	$1.52 \cdot 10^{-5}$	0.195	—	0.236	—
	4-OMe	$3.59 \cdot 10^{-6}$	0.0461	—		

[a] *i.e.*, $f_{4subst.}/f_{2subst.}$. [b] Inaccurate because of side-reactions. Values of k_2 from initial slopes. [c] Approximate, as there is 100° difference in the temperature used in the reactions of the two series. By use of the Arrhenius parameters, the f_p/f_o value obtained is 2.61 at 0°. [d] Rate constants at 50° and 100° are $2.88 \cdot 10^{-1}$ and $1.09_5 \cdot 10^{-1}$ $l \cdot mole^{-1} \cdot sec^{-1}$. [e] Rate constants at 50° and 100° are $7.40 \cdot 10^{-3}$ and $2.88 \cdot 10^{-1}$ $l \cdot mole^{-1} \cdot sec^{-1}$.

nitrobenzenes in reaction with methoxide ion in methanol. While results for an electrically neutral –CH=CH–X group would be preferable, steric effects are likely to be unimportant and the electrostatic interactions are also expected to be weakened by the distance of the CO_2^- group from the reaction centre.

The f_p/f_o value should be less for the corresponding –C≡C–X compounds in which some inductive stabilisation is to be expected However no data to confirm this are yet available.

(*f*) Ortho-*substituent effects in Class 7*, viz. *Fused and attached aromatic rings as substituents*

The situation here is similar to that found with Class 6 substituents. Berliner and his co-workers *[96]* reported that with piperidine as reagent, 2-halogenonaphthalenes are slightly more reactive than 1-halogenonaphthalenes, and both are slightly more reactive than halogenobenzenes. This suggests that the fused ring attached in the 2,3-position is slightly less activating than when attached in the 3,4-position, and may reflect a very weak steric effect. However Amstutz and his co-workers *[97]* have reported data for S_N reactions of chloronaphthalenes and chloroanthracenes in reactions with piperidine as both reagent and solvent at 200°. From their data it may be shown that a fused benzene ring attached in the 2,3-position has an S.R.F. or *f* value of 4.79 as compared with 4.68 when attached in the 3,4-position, from which the two seem to be approximately equal. They also reported that a fused naphthalene ring attached in the 2,3-position has an S.R.F. or *f* value of 3.39, as compared with 6.46 when attached in the 3,4-position, and the value of 4.79 for the fused benzene ring attached in the 2,3-positions. This supports the existence of a weak steric effect thus favouring the 3,4-ring attachment, but the lower *f* value for the 2,3-attached naphthalene than for the benzene ring is surprising, if it is not ascribable to experimental error. A significant result from measurements with 9-chloroanthracene is that *two* fused benzene rings have a marked effect on rate, with an *f* value of 132. Amstutz and his co-workers suggest that it is only in this system that a real conjugative effect is exerted and point to the marked reduction of $\Delta E^{\ddagger}$ in this but not in the other systems. They regard the effect of single rings as largely a weak inductive attraction.

Bamford and Broadbank *[129]* quoted data for the reaction of 1,4-dinitronaphthalene with methoxide ion in methanol. By comparison with the results of Bolto and Miller *[130]* for the reaction with *p*-dinitrobenzene, the *f*-value for the 2,3-fused benzene ring at 50° is 9.3, comparable with that observed in reactions with piperidine.

A much more noticeable steric effect is to be expected if one compares an attached phenyl group as *ortho*- or *para*-substituent. Miller and Williams *[79b]* compared the rates of displacement of chlorine in 1-chloro-4,6-dinitro- 2-phenyl- and 1-chloro-2,6-dinitro-4-phenylbenzene by methoxide ion in methanol at 0°, and obtained an f_p/f_o value of 18.3, thus confirming the suggested steric effect. Whereas the *para*-phenyl group is slightly activating ($f = 1.28$) the *ortho*-phenyl group is definitely deactivating ($f = 0.0788$).

Bunnett *[125,126]* has produced evidence for a rather subtle effect applicable especially to *ortho*-substituents in aromatic substitution, *viz.* that after allowance for normal steric and polar effects, a change at or near the reaction centre to a substituent of higher polarisability causes more acceleration or less deceleration the greater the effective polarisability of the nucleophilic reagent. He identifies the operation of this polarisability effect with Van der Waals (London or dispersion) forces which lower the transition state energy and thus lead to an increase in rate. Evidence, which includes the increasing rate ratio, thiophenoxide ion/methoxide ion, with size of halogen in the reactions of 1-halogeno-2,4-dinitrobenzenes and α-substituted methyl halides, halide exchange reactions of alkyl halides, and in similar changes of iodide ion/hydroxide ion rate ratios in reaction with benzyl halides, is shown in part in Tables 22 and 23. Some results of Miller and Wong *[131]* for reactions with thiomethoxide ion as reagent, which are discussed below, are also included in Table 22.

In commenting on reactions of 2-substituted 1-fluoro-4-nitrobenzenes with five reagents, Bunnett adduced further evidence for such an effect (Table 24).

Sisti and Lowell *[132]* have made similar investigations using reactions of methoxide, thiophenoxide and iodide ions with some *ortho*- and *para*-substituted benzyl halides, which support Bunnett's views only in part. Their conclusions, shown in Table 25 after making

TABLE 22

SOME REACTIONS OF 1-HALOGENO-2,4-DINITROBENZENES IN METHANOL

Halogen	k_2 (0°) ($l \cdot mole^{-1} \cdot sec^{-1}$) for reaction with NaOMe	Piperidine	NaSPh	NaSMe	k_2 (0°) ratios $\frac{SPh^-}{OMe^-}$	$\frac{\text{Piperidine}}{OMe^-}$	$\frac{SMe^-}{OMe^-}$
F	1.76[a]	1.5	$1.03_5 \cdot 10^2$	$1.51_5 \cdot 10^2$	59	0.85	85.1
Cl	$2.00 \cdot 10^{-3}$	$1.95 \cdot 10^{-3}$	3.89	—	1950	0.98	—
Br	$2.38 \cdot 10^{-3}$	$1.97 \cdot 10^{-3}$	6.68	—	4840	1.43	—
I	$3.08 \cdot 10^{-4}$	$4.57 \cdot 10^{-4}$	5.17	$4.07 \cdot 10^{-4}$	16800	1.48	1.32

[a] Miller and Wong's value *[131]* is 1.74.

necessary adjustments to rates, are that the expected trends are observed when considering only a given substituent, but not with a given nucleophile *i.e.* nucleophiles of relatively high polarisability did not give exalted values for the most polarisable of the substituents considered. They consider that field effects (*D*) may cause the anomaly.

The thiophenoxide ion is a key reagent in these comparisons, and if polarisability is the main factor involved, the thiomethoxide ion

TABLE 23

SOME REACTIONS OF α-SUBSTITUTED METHYL BROMIDES AND IODIDES

Halogen	α-Substituent	$10^5 k_2$ (50°) ($l \cdot mole^{-1} \cdot sec^{-1}$) for reaction with NaI/$Me_2CO$	NaOMe/MeOH	NaSPh/MeOH	k_2 ratios PhS^-/OMe^-	I^-/OMe^-
Br	Me	1700	47.2	4120	87	36
Br	F	1350	225	3800	17	6.0
Br	Cl	218	2.36	378	288	92
Br	Br	69	0.370	277	749	186
I	Cl	—	4.41	3770	855	—
I	Br	—	1.02	2100	2060	—
I	I	—	0.536	2800	5230	—

TABLE 24

RATE RATIOS k_2Y/k_2OH^- AT 0° AND ADJUSTED RATE RATIOS $\frac{(k_2Y/k_2OH^-)_{2-R}}{(k_2Y/k_2OH^-)_{2-H}}$ SHOWN IN PARENTHESES FOR SOME 2-SUBSTITUTED 1-FLUORO-4-NITROBENZENES

2-Substituent[a]	Reagent			
	OMe^-	NH_3	Piperidine	SPh^-
H	36 (1)	$6.5 \cdot 10^{-4}$ (1)	26 (1)	48 (1)
Me	23 (0.65)	$8.6 \cdot 10^{-4}$ (1.5)	2.2 (0.084)	208 (4.3)
Br	47 (1.3)	$11 \cdot 10^{-4}$ (1.7)	6.0 (0.23)	298 (6.2)
NO_2	18 (0.50)		16 (0.62)	1090 (23)

[a] The 2-Me group is deactivating and the 2-Br group activating except that with SPh^- the 2-Me group is slightly activating.

should show similar behaviour. However, Miller and Wong *[131]* have compared the reactivity of thiomethoxide and methoxide ions in reactions with 1-halogeno-2,4-dinitrobenzenes in methanol (included in Table 22), and in contrast with Bunnett's results for these compounds with thiophenoxide ion shown in the same Table, found a *decrease* in the RS^-/RO^- rate ratio with the size of halogen displaced, *i.e.* a normal small steric effect. The overall pattern of results quoted

TABLE 25

RATE RATIOS AND ADJUSTED RATE RATIOS FOR SOME SIDE-CHAIN SUBSTITUTION REACTIONS OF *o*- AND *p*-SUBSTITUTED BENZYL HALIDES AT 20°

Reagent in solvent	Rate ratios						Adjusted rate ratios $\frac{(o\text{-R}/p\text{-R})_{reagent}}{(o\text{-R}/p\text{-R})_{OMe^-}}$		
	$\frac{o\text{-Br}}{o\text{-H}}$	$\frac{p\text{-Br}}{p\text{-H}}$	$\frac{o\text{-Me}}{o\text{-H}}$	$\frac{p\text{-Me}}{p\text{-H}}$	$\frac{o\text{-Br}}{p\text{-Br}}$	$\frac{o\text{-Me}}{p\text{-Me}}$	*o*-H	*o*-Me	*o*-Br
OMe^- in MeOH	1.05	1.41	2.87	1.32	0.74	2.17	1.00	1.00	1.00
SPh^- in MeOH	2.84	2.38	5.26	1.50	1.19	3.50	1.00	1.60	1.61
I^- in Me_2CO	4.35	2.58	10.00	1.49	1.68	6.70	1.00	3.11	2.30

by Bunnett *[125,126]* and by Sisti and Lowell *[132]* is clearly significant, but additional results and further study are required and the interpretation of the results may have to be modified.

In the above discussion of *ortho*-substituent effects, there are numerous examples of the substituent effect being less at the *ortho*- than at the *para*-position. In quite a number of cases a kinetic steric retardation is evident, as a consequence of the increase in energy of the transition-state relative to that of the initial state by the onset or increase of non-bonding compression in forming the transition-state.

Examples were also given in which the substituent effect is greater at the *ortho*- than at the *para*-position, but in none of these was the difference ascribed to a kinetic steric acceleration, as a consequence of a decrease in energy of the transition-state relative to that of the initial state by the elimination or decrease of non-bonding compression in forming the transition-state. The likelihood of such an effect in aromatic S_N reactions was however discussed by Miller and Williams *[79a]*. Parker and Read *[133]* have compared the reactivities of five pairs of 1-X-2,4- and 1-X-2,6-dinitrobenzenes in reaction with aniline in ethanol, and found the reactivity order 1-X-2,4- > 1-X-2,6-dinitro compound for X = F, Cl, Br and I, but the reverse order for X = NO_2. In both series the group mobility order is NO_2 > F > Cl, Br, I. They explained the anomalously high reactivity of 1,2,3-trinitrobenzene on the basis of a kinetic steric acceleration, as described.

7. *Meta*-Substituent Effects

Transition states such as have been described earlier (p. 9) can be influenced conjugatively by a *meta*-substituent only by relay from the *meta*-ring atom to which it is attached, to the neighbouring *ortho*- and *para*-positions. Thus true conjugative stabilisation or destabilisation will be absent. However moderate inductive stabilisation or destabilisation will be observed, including the above inducto-electromeric effect *[3]* (as in Fig. 38).

Two positional orders of the operation of inductive effects within the benzene ring have been suggested and there is still controversy. The main arguments have utilised as data *meta*-directing effects in aromatic S_E reactions and effects on acidities of phenols and anilinium

Fig. 38. Stabilisation (A) and destabilisation (B) of nucleophilic substitution by a *meta*-substituent.

ions. One view, expressed by Roberts and his co-workers *[134]*, is that σ-bond relay is in control, leading to the order: *o*- > *m*- > *p*-position. Another view *[135–137]* is that both σ- and π-bond relay are involved leading to the order *o*- > *p*- > *m*-position. All are agreed that the direct effect of a substituent (*D* effect) operating through space is important. It is significant that the important evidence for the former view, *e.g.* the greater acidity of *meta*- than of *para*-trimethylammoniophenols and -anilinium ions, is observed where a *meta*-substituent has a formal charge, and thus the direct or *D* effects are large, and in reactions with fairly small ρ-values in which substituent effects are therefore small.

Liveris and Miller *[108,138a]* in a discussion of reactivity and transmission of the effect of the nitrogen in the pyridinium system, pointed out that the positively charged nitrogen provided a parallel for the electron deficient carbon atom ($C^{\delta+}$) resulting from the attachment of powerful electron-withdrawing substituents such as $-NMe_3^+$. They showed that the reactivity order is nevertheless *o*- > *p*- ≫ *m*-position.

Because of the low intrinsic susceptibility of the benzenoid system to undergo a substitution reaction with nucleophiles, such reactions are generally very slow in simple *m*-C_6H_4XY compounds; and if reaction is observable at all it may involve a mechanism other than the simple activated S_N2 mechanism and may also require the use of catalysts. Reaction may be complicated further by the displacement of both X and Y at comparable rates. Investigators concerned with a range of substituents have therefore generally chosen systems with one or more standard activating groups present *ortho* and/or *para* to the displaced group. The main complications observed then are the occurrence of cross-conjugation and secondary steric effects between

References p. 132

the *meta*-substituent and other substituents present. These may obscure the differences in polar effects from *meta*- and *para*-positions, and are considered as they are seen to occur in the various classes.

In the following discussion the ratio of *para*- and *meta*-substituent rate factors (S.R.F. or f) values is used*. In general an f_p/f_m value differing markedly from unity indicates that the substituent acts mainly by conjugative interactions, whereas a value close to unity indicates the absence of such interactions; but secondary steric effects and cross-conjugation may also lead to large values.

(*a*) Meta-*substituent effects in Classes 1 and 2,* viz. *Substituents attached by charged atoms*

There are experimental difficulties in assessing the activating effect of a positively charged group in the *meta*-position because of the high reactivity of the substrate by alternative paths in which the group is itself displaced, or in which a substituent on the group is displaced *[138b]*. This is a consequence of the much higher mobility of positively charged groups than other substituents. A standard activating substituent *ortho* or *para* to the group to be displaced is also *ortho* or *para* to the *meta* X^+ group and thus activates its displacement too, and the alternative paths are thus still open. This argument is illustrated by Fig. 39. However σ-values of 0.755 and 0.924 at 0° may be estimated for *m*- and *p*-$\overset{+}{N}Me_3$ groups, in different alkaline methanolysis reactions *[51b]*. This clearly indicates the order of inductive effect, *para*- > *meta*-substituent.

Other data available to gauge the positional effects of positively charged groups are the measurements by Liveris and Miller *[108,137a]* of the rates of displacement of chlorine in N-methyl-2-, -3- and -4-chloropyridinium ions, in which ring -N^+ is *ortho*, *meta* or *para* to the leaving group (chlorine).

Their results demonstrate the reactivity order *o*- > *p*- ≫ *m*-substituent, with an f_p/f_m value of $1.62 \cdot 10^6$ at 50°. They pointed out that

* Miller and his co-workers *[71]* have called this the Conjugative Index or C.I., though it might without implications have been called the Meta Index.

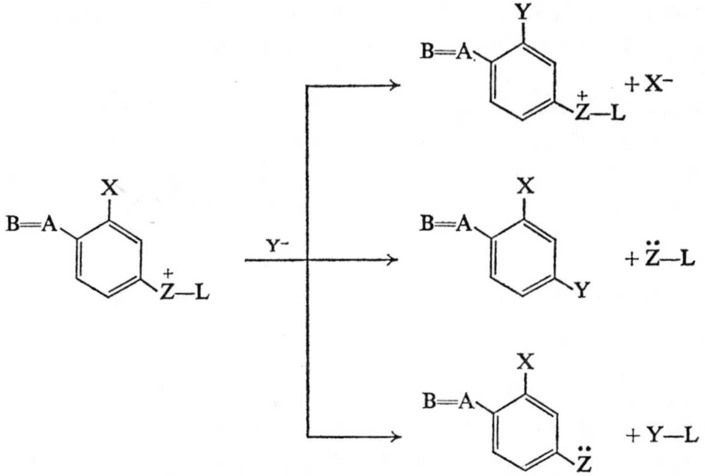

Fig. 39. Illustration of competitive S_N reactions of aromatic substrates, with or without an activating group (– A=B).

activation from the *meta*-position is also large, and recorded the σ^- values 1.584, 2.317 and 2.492 for ring $-N^+$ acting from the *meta-*, *para-* or *ortho*-position respectively. Clearly the influence of an atom more electronegative than the ring-carbon involves the π-electron system, and there seems to be no difference in principle as regards resultant perturbation of the ring as between the changes from aromatic carbon not bearing a substituent to $C^{\delta+}\cdots(\cdots NR_3{}^+)$ and to $N^+\cdots R$.

Clark and Hall *[61]* have considered a negatively charged group ($-O^-$) in comparing the reactivity of chlorine in *m-* with that in *o-* and *p*-chlorophenol in reaction with methoxide ion in methanol. In these conditions the hydroxyl group is largely ionised. After 50–60 hours at 155°C, 12% of the chlorine in the *meta*-compound was found to be replaced, but *no* replacement of chlorine had occurred in either the *ortho-* or *para*-compound. This indicates a large value of f_p/f_m for the oxido group ($-O^-$).

Miller and his co-workers *[71]* have also measured the value for the *m*-oxido group in 5-chloro-2,4-dinitrophenoxide ion. The value is affected by cross-conjugation of the negatively charged oxygen with the nitro group *ortho* and *para* to it, together with a secondary steric effect involving its interaction with the nitro group

References p. 132

ortho to it. (Similar complications with other substituents affected the Hammett plot.) The reported f value for *m*-oxido, $9.51 \cdot 10^{-6}$, is therefore regarded as higher than an unperturbed value. The corresponding σ-value, though affected by cross-conjugation and steric effects is -1.50, as compared with the usual σ-value for *m*-oxido of -0.708, based on acidity measurements, suggesting substantial inductoelectromeric and probably electrostatic effects.

(*b*) Meta-*substituents of Class 3,* viz. *Substituents attached by the positive end of a dipole*

As compared with their *ortho*- and *para*-substituent effects, *meta*-activation by these groups has been much less investigated. Several groups of workers have investigated the nitro group, though there are in some systems complications of the kind mentioned above for Class 1 substituents, and reduction constitutes another competitive reaction. Whereas *o*- and *p*-chloronitrobenzenes react with methoxide ion in methanol to form the corresponding nitroanisoles, *m*-chloronitrobenzene is reduced to the dichloroazoxy compound *[139]*. The nitroso group is even more easily reduced, and Miller and Parker *[66]* reported that even *o*- and *p*-chloronitrosobenzenes undergo reduction instead of substitution in reaction with methoxide ion in methanol.

Bevan and Bye *[51]*, and Miller *[13]* compared the reactivity towards methoxide ion in methanol of fluorobenzene with those of *m*- and *p*-fluoronitrobenzene and obtained concordant results. Values are $f_p^{NO_2}$, $4.73 \cdot 10^8$; $f_m^{NO_2}$, $4.70 \cdot 10^4$; $f_p^{NO_2}/f_m^{NO_2}$, $1.00_5 \cdot 10^4$ at 49.55° *[51]*; $f_p^{NO_2}$, $1.33 \cdot 10^9$; $f_m^{NO_2}$, $7.22 \cdot 10^4$; $f_p^{NO_2}/f_m^{NO_2}$, $1.84 \cdot 10^4$ at 50° *[13]*. The absolute values of f are in slight doubt, because for both workers the results for the parent compound (fluorobenzene) had to be extrapolated over nearly 150° by using the $\Delta E^{\ddagger}$ value, and some inaccuracy inevitably results. Bevan and Bye also measured rates for 3,5 dinitrofluorobenzene and showed that the second nitro group has almost as much effect as the first, whereas they, and Miller *[13]*, showed that the 2,4-dinitro grouping is less effective than would result from additive activation by the *o*- and *p*-nitro group. This seems a distinguishing difference between attached inductive and conjugative activating groups.

The large value of $f_p^{NO_2}/f_m^{NO_2}$, *ca.* 10^4, clearly shows that there is substantial conjugative stabilisation by the *para*-nitro compared with the *meta*-nitro group, which however activates considerably (*cf.* ring-N^+ discussed above).

Brieux and his co-workers *[70]* have compared a number of Class 3 substituents, *viz.*, carboxyl (presumably as carboxylate in the conditions used), ethoxycarbonyl, cyano- and azophenyl, as 4- and 5-substituents in 1-chloro-2-nitrobenzenes. Their results are shown in Table 26 which includes also some results of Miller and his co-workers *[71]* and from which σ-values have been computed.

The same pattern as with the nitro group is seen, with much reduced but still substantial activation from the *meta*-position, and the f_p/f_m values are high. For the carboxylate group the ratio is less, as expected, despite some magnification arising from weak deactivation from the *meta*-position. This may be a field effect. The similar but much larger difference with the sulphonate group probably arises because there is also then a secondary steric effect forcing the 4-nitro group out the plane of the ring and reducing its activating power. This effect

TABLE 26

REACTIONS OF (*a*) 1-CHLORO-2-NITRO- AND 1-CHLORO-4-NITRO-5-X-BENZENES WITH PIPERIDINE IN C_6H_6 AT 45° ($\rho = 4.08$) *[70]*; (*b*) 1-CHLORO-2,6-$(NO_2)_2$-4-X- AND 1-CHLORO-2,4-$(NO_2)_2$-5-X-BENZENES WITH OMe^- IN MeOH AT 50° ($\rho = 3.36$) *[71]*

Series	Substituent	Rate constant k_2 ($l \cdot mole^{-1} \cdot sec^{-1}$)	Substituent rate factor (S.R.F. or f) (45°)	Substituent constant (σ or σ^-)	f_p/f_m [a]
(*a*)	4- and 5-H	$3.63 \cdot 10^{-6}$	1	0	—
at 45°	4-CO_2^-	$5.20 \cdot 10^{-5}$	14.3	0.283	26.5
	5-CO_2^-	$1.96 \cdot 10^{-6}$	0.54	– 0.0656	
	4-CO_2Et	$3.35 \cdot 10^{-3}$	922	0.729	174
	5-CO_2Et	$1.91 \cdot 10^{-5}$	5.3	0.177	
	4-CN	$2.14 \cdot 10^{-2}$	5890	0.925	101
	5-CN	$2.12 \cdot 10^{-4}$	58.5	0.434	
(*b*)	4-SO_3^- [b]	—	8.11	0.270	279
at 50°	5-SO_3^- [b]	—	0.0291	– 0.457	

[a] *i.e.* $f_{4-subst.}/f_{5-subst.}$. [b] Data estimated from reaction with OH^- in H_2O.

possibly operates more weakly also with the carboxylate group. Illuminati and his co-workers *[140]* have measured the reactivity of the cyano compounds, *inter alia*, in 2-Cl-4-R- and 4-Cl-2-R-quinolines. In this system the usual secondary steric effect consequent on interactions between substituents is largely absent but cross-conjugative effects are present. Steric effects still seem to be present however, leading to differences in the effects of the second ring in the two systems. In one there will be a primary steric effect of the second ring on the replacement of the 4-chlorine atom. In the other a secondary steric effect occurs between the 4-R group and the second ring. The two systems, though similar, are therefore not equivalent and this shows in the substantially differing ρ-values. Results for all the substituents measured by these workers *[140]* are given together in Table 27.

The cyano group is seen to cause substantial activation from the *meta*-position, and the σ-values, 0.650 and 0.620, are fairly comparable with that, *viz.* 0.434, reported by Brieux and his co-workers *[70]*. With this linear electron-withdrawing group composed of small atoms

TABLE 27

REACTIONS OF OMe^- IN MeOH AT 75.2° WITH (*a*) 2-Cl-4-R-QUINOLINES ($\rho = 5.8$) AND (*b*) 4-Cl-2-R-QUINOLINES ($\rho = 4.2$) *[140]*

Series	Substituent	Rate constant $10^4\ k_2$ ($l \cdot mole^{-1} \cdot sec^{-1}$)	Substituent rate factor (S.R.F. or f)	Substituent constant (σ or σ^-)
(*a*)	H	2.22	1	0
	CN	1190	536	0.650
	CH_3	0.877	0.395	−0.096
	OMe	0.486	0.219	−0.157
	OEt	0.478	0.216	−0.159
	SMe	1.94	0.874	−0.014
	Cl	39.4	17.8	0.298
(*b*)	H	2.47	1	0
	CN[a]	9925	4070	0.620
	CH_3	0.776	0.314	−0.086
	OMe	0.143	0.0579	−0.213
	OEt	0.159	0.0644	−0.205
	SMe	1.17	0.473	−0.056
	Cl	747	30.2	0.254

[a] Results affected by imidoester formation *[cf. 104]*.

the usual steric and cross-conjugation effects are not expected, and the σ^--values are similar to those derived from acidities.

The σ-values of 0.559 and 0.617 at 0° may be estimated for *m*- and *p*-$\overset{+}{N}Me_2\overset{-}{O}$ groups in alkaline methanolysis reactions *[51b]*. From another reaction the σ_m value is 0.513. The inductive order $\sigma_m < \sigma_p$ is very similar to that found with *m*- and *p*-$\overset{+}{N}Me_3$ (p. 116), but at a lower level of activating power.

(*c*) Meta-*substituents of Class 4*, viz. *Substituents attached by an electrically neutral atom bearing neither unshared electrons nor forming part of a conjugated side-chain*

As *ortho*- or *para*-substituents, these have been classed as having inductive effects with some indications of weak hyperconjugative effects. Alkyl groups should therefore deactivate less from the *meta*-positions but the difference is likely to be small, and the f_p/f_m value little less than unity. An electron-attracting group such as the trifluoromethyl group should activate more from the *para*- than the *meta*-position but again the difference is not expected to be large, and the f_p/f_m value should be little more than unity. Differences will be accentuated, however, if hyperconjugation effects are present. Some relevant data are given in Table 28.

The slightly greater deactivation by *para*- than by *meta*-alkyl groups is clear for both the methyl and tert-butyl group. The f_p/f_m values are a little less than unity as suggested. The small difference between methyl and tert-butyl groups is evidence for weak hyperconjugation in the former. The results of Illuminati and his co-workers *[140]* also indicate a very weak effect for the *meta*-methyl group. The greater activation of the *para*- than *meta*-trifluoromethyl group is evident and the f_p/f_m value is high enough to suggest there the possibility of a more substantial hyperconjugation.

(*d*) Meta-*substituents of Class 5*, viz. *Substituents attached by an electrically neutral atom bearing unshared electrons, but not forming part of a conjugated side-chain*

At the *para*-position there is for all these substituents, *e.g.* alkoxyl and dialkylamino groups, conflicting inductive stabilisation and

TABLE 28

REACTIONS OF (*a*) 1-Cl-2-NO_2-4- AND -5-X-BENZENES WITH PIPERIDINE IN C_6H_6 AT 45°, FOR WHICH THE ρ-VALUE 4.08 AT 45° IS USED [70]; (*b*) 1-Cl-3- AND 4-X-BENZENES WITH OEt^- IN EtOH AT 150° [141]; (*c*) 1-F-3- AND -4-X-BENZENES WITH OEt^- IN EtOH AT 150° [141]; (*d*) 1-Cl-2,6-$(NO_2)_2$-4-X- AND 1-Cl-2,4-$(NO_2)_2$-5-X-BENZENES WITH OMe^- IN MeOH AT 50°, FOR WHICH THE ρ VALUE 3.35 AT 50° IS USED [71]

Series	Substituent	Rate constant k_2 ($l \cdot mole^{-1} \cdot sec^{-1}$)	Substituent rate factor (S.R.F. or f)	Substituent constant (σ)	f_p/f_m[a]
(*a*)	4- and 5-H	$3.63 \cdot 10^{-6}$	1	0	1
at 45°	4-CH_3	$5.3 \cdot 10^{-7}$	0.15	– 0.195	0.175
	5-CH_3	$3.12 \cdot 10^{-6}$	0.86	– 0.016	
	4-CMe_3	$6.1 \cdot 10^{-7}$	0.17	– 0.189	0.370
	5-CMe_3	$1.67 \cdot 10^{-6}$	0.46	– 0.083	
(*b*)	4-CF_3	$5.83 \cdot 10^{-6}$	—	—	10.0
at 150°	5-CF_3	$5.83 \cdot 10^{-7}$	—	—	
(*c*)	4-CF_3	$4.5 \cdot 10^{-3}$	—	—	41.5
at 150°	5-CF_3	$1.08 \cdot 10^{-4}$	—	—	
(*d*)	4-H	$7.40 \cdot 10^{-3}$	1	0	1
at 50°	5-H	$2.88 \cdot 10^{-1}$	1	0	
	4-CH_3	$1.34 \cdot 10^{-3}$	0.166	– 0.191	0.654
	5-CH_3	$7.30 \cdot 10^{-2}$	0.254	– 0.146	

[a] $f_{4subst.}/f_{5subst.}$.

mesomeric destabilisation with the latter dominant in substituents attached by atoms of Groups V and VI of the Periodic Table. Inductive effects are strong for oxygen, fluorine and chlorine, and moderate for nitrogen, sulphur and bromine. With the inductive order *o*- > *p*- > *m*-position, one would predict for dialkylamino groups the reactivity order *m*- > *p*-, both deactivating; for alkoxy groups *m*- > *p*-, with *para* deactivating and *meta* having little effect; for halogens also *m*- > *p*-, with both activating and with the difference decreasing as follows: F > Cl > Br > I (f_p/f_m value increasing towards unity from F to I), and easily upset by steric effects in systems with several substituents [142]. Some data are given in Table 29 and are consistent with the above qualitative discussion.

TABLE 29

REACTIONS OF (*a*) 1-Cl-2-NO_2-4- AND -5-X-BENZENES WITH PIPERIDINE IN BENZENE AT 45° (ρ-VALUE = 4.08) *[70]*; (*b*) 1-Cl-2,6-$(NO_2)_2$-4-X- AND 1-Cl-2,4-$(NO_2)_2$-5-X-BENZENES WITH OMe^- IN MeOH, AT 0° FOR X = Cl; AT 50° FOR X = OMe; AT 100° FOR X = NH_2 (ρ-VALUES ARE 3.80, 3.35 AND 3.03 RESPECTIVELY) *[71]*

Series	Substituent	Rate constant k_2 ($l \cdot mole^{-1} \cdot sec^{-1}$)	Substituent rate factor (S.R.F. or f)	Substituent constant (σ, σ^-, or σ_N^-)	f_p/f_m [a]
(*a*)	4- and 5-H	$3.63 \cdot 10^{-6}$	1	0	1
at 45°	4-Cl	$2.29 \cdot 10^{-5}$	6.2	0.194	0.192
	5-Cl	$1.17 \cdot 10^{-4}$	32.3	0.370	
	4-Br	$3.49 \cdot 10^{-5}$	9.6	0.241	0.278
	5-Br	$1.25 \cdot 10^{-4}$	34.5	0.377	
	4-I	$4.41 \cdot 10^{-5}$	12.1	0.266	0.485
	5-I	$9.07 \cdot 10^{-5}$	24.9	0.342	
	4-OMe	$9.1 \cdot 10^{-8}$	0.025	−0.393	0.006_0
	5-OMe	$1.52 \cdot 10^{-5}$	4.2	0.153	
	4-OEt	$7.8 \cdot 10^{-8}$	0.022	−0.406	0.006_0
	5-OEt	$1.34 \cdot 10^{-5}$	3.7	0.139	
	4-NH_2	$3 \cdot 10^{-9}$	0.001	−0.735	0.001_2
	5-NH_2	$3.11 \cdot 10^{-6}$	0.86	−0.016	
	4-Ph	$7.58 \cdot 10^{-6}$	2.1	0.079	2.0
	5-Ph	$3.80 \cdot 10^{-6}$	1.05	0.005	
(*b*)	4-H	$4.98 \cdot 10^{-5}$ [b]	1	0	1
at 0°	5-H	$2.00 \cdot 10^{-3}$ [c]	1	0	
	4-Cl	$4.23 \cdot 10^{-4}$	8.49	0.244	1.18
	5-Cl	$2.89 \cdot 10^{-2}$	7.20[d]	0.226	
at 50°	4-OMe	$1.48 \cdot 10^{-4}$	0.0200	−0.507	0.053
	5-OMe	$1.09_5 \cdot 10^{-1}$	0.380	−0.125	
at 100°	4-NH_2[e]	$6 \cdot 10^{-4}$	0.002	-0.8_9	0.09
	5-NH_2	$2.69 \cdot 10^{-1}$	0.0246	−0.053	

[a] $f_{4-subst.}/f_{5-subst.}$. [b] Values at 50° and 100° are $7.40 \cdot 10^{-3}$ and $2.88 \cdot 10^{-1}$. [c] Values at 50° and 100° are $2.88 \cdot 10^{-1}$ and $1.09_5 \cdot 10^{1}$. [d] Half rate ratio because there are two identical chlorine atoms. [e] Inaccurate because of side-reactions. Values of k_2 from initial slopes.

The halogens are all activating in the order *m*- > *p*- as predicted and have f_p/f_m values a little less than unity. The reactivity reversal in series (*b*) is probably a minor secondary steric effect of the chlorine on the coplanarity of the neighbouring nitro group; this has a conspicuous effect where the f_p/f_m value is close to unity.

The rate data and small f_p/f_m value for the methoxyl group are also in accord with predictions. The *meta*-methoxyl group is weakly activating in series (*a*), but slightly deactivating in series (*b*), probably because the same minor secondary steric effect and also cross-conjugation weakens activation by the nitro groups.

The data for the amino group are similar and there is an even smaller value of f_p/f_m as suggested. The *meta*-amino group is only just deactivating in series (*a*). In series (*b*) the same general pattern is seen but the inaccuracy of the results for the *para*-amino compound means that the detailed figures cannot be relied upon in any precise way. Results obtained by Illuminati and his co-workers *[140]* who studied the effects of *meta*-substituents on the methanolysis of 2- and 4-chloroquinoline, agree with those noted above. They report deactivation by the *meta*-thiomethoxyl group, which is surprising in view of the fact that the *para*-thiomethoxyl group is weakly activating *[43]*.

(*e*) Meta-*substituents of Classes 6 and 7*, viz. *Substituents containing carbon–carbon multiple bonds conjugated with the ring; and Fused and attached aromatic rings*

These substituents, *e.g.* –CH=CH–X and –Ar groups, have only weak effects even at the *para*-position and it is uncertain to what extent these are electromeric in origin. Effects of *meta*-substituents should be still weaker and both f_p and f_m values and their ratio should be close to unity.

There is very little experimental evidence, but Brieux and his co-workers *[70]* have compared the effect of 4- and 5-Ph groups and their results, included in Table 29, are in accord with this brief discussion.

8. Substitution of Polyhalogenobenzenes

(*a*) *General*

It has been possible to consider substituent effects so far in relation to a single point of substitution. When considering polyhalogenobenzenes, more particularly perhalogenobenzenes of the general formula C_6Hal_5X, including the parent C_6Hal_6 compounds, substituent effects more nearly parallel those in aromatic S_E reactions in the sense that replacement at alternative points may occur. Substituents may then be classified, for example, as *ortho–para*-directing and *meta*-directing. Associated with substitution at alternative points are experimental complications of isolating and determining the amounts of isomers formed; and this includes cases where the replacement of more than one atom in the *same* ring occurs, as when the entering group is activating.

A difference between substituent effects in electrophilic substitution in benzenes and nucleophilic substitution in perhalogenobenzenes is that, in the latter, hydrogen may be regarded as a substituent. Thus Miller *[40]* predicted that hydrogen in pentachlorobenzene is *ortho–para*-directing with weak deactivation.

Whereas hexafluorobenzene is probably planar, it is likely that other hexahalogenobenzenes exist in a buckled form in order to minimise steric interactions between the halogen atoms *[e.g. 143–147]*.

In 1956, Rocklin *[148]* discussed the reactivity of hexachlorobenzene, mentioning difficulties caused by its low solubility. He showed that it reacts in relatively mild conditions with alkoxide, thioalkoxide and hydrosulphide ions; and with amines of low steric requirements under pressure. Such results are to be expected since the halogens are activating groups, and lead to a substantially higher reactivity of C_6Hal_6 than C_6H_5Hal compounds *[149]*. There is little information on perbromo and periodo compounds. More is known about perchloro compounds but most quantitative work has been carried out with the perfluoro compounds which have become readily available in recent years.

The substantially higher reactivity of perfluoro than perchloro compounds has been determined with first row nucleophiles, and reactive sulphur nucleophiles with which the replaced group mobility order is $F \gg Cl$ *[e.g. 130,131,150]* (see Chapter 5), and it is to this factor alone that the higher reactivity of the perfluoro compounds is due. Ho and Miller *[149]* predicted that total activation at the sixth position by the five chlorines of a pentachlorophenyl group is in fact greater than that of the five fluorines of a pentafluorophenyl group, on the basis that activation by a *para*-chlorine exceeds that of *para*-

fluorine by a margin greater than that by which that of *meta*-fluorine exceeds that of *meta*-chlorine, and that the difference in activating power between *ortho*-chlorine and *ortho*-fluorine is unlikely to upset this; they quoted in support the greater acidity of pentachloro- than of pentafluoro-phenol *[151,152]*. Their prediction has since been confirmed by Miller and Yeung *[153]*.

With much recent interest in perfluoro compounds, the patterns of reactivity in perhalogenobenzenes are now much clearer. In a continuing series of papers *[151,154]* Tatlow and his co-workers have been major contributors to this study, their work being in the main qualitative or semi-quantitative. Their results are summarised in Table 30. Some data on reactivities of perfluoroheterocyclic compounds are also available *[e.g. 155,156]*. Ho and Miller *[149]* have considered quantitatively substituent effects in a series of perfluorophenyl compounds (C_6F_5X), including powerful activating and deactivating substituents, and also compared monochloro- and monofluoro-benzenes with hexachloro- and hexafluoro-benzenes.

They suggested that substituent effects in perhalogeno compounds could be predicted by considering the effect of the change from C_6Hal_6 to C_6Hal_5X as being made up of two components, *viz.*, a change from C_6Hal_6 to C_6Hal_5H and from C_6Hal_5H to C_6Hal_5X. In this way the existing large body of knowledge of substituent effects in aromatic S_N reactions could be utilised and comparisons made more easily with other types of reaction. Their experimental results, shown in Table 31, cover most of the substituents quoted in Table 30, and the agreement with results predicted by use of substituent effects in other S_N reactions and the specified two-stage assessment, supports their concept and suggests that it may be relied upon in all cases.

Based on earlier discussions, the activating power of fluorine is in the positional order *m*- > *o*- > *p*-position. In reaction series with ρ-values about 3–4 the value of f_m^F is about 23 at 25° (σ^- *ca.* 0.35) and of f_p^F about unity at 50° (σ^- *ca.* zero). Data are not available for f_o^F but an intermediate value of about 10 is thought reasonable. The change from C_6F_6 to C_6F_5H thus results in a moderate reduction of reactivity *meta*- to the hydrogen; a moderate but lesser reduction *ortho*- to the hydrogen; and approximately no effect *para* to the hydrogen. In the perfluorobenzene series therefore the hydrogen must be mainly *para*-directing with weak deactivation. Tatlow and his

TABLE 30

SUBSTITUENT EFFECTS IN C_6F_5X COMPOUNDS[a] *[151,154]*

Substituent	Main orientation	Approximate activating power
F	Standard	Standard
H	*para*	Similar to F
Cl, Br, I	*para* (some *ortho*)	Similar to F
NO_2	*para* (anionic reagents)	
	ortho and *para* (neutral reagents)	Powerfully activating
NO	*para* (anionic reagents)	
	ortho and *para* (neutral reagents)	Activating
N_2Ph	*para*	Activating
$\overset{+}{N}_2(O^-)Ph$	*para*	Activating
CF_3[b]	*para*	Activating
C_2F_5	*para*	Activating
SO_3^-	*para*	Similar to F
CO_2^-	*para* (anionic reagents)	
	ortho and *para* (neutral reagents)	Similar to F
SO_2NH^-	*para*	Similar to F
SMe	*para*	Activating
SC_2H_4OH	*para*	Activating
SPh	*para*	Activating
SO_2Ph	*para*	Activating
OMe	all positions[c]	Similar to F
Ph	*para*	Similar to F
NHAc	*para*	Uncertain
NH_2	*meta*	Deactivating
O^-	*meta*	Powerfully deactivating
S^-	*meta*	Powerfully deactivating
$C{\equiv}CR$[d]	*para*	Activating

[a] Varying reagents. In some cases several reagents were used with one compound. [b] *o*- and *p*-$C_6F_4(CF_3)_2$ were shown to be still more reactive. [c] Originally thought to be *para*-directing. [d] Ref. *159*.

co-workers *[154]* have shown that hydrogen is consistently more than 90% *para*-directing with a variety of reagents including methoxide ion, which was used by Ho and Miller. The latter obtained the reactivity ratio C_6F_5H/C_6F_6 as 0.288 in methanol and 0.190 in dioxane/methanol (5:1v/v).

In view of the predominance of *para*-direction the reactivity ratio approximates to the value of the partial rate factor f_p^H which is thus about 0.2.

References p. 132

TABLE 31

COMPARATIVE KINETIC DATA FOR NUCLEOPHILIC SUBSTITUTION OF SOME PERHALOGENOBENZENES *[149]*

Compound	Rate constants k_2 ($l \cdot mole^{-1} \cdot sec^{-1}$)	Main product (known or predicted)	Rate ratios relative to $C_6F_6 = 1$	Rate ratios relative to $C_6F_5H = 1$	Activation parameters $\Delta E^{\ddagger}$ ($kcal \cdot mole^{-1}$)	Activation parameters $\log_{10} B$
	At 160°; 2 moles OH^- per mole of C_6Hal_6 in dioxane/water (9:1 v/v)					
C_6F_6	$1.03 \cdot 10^{-2}$	—	1	—	23.2	9.8
C_6Cl_6	$7.40 \cdot 10^{-5}$	—	0.0072	—	27.3	9.6
	At 50°; equimolar OMe^- and C_6F_5X in dioxane/methanol (5:1 v/v)					
C_6F_6	$2.10 \cdot 10^{-2}$	—	1	5.35	19.7	11.6_5
C_6F_5H	$4.00 \cdot 10^{-3}$	*para* to H	0.190	1	19.0_5	10.5
C_6F_5Br	$6.42 \cdot 10^{-2}$	*para* to Br	3.06	16.1	18.3_5	11.2
C_6F_5I	$3.85 \cdot 10^{-2}$	*para* to I	1.83	9.62	18.6_5	11.2
$C_6F_5CH_3$	$2.95 \cdot 10^{-4}$	*para* to CH_3	0.0141	0.0737	23.5	12.3_5
$C_6F_5OCH_3$	$2.71 \cdot 10^{-4}$	all positions rel. to OCH_3	0.0129	0.0677	23.6	12.4
$C_6F_5NH_2$	$2.15 \cdot 10^{-5}$	*meta* to NH_2	0.00102	0.00538	23.7	11.4
$C_6F_5O^-$	$8.11 \cdot 10^{-8}$	*meta* to O^-	$3.86 \cdot 10^{-6}$	$2.02 \cdot 10^{-5}$	25.6	10.2_5
$C_6F_5CO_2^-$	$1.02 \cdot 10^{-2}$	*para* to CO_2^-	0.485	2.55	20.3_5	11.7_5
	At 50°; equimolar OMe^- and C_6F_5X in methanol					
C_6F_6[a]	$3.01 \cdot 10^{-4}$	—	1	3.46	20.4	$10{\cdot}2_5$
C_6F_5H	$8.66 \cdot 10^{-5}$	*para* to H	0.288	1	20.2	9.6
C_6F_5Br	$1.71 \cdot 10^{-3}$	*para* to Br	5.70	19.9	19.2	10.2
$C_6F_5OCH_3$	$9.92 \cdot 10^{-6}$	all positions rel. to OCH_3	0.0330	0.114	22.1	9.9_5
$C_6F_5CF_3$	$3.70 \cdot 10^{0}$	*para* to CF_3	12300	42800	16.3	11.6
C_6F_5CN	$3.95 \cdot 10^{1}$	*para* to CN	131000	460000	14.8_5	11.6_5

[a] Data from ref. *154* give the values k_2 (50°) $4.66 \cdot 10^{-4}$ ($l \cdot mole \cdot sec^{-1}$); $\Delta E^{\ddagger}$ 26.1 ($kcal \cdot mole^{-1}$); $\log_{10} B$ 14.3; $\Delta S^{\ddagger}$ (50°) 4.8 e.u.; $T\Delta S$ (50°) 1.5 ($kcal \cdot mole^{-1}$); $\Delta G^{\ddagger}$ (50°) 24.6 ($kcal \cdot mole^{-1}$).

The predominance of *para*-direction is weaker in the comparison of C_6Cl_5H and C_6Cl_6 because chlorine activates all positions, whereas fluorine activates the *ortho*- and *meta*- but not *para*-position. In pentachlorobenzene therefore, hydrogen should be mainly *ortho*–*para*-directing and deactivating *[40]*, but some *meta*-product should be formed. These conclusions have still to be confirmed.

(*b*) *Substituent effects of activating groups in perfluorobenzenes*

The majority of activating groups exert their influence most effectively at the *para*-position, the main exceptions being the halogens and cationic groups; and dipolar groups when reacting with neutral nucleophiles. Correspondingly in C_6F_5X compounds the substituent effect of an activating substituent X is generally to reinforce *para*-direction, and this is evident from the results of Tatlow and his co-workers *[154]*. *Para*-direction is however weakened in the groups which are listed above as exceptions.

The kinetic results of Ho and Miller *[149]* show that the magnitude of the substituent effects of these, and other groups considered below, in C_6F_5X compounds is satisfactorily related to numerical values obtained in the common activated aromatic S_N reactions in which the Hammett reaction constants (ρ) have values of about 4 to 5.

For bromine as substituent the rate ratios in reaction with methoxide ion in dioxane/methanol (5:1v/v) at 50° are 3.06 relative to C_6F_6 and 16.1 relative to C_6F_5H. In methanol as solvent, values are 5.70 and 19.9 compared to a ratio of 15.4 at 50° in comparing *o*-chloronitrobenzene and 1-chloro-4-bromo-2-nitrobenzene, a result from a series with a ρ-value of 3.90. Values of rate ratios for the trifluoromethyl and cyano groups are a little high compared with values estimated in series with ρ-values about 4, but this is due to an entropy effect of uncertain origin.

The very powerful activating effect of the cyano group (S.R.F. or $f > 10^5$) has been illustrated recently by the report of Wakefield *[158]* that all six fluorine atoms of hexafluorobenzene are replaced by treatment with sodium cyanide in methanol. The product, 1,4-dicyano-2,3,5,6-tetramethoxybenzene, was also shown to be formed by base-catalysed methanolysis of 1,4-dicyano-2,3,5,6-tetrafluorobenzene, the expected intermediate disubstitution product.

Wiles and Massey *[159]* have treated lithium alkynes (LiC≡CR) with hexafluorobenzene in tetrahydrofuran under nitrogen, and obtained 1,4-dialkynyltetrafluorobenzenes as the major product. This supports the theoretical prediction of the activating power of acetylenic groups (Section 5, p. 9).

The carboxylate group is a very weak activating group in the *para*-position, whereas it is weakly deactivating in the *ortho*-position in reactions with anionic nucleophiles and weakly activating in reactions with neutral nucleophiles. It is deactivating in the *meta*-position but probably has a secondary steric effect in the series studied. In C_6F_5X compounds reactivity is thus expected to fall between that of penta- and hexa-fluorobenzenes but to be predominantly *para*-directing. The latter has been confirmed by Tatlow and his co-workers *[154]* and Ho and Miller *[149]* have confirmed the predicted level of reactivity.

(*c*) *Substituent effects of weak deactivating groups in perfluorobenzenes*

Weak deactivating groups characteristically deactivate more at *para*- than at *meta*-position, with the effect at the *ortho*-position intermediate except where steric effects intervene. The general result is therefore deactivation, with the predominance of *para*-direction weakened or lost.

The methyl group is a weak deactivating group with the positional reactivity order *o*-, *p*- < *m*-position, but the differences are only small. One should therefore observe weak deactivation with substantial retention of *para*-direction. The orientation has been confirmed for the methyl group in *ar*-pentafluorotoluene *[154]*. In reaction with methoxide ion in dioxane/methanol at 50° the rate ratio is 0.0737 relative to C_6F_5H as compared with an f_p^{Me} value of 0.119 in the 1-chloro-2-nitrobenzene-4-X-series *[149]*.

The substituent effect of the methoxyl group is known to vary markedly with position, with the reactivity order *p*- < *o*- < *m*-position. In the latter position there is little effect on reactivity compared with hydrogen. In C_6F_5X compounds therefore it should be weakly deactivating and the marked predominance of *para*-direction lost: all three isomers being produced in comparable amounts. The original report by Tatlow and his co-workers that this group is *para*-directing has been amended by them in a recent paper *[154]*. With methoxide ion as reagent, reported percentages are *para*, 57 %, *meta*, 31 %, *ortho*, 16 %. In reactions with methoxide ion at 50° the rate ratios relative to C_6F_5H are 0.0677 in dioxane/methanol (5:1 v/v) and 0.114 in methanol *[149]*. Comparison results are not available with methoxide ion in the monochloronitro series used above but in 4- and 5-substituted chlorodinitro series in which the ρ-value is about 0.6 less than in the chloromononitro series, the *f*-values at 50° for the methoxyl group are *para*, 0.02; *ortho*, 0.16, *meta*, 0.38 (a higher value of about unity would be expected for the latter in the absence of a secondary steric effect and cross-conjugation). Such values would lead to the change in orientation pattern and the level of reactivities actually found *[145,154]*.

(*d*) *Substituent effects of powerful deactivating groups in perfluorobenzenes*

Powerful deactivating groups (*e.g.* amino and oxido groups) characteristically exert their substituent effects in the order *p*- > *o*- > *m*-position, so that substrate reactivity is in the reverse order. In C_6F_5X compounds where X is such a substituent, the result should be powerful deactivation and a switch from predominant *para*- to predominant *meta*-direction. Tatlow and his co-workers *[154]* have confirmed that the amino and oxido groups are *meta*-directing and deactivating and that the sulphido group is deactivating. Ho and Miller *[149]* have shown that in reaction with methoxide ion in dioxane/methanol at 50° the rate ratio C_6F_5X/C_6F_5H is $5.38 \cdot 10^{-3}$ for the amino group, and $2.02 \cdot 10^{-5}$ for the oxido group. For comparison, values of f_m^X in the 1-chloro-2,4-dinitro-5-X-benzenes are $1.33 \cdot 10^{-2}$ for the amino group and $9.31 \cdot 10^{-6}$ for the oxido group.

Tatlow and his co-workers *[154]* have demonstrated the qualitative reactivity order $C_6F_5NH_2 < C_6F_5NHMe < C_6F_5NMe_2$ with a shift away from *meta*- to *para*-direction. They ascribe these changes to a secondary steric effect, *viz.* of *ortho*-fluorines on the substituted amino group. This results in the inability of the substituted amino group to take up a position in the ring plane. Consequently its deactivating and *meta*-directing power is lessened.

References p. 132

The kinetic data of Ho and Miller *[149]* also show that rates of substitution by methoxide ion are substantially higher in admixtures of dioxane with a protic solvent than in the latter alone. This is an example of a general effect on reagent reactivities, which is discussed in Chapter 6.

REFERENCES

1 J. MILLER, *J. Chem. Soc.*, (1952) 3550.

2 C. K. INGOLD, *Structure and Mechanism in Organic Chemistry*, Cornell Univ. Press, Ithaca, N.Y., 1953, (a) p. 88; (b) p. 105; (c) p. 60, p. 400–403; (d) p. 77, p. 89; (e) p. 266; (f) p. 75, p. 240–241; (g) p. 108; (h) p. 267.

3 M. J. S. DEWAR, *J. Am. Chem. Soc.*, 84 (1962) 3539.

4 E. HÜCKEL, *Z. Phys.*, 60 (1930) 423; 70 (1931) 204; 72 (1931) 310; 76 (1932) 628.

5 G. W. WHELAND, *J. Am. Chem. Soc.*, 64 (1942) 900.

6 R. D. BROWN, *Quart. Rev.*, 6 (1952) 63.

7 Symposium on "Status of Quantum Chemistry in the Interpretation of Organic Chemical Phenomena", *Tetrahedron*, 19 (1963) Suppl. 2, p. 1–477.

8 S. NAGAKURA, *Tetrahedron*, 19 (1963) Suppl. 2, p. 36.

9*a* J. MILLER, *J. Am. Chem. Soc.*, 85 (1963) 1628; *b* D. L. HILL, K. C. HO AND J. MILLER, *J. Chem. Soc.*, *B*, (1966) 299.

10 T. ABE, *Bull. Chem. Soc. Japan*, 37 (1694) 508.

11 G. D. LEAHY, M. LIVERIS, J. MILLER AND A. J. PARKER, *Australian J. Chem.*, 9 (1956) 382.

12 J. R. KNOWLES, R. O. C. NORMAN AND J. H. PROSSER, *Proc. Chem. Soc.*, (1961) 341.

13 J. MILLER, *Australian J. Chem.*, 9 (1956) 61.

14*a* L. P. HAMMETT, *J. Am. Chem. Soc.*, 59 (1937) 96; *b* G. N. BURKHARDT, *Nature*, 136 (1935) 684.

15 H. H. JAFFÉ, *Chem. Rev.*, 53 (1953) 191.

16 J. E. LEFFLER, *J. Org. Chem.*, 20 (1955) 1202.

17*a* R. W. TAFT, *Steric Effects in Organic Chemistry*, Wiley, New York, 1956, Chapter 13; *b* R. W. TAFT, *J. Am. Chem. Soc.*, 74 (1952) 3120; 75 (1953) 4231.

18 R. W. TAFT AND I. C. LEWIS, *J. Am. Chem. Soc.*, 80 (1958) 2436; 81 (1959) 5343; *Tetrahedron*, 5 (1959) 210.

19 H. VAN BEKKUM, P. E. VERKADE AND B. M. WEPSTER, *Rec. Trav. Chim.*, 78 (1959) 815.

20 A. C. FARTHING AND B. NAM, *Steric Effects in Conjugated Systems*, Chemical Society, London, 1958, p. 319 *et seq.*

21 I. J. SOLOMON AND R. FILLER, *J. Am. Chem. Soc.*, 85 (1963) 3492.

22 D. E. PEARSON, J. F. BAXTER AND J. C. MARTIN, *J. Org. Chem.*, 17 (1952) 1511.

23 P. B. D. DE LA MARE, *J. Chem. Soc.*, (1954) 4450.

24 E. BERLINER AND L. C. MONACK, *J. Am. Chem. Soc.*, 74 (1952) 1574.

25 H. C. BROWN AND K. L. NELSON, *J. Am. Chem. Soc.*, 75 (1953) 6296.

26 H. C. BROWN AND Y. OKAMOTO, *J. Am. Chem. Soc.*, 79 (1957) 1913.

27 H. C. BROWN, *Steric effects in Conjugated Systems*, Chemical Society London, 1958, p. 100 *et seq.*
28 L. M. STOCK AND H. C. BROWN, *Advances in Physical Organic Chemistry*, Vol. 1, Academic Press, New York, 1963, p. 35 *et seq.*
29*a* E. GRUNWALD AND S. WINSTEIN, *J. Am. Chem. Soc.*, 70 (1948) 846; *b* A. H. FAINBERG AND S. WINSTEIN, *J. Am. Chem. Soc.*, 78 (1956) 2770.
30 C. G. SWAIN AND C. B. SCOTT, *J. Am. Chem. Soc.*, 75 (1953) 141.
31 J. O. EDWARDS, *J. Am. Chem. Soc.*, 76 (1954) 1541; 78 (1956) 1819.
32 P. R. WELLS, *Chem. Rev.*, 63 (1963) 171.
33 G. S. HAMMOND, *J. Am. Chem. Soc.*, 77 (1955) 334.
34*a* J. MILLER AND K. Y. WAN, *J. Chem. Soc.*, (1963) 3492; *b* J. MILLER AND K. Y. WAN, unpublished work.
35 B. A. BOLTO, J. MILLER AND V. A. WILLIAMS, *J. Chem. Soc.*, (1955) 2926.
36 L. P. HAMMETT AND M. A. PAUL, *J. Am. Chem. Soc.*, 56 (1934) 827.
37 G. W. WHELAND, R. M. BROWNELL AND E. C. MAYO, *J. Am. Chem. Soc.*, 70 (1948) 2492.
38*a* B. M. WEPSTER, *Rev. Trav. Chim.*, 76 (1957) 335; *b* B. M. WEPSTER, *Steric Effects in Conjugated Systems*, Chemical Society, London, 1958, p. 82 *et seq.*
39 J. F. BUNNETT AND R. E. ZAHLER, *Chem. Rev.*, 49 (1951) 273.
40 J. MILLER, *Rev. Pure and Appl. Chem.* (*Australia*), 1 (1951) 171.
41 J. F. BUNNETT, *Quart. Rev.*, 12 (1958) 1.
42 S. D. ROSS, *Progress in Physical Organic Chemistry*, Vol. 1, Interscience, New York, 1963, p. 31, *et seq.*
43 N. J. DALY, G. KRUGER AND J. MILLER, *Australian J. Chem.*, 11 (1958) 290.
44 J. F. BUNNETT, F. DRAPER, P. R. RYASON, P. NOBLE, R. G. TONKYN AND R. E. ZAHLER, *J. Am. Chem. Soc.*, 75 (1953) 642.
45*a* F. G. BORDWELL AND P. J. BOUTAN, *J. Am. Chem. Soc.*, 78 (1956) 87; *b* J. H. RIDD AND J. H. P. UTLEY, *Proc. Chem. Soc.*, (1964) 24.
46 C. K. INGOLD, F. R. SHAW AND I. S. WILSON, *J. Chem. Soc.*, (1928) 1280.
47 J. W. BAKER AND W. G. MOFFITT, *J. Chem. Soc.*, (1930) 1722.
48 L. PAULING, *The Nature of the Chemical Bond*, 3rd Ed., Cornell Univ. Press, Ithaca, N.Y., 1960, p. 88 *et seq.*
49 J. S. LITTLER, *Trans. Faraday Soc.*, 59 (1963) 2296.
50*a* B. A. BOLTO AND J. MILLER, *Chem. and Ind.*, (1953) 640; *b* B. A. BOLTO, M. LIVERIS AND J. MILLER, *J. Chem. Soc.*, (1956) 760; *c* B. A. BOLTO AND J. MILLER, unpublished work.
51*a* C. W. L. BEVAN AND G. C. BYE, *J. Chem. Soc.*, (1954) 3091; *b* C. W. L. BEVAN, T. A. EMOKPAE AND J. HIRST, *J. Chem. Soc., B*, (1968) 238.
52 W. F. SAGER AND C. D. RITCHIE, *J. Am. Chem. Soc.*, 83 (1961) 3498.
53 C. D. RITCHIE, J. D. SALTIEL AND E. S. LEWIS, *J. Am. Chem. Soc.*, 83 (1961) 4603.
54 B. LAMM, *Acta Chem. Scand.*, 16 (1962) 769.
55 J. MILLER AND A. J. PARKER, *J. Am. Chem. Soc.*, 83 (1961) 117.
56 C. K. BANKS, *J. Am. Chem. Soc.*, 66 (1944) 1127.
57*a* R. R. BISHOP, E. S. CAVELL AND N. B. CHAPMAN, *J. Chem. Soc.*, (1952) 437; *b* N. B. CHAPMAN AND C. W. REES, *J. Chem. Soc.*, (1954) 1190.
58 J. F. BUNNETT, E. BUNCEL AND K. V. NAMBEKIAN, *J. Am. Chem. Soc.*, 84 (1962) 4136.
59 R. MELDOLA AND E. G. C. STEPHENS, *J. Chem. Soc.*, 87 (1905) 1199.
60 J. RIBKA, *Angew. Chem.*, 70 (1958) 241.

61 R. H. CLARK AND R. H. HALL, *Trans. Roy. Soc. Canada*, [3] 21, Sect. 3 (1927) 311.
62 T. DE CRAUW, *Rec. Trav. Chim.*, 50 (1931) 753.
63 J. MILLER, *J. Am. Chem. Soc.*, 76 (1954) 448.
64*a* R. L. HEPPOLETTE AND J. MILLER, *J. Chem. Soc.*, (1956) 2329; *b* R. L. HEPPOLETTE AND J. MILLER, unpublished work.
65 J. MILLER, A. J. PARKER AND B. A. BOLTO, *J. Am. Chem. Soc.*, 79 (1957) 93.
66 J. MILLER AND A. J. PARKER, *Australian J. Chem.*, 11 (1958) 302.
67 T. L. CHAN, J. MILLER AND F. STANSFIELD, *J. Chem. Soc.*, (1964) 1213.
68 J. F. BUNNETT, H. MOE AND D. KNUTSON, *J. Am. Chem. Soc.*, 76 (1954) 3936.
69 E. ELIEL AND K. W. NELSON, *J. Org. Chem.*, 20 (1955) 1657.
70 W. GREIZERSTEIN, R. A. BONELLI AND J. A. BRIEUX, *J. Am. Chem. Soc.*, 84 (1962) 1026.
71 M. LIVERIS, P. G. LUTZ AND J. MILLER, *J. Am. Chem. Soc.*, 78 (1956) 3375.
72 J. MILLER, *J. Am. Chem. Soc.*, 77 (1955) 180.
73*a* W. A. SHEPPARD, *J. Am. Chem. Soc.*, 84 (1962) 3072; *b* D. R. EATON AND W. A. SHEPPARD, *J. Am. Chem. Soc.*, 85 (1963) 1310; *c* W. A. SHEPPARD, *J. Am. Chem. Soc.*, 85 (1963) 1314.
74 M. J. JANSSEN, *Rec. Trav. Chim.*, 79 (1960) 464, 1066; 81 (1962) 650; 82 (1963) 931.
75 F. G. BORDWELL AND P. J. BOUTAN, *J. Am. Chem. Soc.*, 79 (1957) 717.
76 E. ROTHSTEIN, *J. Chem. Soc.*, (1953) 3991.
77 J. J. LAGOWSKI, *Quart. Rev.*, 13 (1959) 233.
78*a* D. T. DOWNING, R. L. HEPPOLETTE AND J. MILLER, *Chem. and Ind.*, (1953) 1260; *b* R. L. HEPPOLETTE, J. MILLER AND V. A. WILLIAMS, *J. Chem. Soc.*, (1955) 2929.
79*a* J. MILLER AND V. A. WILLIAMS, *J. Chem. Soc.*, (1953) 1475; *b* J. MILLER AND V. A. WILLIAMS, unpublished work.
80 G. D. MELROSE AND J. MILLER, unpublished work.
81 P. J. C. FIERENS AND A. HALLEUX, *Bull. Soc. Chim. Belges*, 64 (1955) 696, 704, 709, 717.
82*a* W. A. SWEENEY AND W. M. SCHUBERT, *J. Am. Chem. Soc.*, 76 (1954) 4625; *b* W. M. SCHUBERT, J. M. CRAVEN, R. G. MINTON AND R. B. MURPHY, *Tetrahedron*, 5 (1959) 194.
83 V. J. SHINER, *J. Am. Chem. Soc.*, 76 (1954) 1603.
84 R. A. CLEMENT, J. N. NAGHIZADEH AND M. R. RICE, *J. Am. Chem. Soc.*, 82 (1960) 2449.
85 R. A. BENKESER, T. V. LISTON AND G. M. STANTON, *Tetrahedron Letters*, No. 15 (1960), p. 1.
86 J. D. ROBERTS, R. L. WEBB AND E. A. MCELHILL, *J. Am. Chem. Soc.*, 72 (1950) 408.
87 G. CILENTO, *Chem. Rev.*, 60 (1960) 146.
88 C. EABORN AND S. J. PARKER, *J. Chem. Soc.*, (1955) 126.
89 P. B. D. DE LA MARE AND C. A. VERNON, *J. Chem. Soc.*, (1951) 1764 and subsequent papers.
90 E. A. KRYUGER AND M. S. BEDNOVA, *J. Gen. Chem. (USSR)*, 3 (1933) 67.
91 G. ILLUMINATI, *J. Am. Chem. Soc.*, 80 (1958) 4945.
92 F. G. BORDWELL AND K. ROHDE, *J. Am. Chem. Soc.*, 70 (1948) 1191.
93 R. L. HEPPOLETTE AND J. MILLER, *Chem. and Ind.*, (1954) 904.
94 K. BOWDEN, *Canad. J. Chem.*, 41 (1963) 2781.
95 R. FUCHS, *J. Org. Chem.*, 28 (1963) 3209.

96 E. BERLINER, M. J. QUINN AND P. J. EDGERTON, *J. Am. Chem. Soc.*, 72 (1950) 5305.
97 A. RICHARDSON, K. R. BROWER AND E. AMSTUTZ, *J. Org. Chem.*, 21 (1956) 890.
98 P. VAN BERK, P. E. VERKADE AND B. M. WEPSTER, *Rec. Trav. Chim.*, 76 (1957) 286.
99 E. BERLINER, B. NEWMAN AND T. M. RIABOFF, *J. Am. Chem. Soc.*, 77 (1955) 478.
100 A. STREITWIESER, *Molecular Orbital Theory for Organic Chemists*, Wiley, New York, 1961, p. 242–243.
101 B. NICHOLLS AND M. C. WHITING, *J. Chem. Soc.*, (1959) 551.
102*a* D. A. BROWN, *J. Chem. Soc.*, (1963) 4389; *b* D. A. BROWN AND J. R. RAJU, *J. Chem. Soc., A*, (1966) 40.
103 A. L. GREEN, *J. Chem. Soc.*, (1954) 3538.
104*a* R. L. HEPPOLETTE, J. MILLER AND V. A. WILLIAMS, *J. Am. Chem. Soc.*, 78 (1956) 1975; *b* N. S. BAYLISS, R. L. HEPPOLETTE, L. H. LITTLE AND J. MILLER, *J. Am. Chem. Soc.*, 78 (1956) 1978.
105 M. S. NEWMAN, *J. Am. Chem. Soc.*, 77 (1955) 5552.
106 J. D. ROBERTS, *J. Am. Chem. Soc.*, 77 (1955) 5554.
107 M. CHARTON, *Canad. J. Chem.*, 38 (1960) 2943.
108 M. LIVERIS AND J. MILLER, *J. Chem. Soc.*, (1963) 3486.
109 J. MILLER AND V. A. WILLIAMS, *J. Am. Chem. Soc.*, 76 (1954) 5482.
110*a* J. F. BUNNETT, R. J. MORATH AND T. OKAMOTO, *J. Am. Chem. Soc.*, 77 (1955) 5055; *b* J. F. BUNNETT AND R. J. MORATH, *J. Am. Chem. Soc.*, 77 (1955) 5051.
111 G. MARINO, *Ricerca. Sci.*, 30 (1960) 2094.
112 G. P. BRINER, J. MILLER, M. LIVERIS AND P. G. LUTZ, *J. Chem. Soc.*, (1954) 1265.
113 W. GREIZERSTEIN AND J. A. BRIEUX, *J. Am. Chem. Soc.*, 84 (1962) 1032.
114 R. R. BISHOP, E. A. S. CAVELL AND N. B. CHAPMAN, *J. Chem. Soc.*, (1952) 437.
115 M. F. HAWTHORNE, *J. Am. Chem. Soc.*, 76 (1954) 6358.
116 S. D. ROSS AND M. FINKELSTEIN, *J. Am. Chem. Soc.*, 85 (1963) 2603.
117 A. M. KRISTJANSON AND C. A. WINKLER, *Canad. J. Chem.*, 29 (1951) 154.
118 Y. OGATA AND M. TSUCHIDA, *J. Org. Chem.*, 20 (1955) 1631.
119 H. KLOOSTERZIEL AND H. J. BACKER, *Rev. Trav. Chim.*, 72 (1953) 185.
120 J. F. BUNNETT AND J. Y. BASSETT, *J. Org. Chem.*, 27 (1962) 2345.
121 P. B. D. DE LA MARE, E. D. HUGHES AND C. K. INGOLD, *J. Chem. Soc.*, (1948) 17.
122 F. SMITH AND L. M. TURTON, *J. Chem. Soc.*, (1951) 1701.
123 H. C. BROWN, *J. Am. Chem. Soc.*, 78 (1956) 5387.
124 P. B. D. DE LA MARE, *J. Chem. Soc.*, (1957) 131.
125 J. F. BUNNETT, *J. Am. Chem. Soc.*, 79 (1957) 5969.
126 J. D. REINHEIMER AND J. F. BUNNETT, *J. Am. Chem. Soc.*, 81 (1959) 315.
127 J. A. BRIEUX AND V. DEULOFEU, *J. Chem. Soc.*, (1954) 2519.
128 R. B. SANDIN AND M. LISKEAR, *J. Am. Chem. Soc.*, 57 (1935) 1304.
129 A. W. BAMFORD AND R. W. C. BROADBANK, *Tetrahedron*, 3 (1958) 321.
130 B. A. BOLTO AND J. MILLER, *Australian J. Chem.*, 9 (1956) 74, 304.
131 J. MILLER AND K. W. WONG, *J. Chem. Soc.*, (1965) 5454.
132 A. J. SISTI AND S. LOWELL, *J. Org. Chem.*, 29 (1964) 1635.
133 R. E. PARKER AND T. O. READ, *J. Chem. Soc.*, (1962) 3149.
134*a* J. D. ROBERTS, R. A. CLEMENT AND J. J. DRYSDALE, *J. Am. Chem. Soc.*, 73 (1951) 2181; *b* J. D. ROBERTS AND W. T. MORELAND, *J. Am. Chem. Soc.*,

75 (1953) 2167; *c* J. D. ROBERTS AND R. A. CARBONI, *J. Am. Chem. Soc.*, 77 (1955) 5554.
135 J. ALLAN, A. E. OXFORD, R. ROBINSON AND J. C. SMITH, *J. Chem. Soc.*, (1926) 401.
136 C. K. INGOLD, *Chem. Rev.*, 15 (1934) 225.
137 H. C. LONGUET–HIGGINS, *Proc. Chem. Soc.*, (1957) 157.
138*a* M. LIVERIS AND J. MILLER, *Australian J. Chem.*, 11 (1958) 297; *b* B. A. BOLTO AND J. MILLER, *J. Org. Chem.*, 20 (1955) 558.
139 A. F. HOLLEMAN AND W. J. DE MOOY, *Rec. Trav. Chim.*, 35 (1916) 1.
140 M. L. BELLI, G. ILLUMINATI AND G. MARINO, *Tetrahedron*, 19 (1963) 345.
141 J. MILLER AND J. M. WRIGHTSON, *Abstracts, 112th meeting, Am. Chem. Soc.*, (1947) 16J.
142 R. L. HEPPOLETTE, M. LIVERIS, P. G. LUTZ, J. MILLER AND V. A. WILLIAMS, *Australian J. Chem.*, 8 (1955) 454.
143 G. FERGUSON AND J. M. ROBERTSON, in V. GOLD (Ed.), *Advances in Physical Organic Chemistry*, Vol. 1, Academic Press, New York, 1963, p. 233.
144 C. A. COULSON AND D. STOCKER, *Mol. Phys.*, 2 (1959) 397.
145 O. BASTIANSEN AND O. HASSEL, *Acta. Chem. Scand.*, 1 (1947) 489.
146 J. DUCHESNE AND A. MONFILS, *J. Chem. Phys.*, 22 (1954) 562.
147 T. SAITO, *Bull. Chem. Soc. Japan*, 33 (1960) 343.
148 A. L. ROCKLIN, *J. Org. Chem.*, 21 (1956) 1478.
149 K. C. HO AND J. MILLER, *Australian J. Chem.*, 19 (1966) 423.
150 K. C. HO, J. MILLER AND K. W. WONG, *J. Chem. Soc., B*, (1966) 310.
151 E. J. FORBES, R. D. RICHARDSON, M. STACEY AND J. C. TATLOW, *J. Chem. Soc.*, (1959) 2019.
152 G. J. TIESSENS, *Rec. Trav. Chim.*, 48 (1929) 1068.
153 J. MILLER AND H. W. YEUNG, *Australian J. Chem.*, 20 (1967) 379.
154 J. C. TATLOW, WITH J. BURDON, C. PATRICK, P. L. COE and other workers, *J. Chem. Soc.*, (1959) 166; (1960) 1768, 4754; (1961) 802; (1962) 1801, 3253; (1963) 3692, 4281; (1964) 763, 1777, 2975; (1965) 1045, 2088, 2094, 2621, 2658, 5152, 6326, 6329, 6336; (1966) 597, 2020, 2323; *Endeavour*, 22 (1963) 89; *Nature*, 178 (1956) 199; *Tetrahedron*, 8 (1960) 38; 9 (1960) 240; 22 (1966) 1183, 2389, 3373; 23 (1967) 505.
155 M. BELLAS AND H. SUSCHITZSKY, *J. Chem. Soc.*, (1963) 4007.
156 R. D. CHAMBERS, J. HUTCHISON AND W. K. R. MUSGRAVE, *J. Chem. Soc.*, (1964) 3736, 5634.
157 R. E. BANKS, J. E. BURGESS, W. M. CHENG AND R. N. HASZELDINE, *J. Chem. Soc.*, (1965) 576.
158 B. J. WAKEFIELD, *J. Chem. Soc., C*, (1967) 72.
159 M. R. WILES AND A. G. MASSEY, *Chem. and Ind.*, (1967) 663.

Chapter 5

VARIATION OF LEAVING GROUPS

1. Introduction

The effects of varying the leaving groups in many cases depend on the nucleophilic reagent also. This is especially noticeable for example with the halogens. With many common reagents such as alkoxide ions or amines, fluorine in activated aromatic substrates is up to a thousand times more mobile than the other halogens, whereas with many other reagents, *e.g.* thiocyanate and iodide ions it is up to a thousand times *less* mobile *[1–3]*. With one reagent, *viz.* thiomethoxide*, it is now known *[3b]* that fluorine can have a mobility up to about 10^5 times that of other halogens. Apart from the marked dependence such as that described above, the roles of nucleophile and leaving groups are obviously interchangeable in reversible reactions, such as halogen exchange (Cl, Br or I) in 1-halogeno-2,4-dinitrobenzenes *[4]*. While there is ample evidence for interdependence, the reasons for this are sufficiently well known, and reactivity patterns have been sufficiently investigated to allow largely separate consideration to be given to the effects of varying the leaving group.

2. Comparison of the Mobility of Leaving Groups in Aliphatic and Aromatic Systems

In saturated aliphatic S_N reactions, the effect of varying the leaving group has received surprisingly little attention *[5–7]*, although a general pattern is known. Still less is known in unsaturated aliphatic S_N reactions *[8–10]* of which little more than preliminary investigations have been made. In aromatic S_N reactions however, greater attention has been paid to this aspect of substitution *[11–14]*.

For saturated aliphatic S_N reactions, Hine *[6]* quoted the order of

* Methylthiolide is an alternative name.

References p. 176

mobility: $N_2^+ > OSO_2R > I > Br > ONO_2 \sim Cl > OH_2^+ \sim SMe_2^+ > F > OSO_3^- > NR_3^+ > OR > NR_2$(R = alkyl). He gave little of the supporting quantitative results, but for example quoted the results of Glew and Moelwyn-Hughes (relative rate coefficients) *[15]* for the S_N2 hydrolysis in water of methyl halides at 100°, 1:25:300:100 for the series F to I; and of Tronov and Krueger *[16]* for the S_N2 reaction with piperidine (in the same order): 1:16:17800:50500. He quoted also the much lower F/Cl ratio of about $1:10^6$ demonstrated by Swain and Scott *[17]* in the S_N1 solvolysis of triphenylmethyl halides in 85% acetone/water at 25°.

In activated (addition–elimination) aromatic S_N2 reactions, Bunnett and Zahler *[11]* quoted the order: $F > NO_2 > Cl, Br, I > N_3 > OSO_2R > NR_3^+ > OAr > OR > SR, SAr > SO_2R > NR_2$ (R = alkyl); and commented that this order varies with the reagent. Some quantitative and semiquantitative results were reported. Bunnett has since amended the order somewhat *[13]*. In his review, Miller *[12]* suggested that in aromatic S_N reactions the electron-attracting power of the replaceable group usually exerts the major influence on mobility, and thus suggested the influence of electrical charge to be in the order $X^+ > X^0 > X^-$ *e.g.* $SR_2^+ > SR > S^-$; and that within a polar category mobility occurred according to electronegativity, *e.g.* $F > OR > NR_2$ and $F > Cl > Br > I$. High polarizability was regarded as an additional favourable influence. A low bond-dissociation energy was also regarded as favourable, but as a subsidiary factor important only in cases where electronegativity is unimportant.

Before proceeding further with this discussion it is necessary to give some details of group mobility. In this respect, compounds with a replaceable halogen form well investigated series in which the replaced groups differ substantially in such factors as electronegativity, polarisability and bond dissociation energy. It is most instructive therefore to consider first patterns of halogen mobility, especially the differences between fluorine and the other halogens, since these differences are much larger than those between the heavier halogens: the mobility of fluorine relative to the other halogens is undoubtedly the key to the assessment of the factors on which mobility depends.

In uncatalysed reactions the characteristic order of mobility in saturated aliphatic S_N2 reactions is $F \ll Cl < Br < I$. The same order but with enhanced differences is found in saturated aliphatic S_N1

reactions *[5–7]*. Halogen mobility in unsaturated aliphatic S_N reactions has been investigated a little *[8–10]* and appears to be intermediate between those in saturated aliphatic and in aromatic reactions. It seems likely that marked variations, dependent on the nucleophile used, will be found.

In activated aromatic S_N2 reactions the order depends markedly on the reagent. In many of the common reagents the nucleophilic (or bond-forming) atom belongs to the first horizontal row of the Periodic Table, and for such reagents, particularly in a protic solvent the normal mobility order is: F > Cl > Br > I (commonly F ≫ Cl). With the majority of other nucleophiles the order is the reverse, thus following essentially the saturated aliphatic S_N pattern. There is evidence also for reagents showing a borderline pattern *[3,18]*.

It is noteworthy that Miller and his co-workers *[19]* have suggested and given supporting evidence that in inorganic S_N2 reactions "aliphatic" (one step) and "aromatic" (two step addition–elimination via a relatively stable intermediate complex) type substitution will be found, depending on whether or not the lowest unfilled electronic energy level of the central atom is substantially above the highest filled one, and thus on whether covalent bond formation by the entering nucleophile can proceed without synchronous breaking of the bond to the replaceable group.

3. Halogen Mobility in Addition–Elimination Aromatic S_N2 Reactions

(*a*) *General*

In order to discuss halogen mobility fully, it is necessary to have available data from a wide variety of reactions. Details for many aromatic S_N reactions are given in Tables 32–35, including additional references *20* to *55*.

For comparison and for later discussion (Section 5) some data for elimination–addition reactions are included in Table 32.

The data in Table 32(*a*) and in Tables 33–35 illustrate characteristic mobilities of halogen in activated aromatic S_N2 reactions, and though

TABLE 32

HALOGEN MOBILITY IN HALOGENOBENZENES (*a*) WITH OMe^- IN MeOH AT 200° *[20–22]*; (*b*) WITH PIPERIDINE AT 165° *[23]*; (*c*) WITH NH_2^- IN NH_3 (liq.) AT −33° *[24]*; (*d*) WITH PIPERIDIDE ION IN Et_2O AT 20° *[25]*

Series	Halogen	Rate constant (k_2) ($l \cdot mole^{-1} \cdot sec^{-1}$)	Mobility: rel. to Cl = 1 (G.R.F.)	Mobility: rel. to I = 1	$\Delta E^{\ddagger}$ (kcal·mole^{-1})	$\log_{10} B$
(*a*)	F[a]	$6.65 \cdot 10^{-5}$	1960	—	34.9_5	11.9_5
		$(1.5_5 \cdot 10^{-4})$	(4570)	—	(36.4)	(13.0)
	Cl[b]	$3.39 \cdot 10^{-8}$	1	—	39.9_5	11.1
(*b*)	Cl	$1.78 \cdot 10^{-9}$	1	0.0682	—	—
	Br	$1.44 \cdot 10^{-8}$	8.09	0.552	—	—
	I	$2.61 \cdot 10^{-8}$	14.7	1	23.6	8.75
(*c*)	F	very slow	very small	—	—	
	Cl	fast	1	—	—	
	Br	fast	21	—	—	
	I	fast	$8._3$	—	—	
(*d*)	F	$1.10 \cdot 10^{-3}$	4.02	—	—	
	Cl	$2.74 \cdot 10^{-4}$	1	—	—	
	Br	$4.46 \cdot 10^{-4}$	1.63	—	—	
	I	$1.72 \cdot 10^{-4}$	0.628	—	—	

[a] Values in parentheses from ref. *21*, which its authors regard as subject to some doubt. [b] By extrapolation of kinetic data for a series of 1-chloro-4-X-benzenes *[22a, cf. 22b]*.

results for a still wider range of nucleophiles are desirable, an interesting and informative pattern emerges for these reactions and is seen most clearly with the anionic nucleophiles.

(*b*) *Mobility in reactions with anionic nucleophiles*

For anionic nucleophiles in which the nucleophilic atom is in the first horizontal row of the Periodic Table, and for protic solvents particularly, the characteristic mobility pattern is F ≫ Cl > Br > I. This is the opposite order to that found in saturated aliphatic S_N

TABLE 33A

HALOGEN MOBILITY IN SOME *para*-SUBSTITUTED HALOGENOBENZENES (*a*) *p*-HALOGENONITROBENZENES WITH N_3^- IN MeOH AT 50° *[1,26,27]*; (*b*) *p*-HALOGENONITROBENZENES WITH N_3^- IN DMF AT 50° *[1,26,27]*; (*c*) *p*-HALOGENONITROBENZENES WITH OMe^- IN MeOH AT 50° *[28]*; (*d*) *p*-HALOGENONITROBENZENES WITH OEt^- IN EtOH AT 90.8° *[29]*; (*e*) *p*-HALOGENONITROBENZENES WITH SMe^- IN MeOH AT 0° *[3]*; (*f*) *p*-HALOGENONITROBENZENES WITH SPh^- IN MeOH AT 0° *[18,30]*; (*g*) *p*-HALOGENONITROBENZENES WITH PIPERIDINE IN EtOH AT 90° *[31]*; (*h*) *p*-HALOGENONITROBENZENES WITH PIPERIDINE IN C_6H_6 AT 90° *[31]*

Series	Halogen	Rate constant (k_2) ($l\cdot mole^{-1}\cdot sec^{-1}$)	Mobility rel. to Cl = 1 (or G.R.F.)	Mobility rel. to I = 1	$\Delta E^{\ddagger}$ (kcal·mole^{-1})	$\log_{10}B$ or ($\Delta S^{\ddagger}$ in e.u.)
(*a*)	F	$3.73\cdot10^{-6}$	—	291	24.1	10.4_5
	I	$1.28\cdot10^{-8}$	—	1	27.8	10.9
(*b*)	F	$1.73\cdot10^{-2}$	—	152	18.8	10.9_5 [a]
					19.3_5	11.4 [a]
	I	$1.14\cdot10^{-4}$	—	1	22.1_5	11.0_5
(*c*)	F	$2.64\cdot10^{-3}$	312	866	21.1	11.7_5
	Cl	$8.47\cdot10^{-6}$	1	2.78	24.0_5	11.2
	Br	$7.16\cdot10^{-6}$	0.845	2.35	24.6_5	11.5
	I	$3.05\cdot10^{-6}$	0.361	1	25.0	11.4
(*d*)	F	$2.2\cdot10^{-1}$	228	3100	19.0	10.7
	Cl	$9.63\cdot10^{-4}$	1	13.6	20.1	9.0
	Br	$8.38\cdot10^{-4}$	0.870	11.7	20.3	9.2
	I	$7.09\cdot10^{-5}$	0.0736	1	—	—
(*e*)	F	$1.29\cdot10^{-4}$	—	3720	17.9	(−12.3)
	I	$3.47\cdot10^{-8}$	—	1	19.7	(−22.4)
(*f*)	F	$7.79\cdot10^{-6}$	—	2.29	19.8	(−10.5)
	I	$3.40\cdot10^{-6}$	—	1	19.4	(-13.4_5)
(*g*)	F	$2.25\cdot10^{-3}$	208	563	13.2	5.3
	Cl	$1.08\cdot10^{-5}$	1	2.70	17.1	5.3
	Br	$1.39\cdot10^{-5}$	1.29	3.48	16.8	5.2
	I	$4.0\cdot10^{-6}$	0.37	1	18.0	5.5
(*h*)	Cl	$1.1\cdot10^{-6}$	1	0.917	13.7	2.3
	Br	$2.0\cdot10^{-6}$	1.82	1.67	15.5	3.6 [b]
		$2.6\cdot10^{-6}$	2.36	2.17	13.6	2.6 [b]
	I	$1.2\cdot10^{-6}$	1.09	1	19.5	5.8

[a] Values from different procedures for making rate measurements. [b] Alternative values.

TABLE 33B

HALOGEN MOBILITY IN SOME *ortho*-SUBSTITUTED HALOGENOBENZENES (*a*) *o*-HALOGENONITROBENZENES WITH N_3^- IN MeOH AT 100° *[1]*; (*b*) *o*-HALOGENONITROBENZENES WITH OMe^- IN MeOH AT 50° *[32]*; (*c*) *o*-HALOGENONITROBENZENES WITH PIPERIDINE IN EtOH AT 90° *[31]*; (*d*) *o*-HALOGENOBENZENEDIAZONIUM IONS WITH OH^- IN H_2O AT 0° *[33,34]*

Series	Halogen	Rate constant (k_2) ($l \cdot mole^{-1} \cdot sec^{-1}$)	Mobility rel. to Cl = 1 (or G.R.F.)	Mobility rel. to I = 1	$\Delta E^{\ddagger}$ (kcal·mole^{-1})	$\log_{10} B$ (or ($\Delta S^{\ddagger}$ in e.u.)
(*a*)	F	$2.66 \cdot 10^{-4}$	—	36.1	22.8_5	9.8
	I	$7.35 \cdot 10^{-6}$	—	1	24.7_5	9.3_5
(*b*)	F	$1.81 \cdot 10^{-3}$	722	2180	19.7	10.4
	Cl	$2.52 \cdot 10^{-6}$	1	3.04	23.6_5	10.4
	Br	$1.77 \cdot 10^{-6}$	0.702	1.14	25.5_5	11.5
	I	$8.29 \cdot 10^{-7}$	0.330	1	26.5	11.9
(*c*)	F	$1.23_5 \cdot 10^{-2}$	415	532	14.8	7.0
	Cl	$2.98 \cdot 10^{-5}$	1	1.28_5	18.2	6.4
	Br	$5.60 \cdot 10^{-5}$	1.88	2.42	18.3	6.8
	I	$2.32 \cdot 10^{-5}$	0.779	1	19.2	6.9
d)	F	$1.6 \cdot 10^{-2}$	53	—	—	—
	Cl	$3 \cdot 10^{-4}$	1	—	—	—

reactions and is ascribed *[3,11,12,36,53]* to the effects of differing electronegativity, combined with the unimportance of bond breaking, in an addition–elimination mechanism via an intermediate complex (see Chapter 1, p. 8), in which formation of the first transition state involving bond formation, without concurrent bond rupture and associated factors, is rate-limiting. This is supported by the parallel observations of Fainberg and Miller *[57]* that in the S_N2 reaction of γ-fluoroallyl halides, the γ-fluorine substituent accelerates displacement of halogen, though it is displaced more slowly itself.

With heavy nucleophiles typically the mobility of fluorine relative to the other halogens is strikingly reversed as a consequence of transition-state 2 formation being rate-limiting. Comparative details of Cl, Br and I mobilities are lacking in these cases so that a reversal also among the heavier halogens, which in any case have similar mobilities, is not certain. For example, from Table 34A (*b*) and (*d*) the mobility ratio, fluorine/iodine, in 1-halogeno-2,4-dinitrobenzenes is $5.65 \cdot 10^3$ in reactions with methoxide in methanol at 0°, and may be computed as $3.80 \cdot 10^2$ at 100°; compared with $1.11 \cdot 10^{-3}$ at 100° in reactions with thiocyanate in methanol. This is a change by a factor of $3.42 \cdot 10^5$ (the change would be larger still at lower temperatures) and is ascribed to a shift from formation of the first to formation of the second transition state (involving rupture of the bond to the displaced group and associated factors) as rate-limiting. A similar large change is found in comparing the reactions of methoxide and iodide ion with these two substrates (Table 34A (*b*) and 34B (*i*)).

Whereas thiocyanate ion as a reagent behaves like the heavy halogenide ions in relation to the halogen mobility order, it is of special interest that the structurally simpler and more typical second row reagent thiomethoxide ion appears to be on the borderline with respect to its influence on the mobility order of the displaced groups *[2,3]*. Thiophenoxide ion is somewhat similar, but behaves much more nearly as a heavy nucleophile, though with some differences from other heavy nucleophiles because it is so reactive *[18]*. From Table 33A (*e*) and 34B (*g*) the F/I mobility ratio with thiomethoxide ion is about normal in the mononitro-series and very high in the dinitro-series. In the mononitro series it is however due to an unusually small difference in $\Delta E^{\ddagger}$, augmented by the depression of mobility of iodine in its reactions with a heavy reagent, appearing as a reduction in $\Delta S^{\ddagger}$. In the dinitro series the normal $\Delta E^{\ddagger}$ difference occurs, and the $\Delta S^{\ddagger}$ depression then leads to a very high F/I mobility ratio. From Table 33A (*f*) and 34B (*h*) a similar difference between the mono- and dinitro-series is seen in reactions of thiophenoxide ion, but at a relatively lower level of fluorine mobility consequent on a shift towards the more typical heavy nucleophile behaviour of thiophenoxide than thiomethoxide ion. Parker and Read *[55]* prefer for many aromatic S_N2 reactions a one-stage mechanism and explain the differences in fluorine mobility as a consequence of a spectrum of transition states with varying extent of bond-making and bond-breaking. However, this is inconsistent with other aspects of aromatic S_N2 reactions and some reasons for the rejection of their views have been given in Chapter 1, p. 20.

The mobility patterns discussed above have been considered fully by Miller and his co-workers *[2,18]* in the discussion of a series of semi-

TABLE 34A

HALOGEN MOBILITY IN SOME DISUBSTITUTED HALOGENOBENZENES (*a*) 1-HALOGENO-2,4-DINITROBENZENES WITH N_3^- IN MeOH AT 0° *[18,35]*; (*b*) 1-HALOGENO-2,4-DINITROBENZENES WITH OMe^- IN MeOH AT 0° *[3,36]*; (*c*) 1-HALOGENO-2,4-DINITROBENZENES WITH p-$NO_2C_6H_4O^-$ IN MeOH AT 25° AND 100° *[1]*; (*d*) 1-HALOGENO-2,4-DINITROBENZENES WITH SCN^- IN MeOH AT 100° *[33,35,37]*; (*e*) 1-HALOGENO-2,4-DINITROBENZENES WITH SCN^- IN Me_2CO AT 25° *[1,2]*; (*f*) 1-HALOGENO-2,4-DINITROBENZENES WITH NH_3 IN MeOH AT 97.1° *[38]*; (*g*) 1-HALOGENO-2,4-DINITROBENZENES WITH $PhNH_2$ IN EtOH AT 50° *[39,40]*; (*h*) 1-HALOGENO-2,4-DINITROBENZENES WITH $PhNH_2$ IN ACETONE AT 50° *[41]*; (*i*) 1-HALOGENO-2,4-DINITROBENZENES WITH PhNHMe IN $PhNO_2$ AT 131.5° *[42]*; (*j*) 1-HALOGENO-2,4-DINITROBENZENES WITH PhNHMe IN EtOH AT 67.2° *[43]*

Series	Halogen	Rate constant (k_2) ($l \cdot mole^{-1} \cdot sec^{-1}$)	Mobility rel. to Cl = 1 (or G.R.F.)	Mobility rel. to I = 1	$\Delta E^{\ddagger}$ (kcal·mole^{-1})	$\log_{10} B$ or ($\Delta S^{\ddagger}$ in e.u.)
(*a*)	F	$2.00 \cdot 10^{-2}$	—	1150	16.6_5	11.6
	I	$1.74 \cdot 10^{-5}$	—	1	18.9_5	10.4
(*b*)	F	1.74	890	5650	13.5	11.0_5
	Cl	$2.00 \cdot 10^{-3}$	1	6.50	17.4_5	11.2_5
	Br	$1.38 \cdot 10^{-3}$	0.690	4.49	17.0_5	10.8
	I	$3.08 \cdot 10^{-4}$	0.154	1	18.9_5	11.7
(*c*)	F	$4.61 \cdot 10^{-3}$(25°)	3160	—	—	—
	Cl	$1.46 \cdot 10^{-6}$(25°)	1	—	22.0_5	10.3_5
		$2.60 \cdot 10^{-3}$(100°)	1			
	I	$1.29 \cdot 10^{-3}$(100°)	0.496	—	—	—
(*d*)	F	$3.68 \cdot 10^{-6}$	—	0.00111	29.1	11.6_5
	I	$3.31 \cdot 10^{-3}$	—	1	20.0_5	9.2_5
(*e*)	F	$4.59 \cdot 10^{-4}$ [a]	—	0.176	19.4_5	10.3
	I	$2.60 \cdot 10^{-3}$ [a]	—	1	18.1	10.1_5
(*f*)	F	$2.10 \cdot 10^{-1}$	460	1290	12.1	(−28.2)
	Cl	$4.57 \cdot 10^{-4}$	1	2.86	16.6	(−31.1)
	Br	$4.28 \cdot 10^{-4}$	0.936	2.62	16.7	(−31.2)
	I	$1.63 \cdot 10^{-4}$	0.356	1	17.0	(−31.3)
(*g*)	F	$1.68 \cdot 10^{-2}$	62.5	128	6.4	2.6
	Cl	$2.69 \cdot 10^{-4}$	1	2.05	11.2	4.0
	Br	$4.05 \cdot 10^{-4}$	1.51	3.09	11.2	4.2
	I	$1.31 \cdot 10^{-4}$	0.487	1	—	—

continued on page 145

TABLE 34A (*continued*)

Series	Halogen	Rate constant (k_2) ($l \cdot mole^{-1} \cdot sec^{-1}$)	Mobility: rel. to Cl = 1 (or G.R.F.)	Mobility: rel. to I = 1	$\Delta E^{\ddagger}$ (kcal · mole^{-1})	$\log_{10} B$ or ($\Delta S^{\ddagger}$ in e.u.)
(*h*)	F	$2.20 \cdot 10^{-4}$	12.5	—	F compound 3 kcal · mole^{-1} less than Cl compound	F compound 1.3 kcal · mole^{-1} less than Cl compound
	Cl	$1.75 \cdot 10^{-5}$	1	—		
(*i*)	F	$2.67 \cdot 10^{-5}$	0.0615	—	10	(−56)
	Cl	$4.40 \cdot 10^{-4}$	1	—	12, 10 [b]	(−48)
	Br	$1.30 \cdot 10^{-3}$	1.96	—	11, 10 [b]	(−44)
(*j*)	F	$7.30 \cdot 10^{-5}$	1.00	—	—	—
	Cl[c]	$7.30 \cdot 10^{-5}$	1	—	—	—
	Br[c]	$2.03 \cdot 10^{-4}$	2.78	—	—	—
(Plus	F	$1.02 \cdot 10^{-4}$	0.99	—	—	—
0.104 *M*	Cl	$1.03 \cdot 10^{-4}$	1	—	—	—
$NaClO_4$)	Br	$2.49 \cdot 10^{-4}$	2.42	—	—	—
(Plus	F	$1.10 \cdot 10^{-3}$	12.1	—	—	—
0.104 *M*	Cl	$9.1 \cdot 10^{-5}$	1	—	—	—
KOAc)	Br	$2.32 \cdot 10^{-4}$	2.55	—	—	—

[a] k_2 values at 25° allow for incomplete dissociation of KSCN in acetone, but Arrhenius parameters are based on uncorrected values at all temperatures involved. [b] Alternative values. [c] Rheinlander *[40]* quotes data for replacement of Cl, Br, I at 50° giving G.R.F. values: 1, 2.94, 0.720.

empirical theoretical calculations leading to the complete potential energy–reaction coordinate profiles for reaction of various anionic reagents in activated aromatic S_N2 reactions, which agree well with experiment. Some relevant curves based on these calculations are given in Figs. 40–53 below. The essentials of this procedure are: (*i*) to calculate the energy differences between the intermediate complex (I.C.) and the initial and the final state (I.St. and F.St.) by utilising electron affinities, heats of solvation and bond dissociation energies

TABLE 34B

HALOGEN MOBILITY IN SOME DISUBSTITUTED HALOGENOBENZENES (*a*) 1-HALOGENO-2,4-DINITROBENZENES WITH OH^- IN DIOXANE-H_2O (60/40 v/v) AT 50° *[44]*; (b) 1-HALOGENO-2,4-DINITROBENZENES WITH SO_3^{2-} IN 60% EtOH-H_2O AT 30° *[45]*; (*c*) 1-HALOGENO-2,4-DINITROBENZENES WITH *p*-$NO_2C_6H_4SO_2^-$ IN MeOH AT 50° *[1,46,47]*; (*d*) 1-HALOGENO-2,4-DINITRO BENZENES WITH $S_2O_3^{2-}$, SO_3^{2-}, SH^- IN MeOH AT 25° *[1,46,47]*; (*e*) 1-HALOGENO-2,4-DINITROBENZENES WITH THIOUREA IN MeOH AT 100° *[48]*; (*f*) 1-HALOGENO-2,4-DINITROBENZENES WITH SELENOUREA IN MeOH AT 100° *[48]*; (*g*) 1-HALOGENO-2,4-DINITROBENZENES WITH SMe^- IN MeOH AT 0° *[3,49]*; (*h*) 1-HALOGENO-2,4-DINITROBENZENES WITH SPh^- IN MeOH AT 0° *[18,50]*; (*i*) 1-HALOGENO-2,4-DINITROBENZENES WITH I^- IN MeOH AT 100° *[1,2,51,52]*; (*j*) 1-HALOGENO-2,4-DINITROBENZENES WITH PIPERIDINE IN MeOH AT 0° *[53]*; (*k*) 1-HALOGENO 2-,4-DINITROBENZENES WITH PYRIDINE IN ACETONE AT 50° *[41]*; (*l*) 1-HALOGENO-2,6-DINITROBENZENES WITH OMe^- IN MeOH AT 50° *[41,49]*; (*m*) 1-HALOGENO-2,6-DINITROBENZENES WITH $PhNH_2$ IN EtOH AT 50° *[55]*

Series	Halogen	Rate constant (k_2) ($l \cdot mole^{-1} \cdot sec^{-1}$)	Mobility rel. to Cl = 1 (or G.R.F.)	Mobility rel. to I = 1	$\Delta E^{\ddagger}$ (kcal · $mole^{-1}$)	$\log_{10} B$ or ($\Delta S^{\ddagger}$ in e.u.)
(*a*)	F	Fast	≫ 1	≫ 3.5	—	—
	Cl	$7.47 \cdot 10^{-3}$	1	3.50	16.1	8.7_5
	Br	$5.36 \cdot 10^{-3}$	0.718	2.50	16.4	8.8
	I	$2.14 \cdot 10^{-3}$	0.286	1	17.4_5	9.1_5
(*b*)	F	4.25	35.4	48.6	10.7	8.3_5
	Cl	$1.20 \cdot 10^{-1}$	1	1.37	15.1	9.9_5
	Br	$1.33 \cdot 10^{-1}$	1.11	1.52	13.3	8.7_5
	I	$8.75 \cdot 10^{-2}$	0.730	1	15.0	9.7_5
(*c*)	F	Very slow	Very small	—	—	—
	Cl	$1.66 \cdot 10^{-4}$	1	—	—	—
(*d*)	F	Similar rates	Close to 1	—	—	—
	Cl		1	—	—	—
(*e*)	F	$7.5 \cdot 10^{-3}$	11	—	—	—
	Cl	$6.8 \cdot 10^{-4}$	1	—	—	—
(*f*)	F	$1.6 \cdot 10^{-2}$	2.5	—	—	—
	Cl	$6.4 \cdot 10^{-3}$	1	—	—	—
(*g*)	F	$1.51_5 \cdot 10^{2}$	41500	372000	10.4	10.5_5
	Cl	$3.66 \cdot 10^{-3}$	1	8.98	14.1	8.8_5
	I	$4.07 \cdot 10^{-4}$	0.111	1	14.3	8.0_5

continued on p. 147

TABLE 34B (*continued*)

Series	Halogen	Rate constant (k_2) (l·mole^{-1}·sec^{-1})	Mobility: rel. to Cl = 1 (or G.R.F.)	Mobility: rel. to I = 1	$\Delta E^{\neq}$ (kcal·mole^{-1})	$\log_{10} B$ or ($\Delta S^{\neq}$ in e.u.)
(*h*)	F	$1.29 \cdot 10^2$	33.2	25.0	10.7	(−12.5)
	Cl	3.89	1	0.753	10.3	(−20.0)
	Br	6.68	1.72	1.29	9.8	(−20.6)
	I	5.17	1.33	1	10.7	(−17.9)
(*i*)	F	Very slow (*ca.* 10^{-9})[a]	—	$\ll 1$ (*ca.* 10^{-4})[a]	—	—
			—			
	Br	$1.5 \cdot 10^{-5}$	—	1	24.9_5	9.8
	I	$4.7 \cdot 10^{-5}$	—	3.1	22.8	9.1
(*j*)	F	1.5	770	3280	—	—
	Cl	$1.95 \cdot 10^{-3}$	1	4.27	11.6	(−30.2)
	Br	$1.97 \cdot 10^{-3}$	1.01	4.31	11.8	(−29.5)
	I	$4.57 \cdot 10^{-4}$	0.234	1	12.0	(−31.7)
(*k*)	F	$2.13 \cdot 10^{-7}$	3.26	—	about same for both compounds	—
	Cl	$6.54 \cdot 10^{-8}$	1	—		
(*l*)	Cl	$7.40 \cdot 10^{-3}$	1	12.4	17.5_5	9.7_5
	Br	$4.29 \cdot 10^{-3}$	0.580	7.20	14.4	7.4
	I	$5.96 \cdot 10^{-4}$	0.0806	1	22.2	11.8
(*m*)	F	$8.77 \cdot 10^{-3}$	44.4	$105._5$	9.4	4.3
	Cl	$1.98 \cdot 10^{-4}$	1	2.38	14.2	5.9
	Br	$3.66 \cdot 10^{-4}$	1.85	4.40	14.1	6.1
	I	$8.32 \cdot 10^{-5}$	0.420	1	14.9	6.0

[a] The numerical value is calculated by Miller's procedure *[2]*.

(B.D.E.), including in this differences between the Ar–X bond in the I.St. and F.St. and the Alph–X bond in the I.C.; (*ii*) to calculate the energy levels of the two transition states by using a semi-empirical relationship between a percentage of the relevant B.D.E. and the degree of exo- or endo-thermicity of energy changes from I.C. to I.St. and F.St.; (*iii*) to reduce the T.St. level when the other group attached at the reaction centre is highly electronegative, since this should facilitate bond formation (the α-substituent effect). Also by consideration of

TABLE 35

MOBILITY OF HALOGEN IN PICRYL HALIDES (*a*) WITH OH^- IN H_2O AT 25° *[54]*; (*b*) WITH $PhNH_2$ IN EtOH AT 25° *[55]*; (*c*) WITH PhNHMe IN EtOH AT 50° *[40]*[a]; (*d*) WITH MeOH AS REAGENT AND SOLVENT AT 100° *[18,30]*

Series	Halogen	Rate constant k_2 ($l \cdot mole^{-1} \cdot sec^{-1}$)	Mobility rel. to Cl = 1 (G.R.F.)	Mobility rel. to I = 1	$\Delta E^{\ddagger}$ ($kcal \cdot mole^{-1}$)	$\log_{10} B$
(*a*)	F	$7 \cdot 10^2$	1400	—	—	—
	Cl	$5 \cdot 10^{-1}$	1	—	—	—
(*b*)	F	$2.10 \cdot 10^2$	195	755	4.4	5.5
	Cl	1.08	1	3.88	7.8	5.8
	Br	1.56	1.44	5.66	8.9	6.7
	I	0.278	0.258	1	9.8	6.6
(*c*)	Cl	$1.37 \cdot 10^{-7}$	1	—	—	—
	Br	$7.48 \cdot 10^{-7}$	5.46	—	—	—
	I	$1.85 \cdot 10^{-7}$	1.36	—	—	—
(*d*)	Cl	$1.31 \cdot 10^{-6}$	1	2.20		
		$3.59 \cdot 10^{-6}$[b]		—	18.1	6.65
	I	$4.86 \cdot 10^{-7}$	0.372	1	—	—

[a] Rates seem very low, being less than double those reported by the same author for the 1-halogeno-2,4-dinitro compounds (see footnote c to Table 34A). [b] By calculation from data in Ref. *29* by using Arrhenius parameters, the highest experimental temperature being 67.7°.

entropy terms, $\Delta G^{\ddagger}$ values can be estimated. Additional details of the procedure are given in Chapter 6, Section 5.

Figures 40–44 show curves for two first row nucleophiles, *viz.* methoxide and azide ion. For the specified direction of reaction, the α-substituent or electronegativity effect of fluorine lowers T.St.1 by 4 $kcal \cdot mole^{-1}$, and that of chlorine lowers it by 1 $kcal \cdot mole^{-1}$. Those of methoxyl and azide are estimated to lower T.St.2 by 3 and 2 $kcal \cdot mole^{-1}$ respectively. Less electronegative groups, *e.g.* iodine, have zero α-substituent effects. The levels of T.St.1 and T.St.2 are seen to be about equal in replacement of fluorine (Figs. 40 and 43), but by comparison with the replacement of iodine for example,

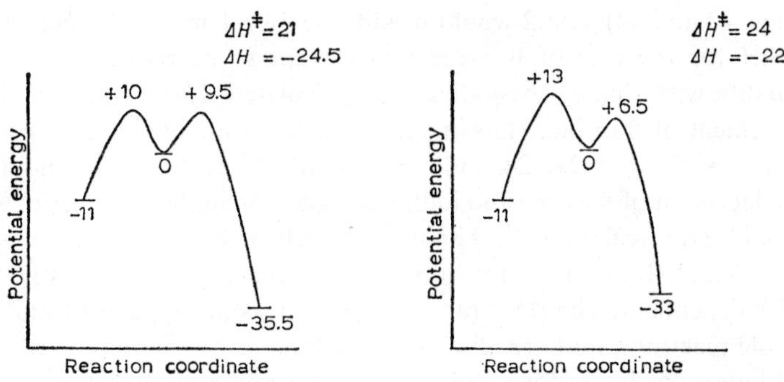

Fig. 40. Potential energy/reaction coordinate profile of $OMe^-/p\text{-}NO_2C_6H_4F$.
Fig. 41. Potential energy/reaction coordinate profile of $OMe^-/p\text{-}NO_2C_6H_4Cl$.

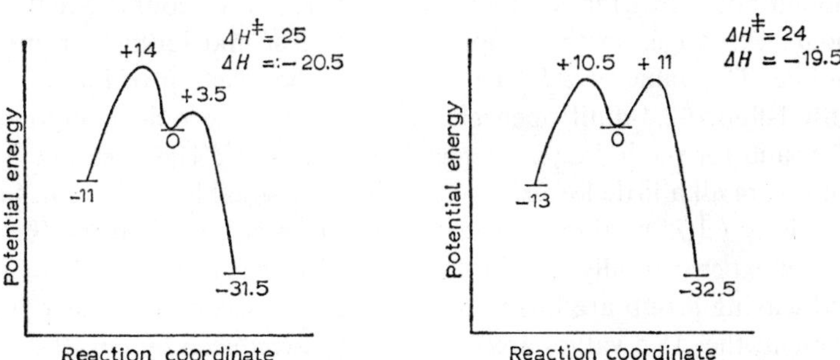

Fig. 42. Potential energy/reaction coordinate profile of $OMe^-/p\text{-}NO_2C_6H_4I$.
Fig. 43. Potential energy/reaction coordinate profile of $N_3^-/p\text{-}NO_2C_6H_4F$.

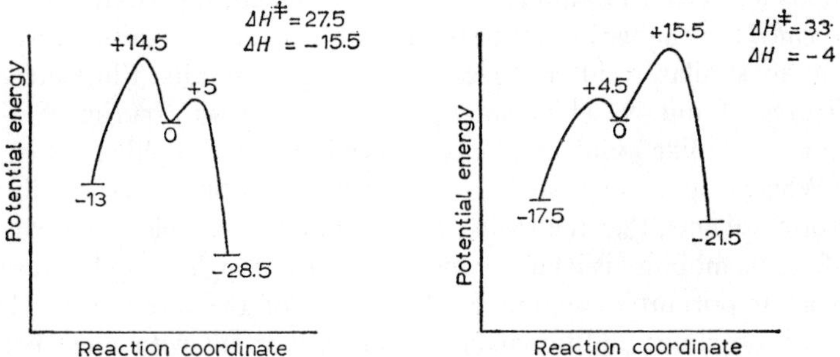

Fig. 44. Potential energy/reaction coordinate profile of $N_3^-/p\text{-}NO_2C_6H_4I$.
Fig. 45. Potential energy/reaction coordinate profile of $I^-/2,4\text{-}(NO_2)_2C_6H_3F$.

References p. 176

(Figs. 42 and 44) T.St.2 would need to be 4 kcal·mole^{-1} higher than T.St.1 for the rate of fluorine replacement to be reduced even to equality with that of the iodine. The calculated $\Delta E^{\ddagger}$ values for displacement of fluorine, chlorine, and iodine by methoxide ion in the *p*-nitro series are 21, 24 and 25 kcal·mole^{-1} respectively and for displacement of fluorine and iodine by azide ion in the same series are 24 and 27.5 kcal·mole^{-1}. These agree well with experiment, which also demonstrates that the mobility differences are essentially all $\Delta E^{\ddagger}$ dependent. The $\Delta S^{\ddagger}$ (and $\log_{10} B$) values are typical of anion-dipole reactions, and are all closely similar.

Figures 45–47 are for iodide ion as an example of a heavy nucleophile. With fluorine as the replaced group, the energy level of T.St.2 is much higher than that of T.St.1. With bromine (and chlorine too, though not shown) it is only a little higher, and of course equal for iodide exchange, so that fluorine is the least and iodine the most mobile. The value of $\Delta E^{\ddagger}$ calculated for the reaction of iodide ion with 1-fluoro-2,4-dinitrobenzene is 33, with the bromo compound 27.5, and for the iodide exchange 26 kcal·mole^{-1}. The experimental values are all a little lower, though still in reasonable agreement, but reactivity ($\Delta G^{\ddagger}$) values are in better agreement, since it has been shown experimentally that in these reactions, when *both* nucleophile and leaving group are linked to the ring by a second or higher row element, the $\Delta S^{\ddagger}$ values are somewhat low (reduced by *ca.* 10 e.u.). This heavy nucleophile interaction is not enough to prevent the reversal of reactivity. While calculations for thiocyanate ion are less satisfactory, because values for both heats of dissociation of C–SCN bonds and electron affinity had to be estimated, (alternative values, *provided* the changes were similar in direction and magnitude, would lead to similar results) the calculations give results, illustrated in Figures 48 and 49, which show, in agreement with experiment, that thiocyanate like iodide ion behaves as a heavy nucleophile.

Whereas it is suggested for first-row nucleophiles, especially in protic solvents, that the main property of a replaceable group which affects its mobility is its electronegativity, for if high enough it results in an important reduction of the energy of the rate-limiting first transition state; additional factors are important with most heavy nucleophiles. Then the strength of the bond to the leaving group, its electron affinity, and its heat of solvation as a displaced group are

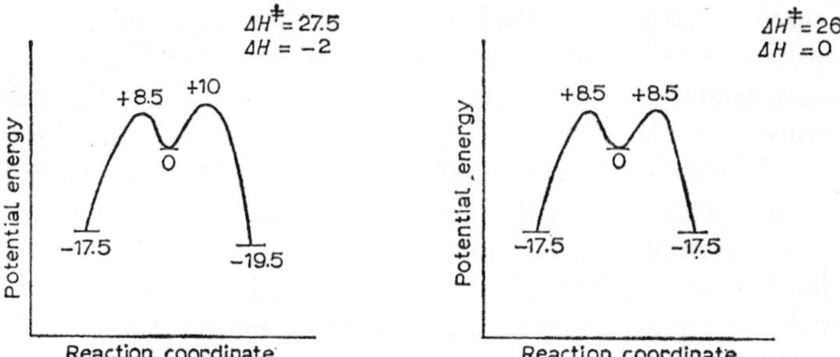

Fig. 46. Potential energy/reaction coordinate profile of I^-/2,4-$(NO_2)_2C_6H_3Br$.
Fig. 47. Potential energy/reaction coordinate profile of I^-/2,4-$(NO_2)_2C_6H_3I$.

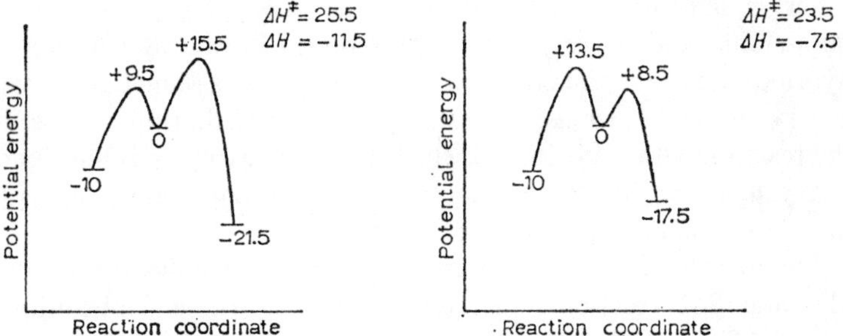

Fig. 48. Potential energy/reaction coordinate profile of SCN^-/2,4-$(NO_2)_2C_6H_3F$.
Fig. 49. Potential energy/reaction coordinate profile of SCN^-/2,4-$(NO_2)_2C_6H_3I$.

major factors. It should be noted that any comparison with aliphatic S_N reactions is then with S_N1 rather than S_N2 reactions, for if the formation of the second transition-state of the aromatic substitution is rate-limiting, this involves bond-rupture without concurrent bond-formation.

These conclusions are strongly supported by the work of Lam and Miller *[37]*. Commenting on the absence of electrophilic catalysis in activated aromatic S_N2 reactions such as is found in the mechanistically one-stage S_N2 reactions of saturated aliphatic compounds, they also pointed out that Miller's calculations *[2,18]* based on the two-stage addition–elimination mechanism for activated aromatic S_N2 reactions, lead to the conclusion that electrophilic catalysis should only be found

References p. 176

in cases where the rate-limiting step is the formation of the second transition state and that fluoro compounds would be likely to exhibit this. It would only be marked where the energy level of the second transition state is substantially higher than that of the first transition state, and it is in these cases that the mobility of fluorine is much less than that of the heavy halogens. They pointed out more specifically that, for example, the reactions of 1-fluoro-2,4-dinitrobenzene with iodide and with thiocyanate ion but not azide ion (although both the latter are pseudo-halogenoid and moderately powerful nucleophiles), should be very markedly increased in rate by addition of electrophilic catalysts.

Experimental results showed that in the reactions of 1-halogeno-2,4-dinitrobenzenes with iodide or thiocyanate ion in which the mobility order is fluorine $<$ heavy halogens, the very slow reaction with potassium iodide is very markedly increased in rate by addition of hydriodic acid, and the very slow reaction with potassium thiocyanate is very markedly increased in rate by addition of thorium nitrate. In the reaction with azide ion, in which the mobility order is fluorine $>$ heavy halogens, the facile reaction with sodium azide is actually markedly *inhibited* by addition of thorium nitrate.

Ingold *[5b]*, in his discussion of electrophile-catalysed saturated aliphatic S_N2 reactions, pictures them as S_N1-like substitutions, whereas Swain had at that time pictured these in terms of a termolecular push-pull mechanism *[58]*, adduced essentially for a solvent acting in this role.

Figures 50–53 show that with thiomethoxide ion the mobilities have a distinctive borderline pattern. With *p*-fluoronitrobenzene the formation of T.St.2 is rate-limiting, but the difference in energy between T.St.2 and T.St.1 is only half the reduction in energy of T.St.1 due to the α-substituent effect of fluorine; so that while T.St.1 formation is rate-limiting for the iodo compound and other halogeno compounds, the $\Delta S^{\ddagger}$ reduction referred to (heavy nucleophile interaction) then increases the F/I mobility ratio. At 0° the ratio is 3720, approximating to that typical of first row nucleophiles, but is partly $\Delta E^{\ddagger}$ and partly $\Delta S^{\ddagger}$ dependent, whereas with first row nucleophiles it is only $\Delta E^{\ddagger}$ dependent.

With 1-fluoro-2,4-dinitrobenzene the level of T.St.1 and of T.St.2 are calculated to be equal, so that the α-substituent effect of fluorine in

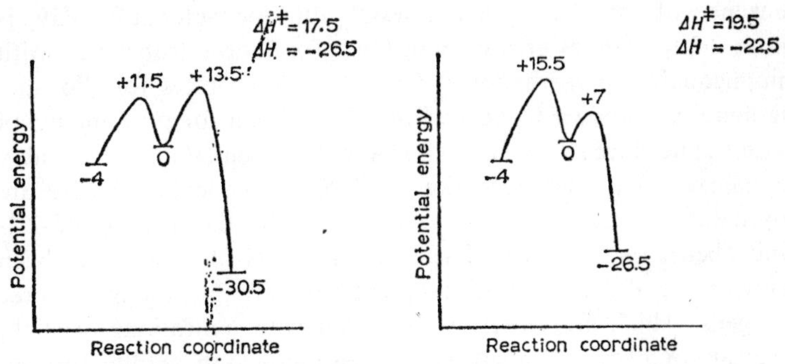

Fig. 50. Potential energy/reaction coordinate profile of SMe^-/p-$NO_2C_6H_4F$.
Fig. 51. Potential energy/reaction coordinate profile of SMe^-/p-$NO_2C_6H_4I$.

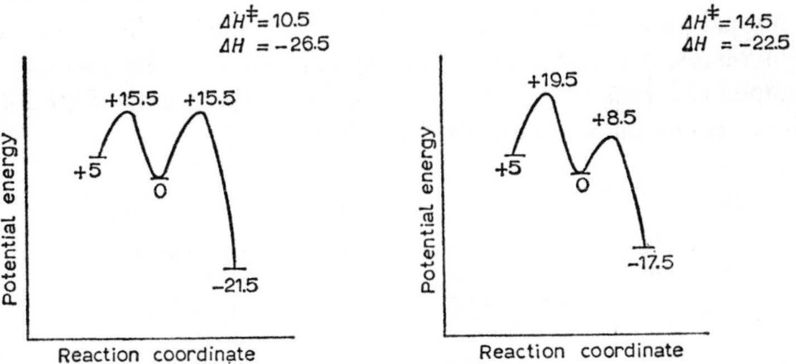

Fig. 52. Potential energy/reaction coordinate profile of $SMe^-/2,4$-$(NO_2)_2C_6H_3F$.
Fig. 53. Potential energy/reaction coordinate profile of $SMe^-/2,4$-$(NO_2)_2C_6H_3I$.

reducing T.St.1 has its full effect on mobility; and since the heavy nucleophile interaction still occurs, an unusually *high* F/I mobility ratio equal to 372000 at 0° is found. The calculated $\Delta E^{\ddagger}$ values for fluoro and iodo compounds sequentially in the *p*-nitro- and 2,4-dinitro-series are 17.5, 19.5, 10.5 and 14.5, and are closely confirmed by experiment *[3]*. Correspondingly $\log_{10}B$ (and $\Delta S^{\ddagger}$) values are 10.4_5 (−12.3), 8.3 (−22.4), 10.5_5 (−12.1), 8.05 (−23.5), *i.e.* near to normal for displacement of fluorine and low for displacement of iodine ($\Delta E^{\ddagger}$ in kcal·mole^{-1}, $\Delta S^{\ddagger}$ in e.v.).

Experimental data with thiophenoxide ion as reagent are somewhat lacking in detail, but it is known that the reactivity of *p*-chloronitrobenzene with thiophenoxide ion in 60% dioxane/water at 25° *[59]* is about 2.5 to 3 times greater than that of *p*-fluoronitrobenzene with thiophenoxide in methanol at 25° *[30]*. Some rates for dioxane/methanol are known to be slightly greater than for pure methanol *[44,60]*, and since this is a similar change in solvent, it may reasonably be inferred that the mobilities of fluorine and chlorine here are approximately equal. In the reactions with the 1-halogeno-2,4-dinitrobenzenes in methanol at 0°, the mobility of fluorine has been shown to be about twenty times greater than that of the other three halogens, which are closely similar, but with indications of slight reversal *[18,50,61]*. Bunnett *[61,62]* has ascribed these differences to polarisability factors favouring displacement of heavier atoms, but Miller and his co-workers *[2,18]* have been able to calculate $\Delta E^{\ddagger}$ for these reactions, and also from other data to suggest that $\Delta S^{\ddagger}$ has a normal value (for these anion-dipole reactions) with the fluoro compounds, and *low* unfavourable values with the other halogeno compounds. Full experimental data are available for the four 2,4-dinitro compounds and are given in Table 36.

TABLE 36

CALCULATED AND EXPERIMENTAL RESULTS FOR THE REACTION OF THE THIOPHENOXIDE ION IN METHANOL WITH 1-HALOGENO-2,4-DINITROBENZENES *[18]*

	Relative energy levels				
Halogen	I.St.	T.St.1	I.C.	T.St.2	F.St.
F	+5	+12.5	0	+15.5	−21.5
Cl	+5	+15.5	0	+10	−19.0
Br	+5	+16.5	0	+10	−19.5
I	+5	+16.5	0	+ 8.5	−17.5

	$\Delta E^{\ddagger}$ (kcal·mole^{-1})		$\Delta S^{\ddagger}$ (0°) (e.u.)		$\Delta G^{\ddagger}$ (0°) (kcal·mole^{-1})
	calcd.	found	predicted	found	found
F	10.5	10.7	Normal	−12.5	14.1
Cl	10.5	10.3	Low	−20.0	15.7_5
Br	11.5	9.8	Low	−20.6	15.4_5
I	11.5	10.7	Low	−17.9	15.6

The relative mobilities with thiophenoxide ion are seen to be a consequence of its being on the heavy nucleophile boundary of a region separating light and heavy nucleophile character, in which thiomethoxide ion shows intermediate behaviour *[18]*. The similarity in mobility of fluorine and the heavy halogens with thiophenoxide ion arises because that of fluorine is low, as with other heavy nucleophiles, and not because the mobilities of the heavy halogens are high. With thiophenoxide ion the low fluorine mobility consequent on formation of T.St.2 being rate-limiting in reactions with heavy anionic nucleophiles, cancels the normally high fluorine mobility due to the electronegativity (α-substituent) effect, whereas with less reactive heavy nucleophiles the still lower values of T.St.1 relative to T.St.2 completely reverse the mobility order resulting from the operation of the α-substituent effect.

We see in detail from Table 36 that with thiophenoxide ion the $\Delta E^{\ddagger}$ values are similar for all the halogeno compounds, but the heavy nucleophile interaction leads to lower values of $\Delta S^{\ddagger}$ and thus $\Delta G^{\ddagger}$ for the chloro, bromo, and iodo than for the fluoro compound. The situation is very similar in the mononitro series, though thiophenoxide ion is a little more clearly a heavy nucleophile there—a corresponding shift is also observed with thiomethoxide ion *[18]*—and the $\Delta G^{\ddagger}$ values for fluoro and iodo compounds are approximately equal. There appears to be no reason here to invoke additional polarisability factors facilitating displacement reactions when large atoms are brought together in forming the transition state *[61]*.

The calculations and resultant mobility patterns may be clarified by re-statement of the calculations in more general terms, recalling however that these reactions, in which a nucleophile forms a bond to carbon, take place in protic solvents.

The energy terms specific to the nucleophile, both as entering or leaving group are its ionisation and solvation energy, and the strength of the bond to the substrate. Large values of bond-strength tend to give high values of the energy of the I.St. or F.St. relative to that of the I.C., whereas large values of each of the other terms tend to give low values of the energy of the I.St. or F.St. For any particular resultant value of the energy of the I.St. or F.St. the corresponding T.St. is lower in energy for a weakly-bonded than for a strongly-bonded nucleophile, *i.e.* for heavy as against light nucleophiles as a class.

References p. 176

For the present discussion the key factor is that the C—F bond is outstandingly strong. The C—H bond approaches it, but other relevant C—X bonds are markedly weaker, decreasing down a column and along a row (left from Group VII). Solvation energy and ionisation energy roughly follow this pattern, except that second-row nucleophiles have somewhat higher values of ionisation energy than corresponding first-row nucleophiles.

The numerical values of the three energy terms are such that I.St. or F.St. energy values are much lower for Group VII nucleophiles than Group VI and so on. Therefore in displacement of halogen the formation of T.St.1 is rate-limiting for reaction with first-row nucleophiles. Even displacement of fluorine is not an exception, but by an extremely small margin. Thus the effect of the electronegativity of the leaving group on bond-formation by the entering group, and not the rupture of the bond to the leaving group, controls the mobility pattern.

At the other extreme the weaker heavy nucleophiles, which form relatively weak bonds but have moderate values of ionisation and solvation energy, have low values of the energy of both I.St. and T.St.1. The energy of T.St.2 is therefore markedly higher for displacement of fluorine, and may be so for other leaving groups, *e.g.* OR. The difference is large enough to counteract the electronegativity factor as well as to lead to bond-rupture as the factor controlling the mobility pattern.

As an intermediate class, the stronger heavy nucleophiles have high values of the energy of the I.St., but since the energy difference, T.St.1 minus I.St., is relatively small, this group of nucleophiles falls in a border region with T.St.1 and T.St.2 at comparable levels. Thus both electronegativity and bond-rupture factors may affect the mobility pattern. Moreover, a quite small difference in the energy terms can then cause a large change in the mobility pattern.

(*c*) *Mobility in reactions with neutral nucleophiles*

With a neutral nucleophile ($\ddot{Y}$) the result of initial coordination by the reagent is a species $\overset{+}{Y}$–$\overset{-}{Bzd}$–X ($\overset{-}{Bzd}$ represents the benzenide system). Since cationic groups have high mobility (see below), the reverse action is relatively facile. In most of the common neutral nucleophiles there is also an ionisable proton attached to the nucleophilic atom, so that the intermediate $H\text{–}\overset{+}{Y}\text{–}\overset{-}{Bzd}\text{–}X$ can transfer the proton to a second molecule of reagent ($\ddot{Y}$–H), then acting as base, to give a new intermediate $Y\text{–}\overset{-}{Bzd}\text{–}X$, which corresponds to that formed by reaction with an anionic nucleophile Y^-. This is equivalent to reduction of the rate of the reverse action. While this involves also the loss of a favourable and substantial α-substituent effect of positively charged nitrogen in reducing the level of T.St.2, proton transfer before elimination, if sufficiently exothermic, would outweigh this. A corresponding transfer of a proton from the F.St., which is an aromatic ammonium ion, would be more exothermic. These relationships are discussed in Chapter 6, p. 226, and illustrated in Fig. 71 of that Chapter.

Since in most cases reactions are nevertheless second order, are not subject to base catalysis, and have the typical aromatic mobility order, fluorine > heavy halogens, the level of T.St.1 relative to the I.St. for reactions of such neutral nucleophiles must be sufficiently high for formation of T.St.1 to be rate-limiting.

Because of lack of fundamental data, such calculations as have been made for anionic reagents *[2,18]* could only be carried out if estimates of the fundamental data required were made (see Chapter 6); but it seems probable that levels of T.St.1 and T.St. 2 are such that with some reagents, in reactions with substrates such as fluoro compounds, where the level of T.St.2 is normally high, the balance might be shifted to the formation of T.St.2 as the rate-limiting step, and the relative mobility of fluorine thus reduced. The study of reactions of 1-halogeno-2,4-dinitrobenzenes in methanol with *N*-methylaniline as nucleophile, made by Bunnett and his co-workers *[43,53,62]*, is particularly relevant to this. They showed that with this reagent addition of base does not affect the rate of displacement of chlorine or bromine, but does affect the rate of displacement of fluorine. Removal of the proton by a base inhibits the reverse step, which with the fluoro compound competes with the forward step, thus suggesting formation of T.St.2 as rate-limiting there. As illustrated in Fig. 54, k_{-1} for the reverse reaction is especially large compared with k_2 (and k_3) for the fluoro but not the other halogeno compounds.

For these reactions, Table 34A (*j*) shows that without a base catalyst the F/Cl mobility ratio (the G.R.F. of fluorine) in ethanol is unity.

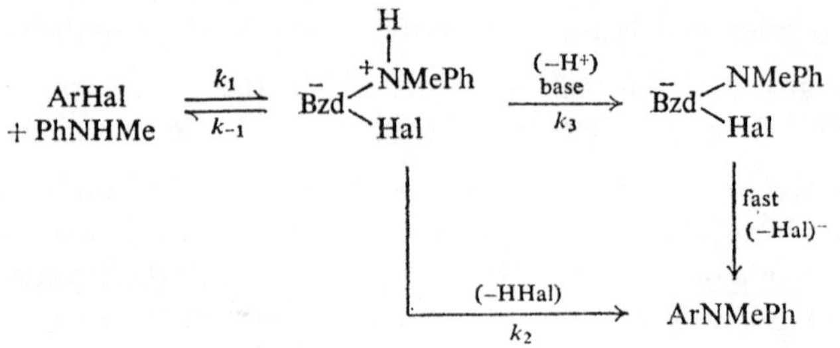

$$\begin{array}{c} \text{ArHal} \\ +\,\text{PhNHMe} \end{array} \underset{k_{-1}}{\overset{k_1}{\rightleftharpoons}} \overset{-}{\text{Bzd}}\!\!\begin{array}{l} \diagup \overset{+}{\text{N}}(\text{H})\text{MePh} \\ \diagdown \text{Hal} \end{array} \xrightarrow[k_3]{(-\text{H}^+)\ \text{base}} \overset{-}{\text{Bzd}}\!\!\begin{array}{l} \diagup \text{NMePh} \\ \diagdown \text{Hal} \end{array} \xrightarrow[\text{fast}]{(-\text{Hal})^-} \text{ArNMePh}$$

$$\overset{-}{\text{Bzd}}\!\!\begin{array}{l} \diagup \overset{+}{\text{N}}(\text{H})\text{MePh} \\ \diagdown \text{Hal} \end{array} \xrightarrow[k_2]{(-\text{HHal})} \text{ArNMePh}$$

Fig. 54. Reaction of *N*-methylaniline with activated halogenobenzenes.

Added potassium acetate inhibits the reverse reaction (k_{-1}) and increases this ratio to 12, so that the relative magnitude of reverse and forward specific reaction rates of the intermediate k_{-1}, and k_2—k_3 is involved in the low mobility of fluorine (see also Chapter 6, p. 225, and Fig. 71). Bunnett and his co-workers also showed that the combination of *N*-methylaniline with the fluoro compound (k_1) is in fact fast compared with that of the corresponding chloro and bromo compounds, and estimated the F/Cl ratio for this as about 70.

Without reliable detailed calculations one cannot predict with certainty the cause of the lower reactivity of tertiary amines than of primary and secondary amines, which allows side reactions with the former at the high temperatures required, and thus whether there is correspondingly a change in the energy levels such that, in some cases at least, T.St.2 formation may be rate-limiting with fluoro compounds. The discussion in Chapter 6 does not suggest that halogen mobility, in protic solvents, has a pattern for aliphatic tertiary amines different from that for secondary or primary amines, but the possibility of a change in this pattern with heterocyclic amines such as pyridine is not excluded. The relative reactivity of the classes of amine is clearly illustrated by the results of Brady, and of Miller and their co-workers *[63,64]*, from which the rate constants for replacement of chlorine in 1-chloro-2,4-dinitrobenzene, in ethanol or methanol at 25° are: $MeNH_2$, $3.16 \cdot 10^{-3}$; Me_2NH, $3.55 \cdot 10^{-2}$; Me_3N, $1.1 \cdot 10^{-6}$ ($l \cdot mole^{-1} \cdot sec^{-1}$). The corresponding ratios are 2870, 32300, 1. Steric factors probably contribute to the low reactivity of tertiary amines, for steric interactions between the reagent and the ring in T.St.1 may reduce the entropy and increase the enthalpy of activation.

A common side reaction is displacement of a nitro group. In the 2,4-dinitroanilinium compound produced, one nitro group is activated by the second nitro group and the NR_3^+ group (Chapter 4, p. 72). A similar displacement of the nitro group in the 2,4-dinitrobenzenediazonium ion has been reported recently *[65]*. Unfortunately there are very few data on halogen mobility in reactions with tertiary amines other than pyridine. The evidence there is conflicting and is discussed below (Section 3(*e*), p. 163). A further possible cause of difference in halogen mobility is hydrogen-bonding (Chapter 8, Section 3(*b*), p. 333). The concept of intramolecular hydrogen-bonding between hydrogen and fluorine in transition states was suggested by Chapman and Parker *[31]* for reaction of primary amines with fluoronitroaromatic compounds, and elaborated more recently by other workers *[41,66–68]*. It is probably a factor contributing to the higher reactivity of primary and secondary than of tertiary amines, especially in replacement of fluorine. Data in Table 34 are illustrative.

With the evidence favouring the view that in protic solvents at least, in the reactions of most neutral nucleophiles the rate-limiting step is that involving bond formation, the mobility pattern should be similar to that with anionic nucleophiles. However in the former type of reaction the highly solvated species are the T.St.'s and I.C. so that the solvation component of the entropy of activation should lead to lower (more negative) entropy for fluoro than the other halogens *[19]*, the order of $\Delta S^{\neq}$ being F < Cl < Br < I. Since this will to some extent counteract the α-substituent effect, the overall range of mobility is generally less. Further, where there is little or no differentiation of mobility due to the α-substituent effect, the solvation entropy effect could reverse the order. This will, for example, affect particularly the relationship of chlorine and bromine mobility, whereas adverse steric contributions might obscure this solvation effect with iodo compounds. The effect on heavy halogen mobility is perhaps more likely to be noticeable with less solvated T.St.'s, as with large and less reactive nucleophiles, *e.g.* more generally in reactions of aromatic than aliphatic amines. These conclusions are supported by the data of Tables 32 to 35.

A generally larger spread of mobility with aliphatic than aromatic amines is also supported by the now very common use of 1-fluoro-2,4-

dinitrobenzene for characterisation of aminoacids *[69]*. The mobility of fluorine relative to other halogens also appears to be high with oxygen nucleophiles *[54]*, though precise kinetic data are lacking. Whalley *[70]* for example discusses the higher reactivity of 1-fluoro- than 1-chloro-2,4-dinitrobenzene with alcohols. Gottlieb *[71]* has also reported that the hydrolysis of 1-fluoro-2,4-dinitrobenzene is facile and proceeds with water at 60°, whereas that of the chloro compound does not. Murto *[54]* has shown that picryl fluoride undergoes solvolysis in water at 25° about 22000 times faster than the chloride, but it does not follow that such a high ratio obtains with larger oxygen nucleophiles.

The experimental pattern with heavy neutral nucleophiles is now becoming clear. Parker *[46]* had reported the qualitative result that 1-fluoro- and 1-chloro-2,4-dinitrobenzene are similarly reactive with thiourea. Miller and Yeung *[48]* have investigated the reactivity of these two substrates with thiourea and selenourea in absolute methanol, and shown that the F/Cl mobility ratio at 100° is about 11 with thiourea and about 2.5 with selenourea. These are both low compared with values usual for oxygen and nitrogen neutral nucleophiles, *e.g.* the above references to water and alcohols, and references to neutral nitrogen nucleophiles in Tables 34A and 34B. Only *N*-methylaniline is comparable, and for this the evidence favours the formation of T.St.2 as rate-limiting. Further, whereas it is normal with light nucleophiles for the value of $\Delta E^{\ddagger}$ for reaction with the fluoro compound to be about 4 kcal·mole^{-1} less than with the chloro compound, it is only about 1–2 kcal·mole^{-1} less for the thiourea reaction and about 1–2 kcal·mole^{-1} *more* for the selenourea reaction. It appears therefore that changes in mobility of fluorine relative to other halogens, such as are found in comparing reactions of light with those of heavy anionic nucleophiles *[2,3]*, are found also in comparing reactions of light with those of heavy neutral nucleophiles *[48]*.

(*d*) *Solvent effect on mobility in reactions with anionic nucleophiles*

For the reactions of anionic nucleophiles with neutral substrates, the discussions of Miller and Parker *[1,2]* are relevant. The gist of these is that a change from protic to an aprotic solvent raises the energy of the initial state relative to transition states and intermediate complex, since the first has a higher solvation energy in protic solvents. When the group eliminated forms an anion there is a corresponding relationship of the final to intermediate state. The main effect of solvent changes is in fact on reagent reactivity (see Chapter 6, p. 196), but mobility may also be affected. Thus the equilibrium in the I/Br exchange of 1-halogeno-2,4-dinitrobenzenes is shifted in favour of the state iodide ion/1-bromo-2,4-dinitrobenzene in acetone as compared with methanol *[1]*, *i.e.* for these reagent-substrate combinations iodine becomes more mobile than bromine. Other relevant data *[1,33]* are that thiocyanate ion reacts faster with both 1-fluoro- and 1-iodo-2,4-dinitrobenzene in acetone than in methanol, but whereas the iodo compound gives good yields of the aryl thiocyanate (though there is always some side-reaction with thiocyanate ion as a reagent), hardly any aryl thiocyanate is produced from the reaction with the fluoro compound. Instead the products correspond to a complex decomposition of the intermediate [F–Bzd–SCN]$^-$, probably initially to ArS$^-$ and F–CN. Such a decomposition, alternative to the normal elimination of thiocyanate ion, suggests that it cannot occur or is very difficult in the aprotic solvent, *i.e. for this path*, reaction does not proceed beyond the intermediate complex. Similarly iodide ion, which does not react appreciably with 1-fluoro-2,4-dinitrobenzene in methanol, does react in acetone though still very slowly, but by a path which does not seem to involve simple displacement of fluorine by iodine ion *[1]*. In contrast and supporting the concept that fluoride is not displaceable directly, fluoride ion in a dipolar aprotic solvent itself displaces other halogens from 1-halogeno-2,4-dinitrobenzenes *[1,26,27,71–74]* and thiocyanate from 1-thiocyanato-2,4-dinitrobenzene *[1]*.

The formation of the well known Meisenheimer complexes, such as [OMe–Bzd–OMe]$^-$ (where Bzd here is picryl), is an example of reactions in which the intermediate complex is stable; and reactions which stop at the intermediate complex are also very familiar when there is no readily displaceable group at activated points, *e.g.* in many colour reactions of *m*-dinitrobenzene and 1,3,5-trinitrobenzene. The Janovsky reaction *[75–82]* is an example, and it is relevant to the present discussion that it and similar reactions are better carried out in aprotic solvents. The only evidence in any way contrary to the discussion is a similarity of F/I mobility ratios in reactions of sodium

References p. 176

azide in methanol and dimethylformamide [Table 33A (*a*) and (*b*)]; whereas one would expect a smaller ratio in dimethylformamide. There is no reliable explanation but it is perhaps relevant that this solvent has the positive end of its dipole more exposed—because of its molecular geometry, and because it is dispersed by internal conjugation of the amide group—than is usual in dipolar aprotic solvents. Further some specific effect of the cation (sodium) may be involved, and it is relevant to this that in the reactions referred to in which fluoride ion displaces other halogens in dipolar aprotic solvents, reasonable yields are obtained with potassium fluoride, still better with rubidium and cesium fluoride, whereas lithium and sodium fluoride are ineffective *[72,83]*.

In contrast to the large solvent effects just discussed the admixture of an aprotic solvent with a protic solvent has much less effect, though in the same direction. The mixtures still appear to behave essentially as protic solvents *[74, cf. 84]*. The effect on relative mobility appears not to have been investigated.

(*e*) *Solvent effect on mobility in reactions with neutral nucleophiles*

In a reaction between a neutral nucleophile and a neutral substrate, a highly polar transition state is formed; and this is expected to be formed more readily in a protic than in an aprotic solvent. The data in Tables 33 and 34 illustrate this, *e.g.* with piperidine at 90°, *p*-chloronitrobenzene reacts faster in ethanol than in benzene. It is perhaps somewhat unexpected at first sight that this does not arise from a lower value of $\Delta E^{\ddagger}$, partly counteracted by a lower value of $\Delta S^{\ddagger}$, in the protic than the aprotic solvent; but instead by a higher value of $\Delta S^{\ddagger}$ partly counteracted by a higher value of $\Delta E^{\ddagger}$. It may be concluded that the polarity and/or polarisability of many aprotic solvents is much higher than is usually assumed. Such a solvent can then substantially lower the activation energy of formation of a polar transition state, but at the expense of a very marked decrease of the entropy of the solvent molecules. A protic, hydrogen-bonded solvent in such circumstances loses hydrogen-bonding energy. The increase in entropy due to this opposes the reduction due to the solvation process and thus modifies the pattern of $\Delta E^{\ddagger}$ and $\Delta S^{\ddagger}$ changes. It is relevant that the addition of benzene to methanol makes it a more nearly ideal solvent for salts *[85,86]*. This is regarded as consequent on just such a depolymerisation of the hydrogen-bonded structure of methanol by benzene. The same explanation was used by Miller and his co-workers *[87]* in their discussion of benzene/methanol as a solvent in aromatic S_N reactions.

As discussed in an earlier section, some neutral nitrogenous nucleophiles give a mobility pattern less strongly favouring fluorine displacement relative to the other halogens than do the anionic first row nucleophiles. Thus reduction, or even reversal in the relative mobility caused by a solvent change, is more likely with these neutral reagents. It is seen in the reaction of *N*-methylaniline, which is also sterically demanding, with 1-halogeno-2,4-dinitrobenzenes in ethanol or nitrobenzene [Table 34A (*i*) and (*j*)]. Both a lower rate and reduction in fluorine mobility in the aprotic compared with protic solvents are shown. With aniline in ethanol or acetone a similar change is noticeable [Table 34A (*g*) and (*h*)] though both are at a higher level of fluorine mobility.

As discussed above, the pattern of halogen mobility with tertiary amines is uncertain, though likely to be fluorine > heavy halogens, with aliphatic amines in a protic solvent. The position in aprotic solvents is uncertain since fluorine mobility is less there. Regrettably data on the relative mobility of fluorine and other halogens with aliphatic tertiary amines are lacking. The pattern is still less predictable with heterocyclic tertiary amines, such as pyridine. Bevan and his co-workers *[41]* [Table 34A (*h*) and 34B (*k*)] have reported that the F/Cl mobility ratio is higher for aniline than pyridine in acetone, though more than unity in both cases. Unpublished work of Lantzke and Miller *[64]* leads to an estimate of k_2 for the reaction of pyridine with 1-chloro-2,4-dinitrobenzene in methanol at 50° as $5.53 \cdot 10^{-6}$, which in comparison with Bevan's value for acetone ($6.54 \cdot 10^{-8}$) confirms the solvent effect, *viz.* a faster reaction in a protic than in an aprotic solvent with tertiary amines also. However, whereas Bevan quotes the k_2 values $2.13 \cdot 10^{-7}$ for 1-fluoro- and $6.54 \cdot 10^{-8}$ ($\text{l} \cdot \text{mole}^{-1} \cdot \text{sec}^{-1}$) for 1-chloro-2,4-dinitrobenzene in acetone at 50° (F > Cl), Bolton and Miller *[27]* were unable to prepare 2,4-dinitrophenylpyridinium fluoride in acetone in conditions in which more than 90% yield of the corresponding chloride was obtained, thus giving the

References p. 176

order F < Cl for an aprotic solvent. They obtained only a little tar as product, and further were able to recover starting materials readily from reaction mixtures. They showed also that under the conditions of Bevan's measurements fluoride ion reacts readily with 2,4-dinitrophenylpyridinium ion to give 1-fluoro-2,4-dinitrobenzene and pyridine. This casts doubts on Bevan's results for the fluoro compound in acetone, except in so far as it may reflect the simple addition of pyridine to the aromatic system, forming an intermediate complex which cannot decompose by direct elimination of fluoride. A comparison of acetone with a protic solvent such as methanol in attempted reaction of pyridine with the fluoro compound is impracticable because in alcohols the fluoride ion produced catalyses alcoholysis, and the product is the ether *[73,83]*. Some further investigation of these reactions is clearly necessary, especially of halogen mobility.

4. Mobility of Groups other than Halogens in Addition–Elimination Aromatic S_N2 Reactions

(*a*) *General*

From the foregoing sections characteristic aromatic mobilities should be found with first row anionic nucleophiles, especially in protic solvents, and also most first row neutral nucleophiles of the type ($\ddot{Y}$–H). Electronegativity is clearly a key factor and although bond breaking is unimportant in these conditions, differences in bond strength between the aromatic bond of the initial state and aliphatic bond of the transition state are likely to be important in some cases.

The question of reversal of mobility in a vertical group of the Periodic Table with heavy nucleophiles, such as is found in halogen displacement, has not been investigated experimentally, though some theoretical predictions have been made for anionic nucleophiles *[2,3,18]*. The theoretical electronegativity orders of mobility: $X^+ > X^+–Y^- > X^0$, and $F > OR > NR_2$ have been investigated by Miller and his co-workers *[1,2,3,18,26,27,33]* (see Table 37).

(b) *Quantitative results*

Data from references already quoted and references *88* to *90* are used. Table 37 is concerned with mobility in 4-nitro-1-X-benzenes. A wider selection of data is available with 2,4-dinitro-1-X-benzenes, given in Table 38. Some kinetic information on picryl compounds is available and is given in Table 39. In discussion, additional qualitative information is also utilised.

TABLE 37

LEAVING GROUP MOBILITY IN SOME 1-X-4-NITROBENZENES (*a*) WITH OMe^- IN MeOH AT 50° *[2,3,33]*; (*b*) WITH N_3^- IN MeOH AT 100° *[1,26,27]*

Series	Leaving group (X)	Rate constant k_2 ($l \cdot mole^{-1} \cdot sec^{-1}$)	Mobility rel. to Cl = 1 (or G.R.F.)	$\Delta E^{\ddagger}$ (kcal · $mole^{-1}$)	$\log_{10} B$
(*a*)	SMe_2^+	$1.77 \cdot 10^0$	$2.09 \cdot 10^5$	24.5_5	16.8_5
	NMe_3^+	$2.00 \cdot 10^{-2}$	$2.36 \cdot 10^3$	20.0	11.8
	NO_2	$3.11 \cdot 10^{-3}$	$1.83 \cdot 10^2$[a]	22.4	$12.6_5(12.3_5)$
	N_3	$8.91 \cdot 10^{-6}$	1.05	25.4_5	12.1_5
	NMe_2	very slow[b]	$\ll 1$	—	—
	OMe	slow[c]	Probably close[c] to unity	25.5[d]	—
	F	$2.64 \cdot 10^{-3}$	$3.12 \cdot 10^2$	21.2	11.7_5
	Cl	$8.47 \cdot 10^{-6}$	1	24.0_5	11.2
(*b*)	NO_2	$5.31 \cdot 10^{-4}$	78.7[a]	—	—
	F	$2.32 \cdot 10^{-4}$	68.8	24.1	10.4_5
	I	$3.38 \cdot 10^{-6}$	1	27.8	10.9

[a] Half the rate ratio for statistical reasons, and correspondingly in series (*a*) a $\log_{10} B$ value reduced by 0.3 ($\log_{10} 2$) is shown in parentheses. [b] Estimated (see text). [c] Refs. *88* and *89*, and *cf.* Table 38. [d] Calculated value *[2]*.

(c) *Factors affecting mobility*

A discussion of the mobility order based on difference in the polar category of the displaced group can be given for a series of nitrogenous groups, *viz.* trimethylammonio (cation), nitro and azido (dipole), and dimethylamino (neutral) groups, towards methoxide ion in methanol.

TABLE 38

LEAVING GROUP MOBILITY IN SOME 1-X-2,4-DINITROBENZENES (*a*) WITH OH^- IN DIOXANE/WATER (60/40 v/v) AT 50° *[38]*; (*b*) WITH OMe^- IN MeOH AT 0° *[3,36,90]*; (*c*) WITH OMe^- IN MeOH/MeOAc (1/1 v/v) AT 0° *[36]*; (*d*) WITH OMe^- IN MeOH/DIOXANE (1/1 v/v) AT 0° *[36]*; (*e*) WITH KOH IN MeOH AT 20° *[91]*; (*f*) WITH PIPERIDINE IN MeOH AT 0° *[53]*; (*g*) WITH OH^- IN H_2O AT 50° *[92]*; (*h*) WITH SPh^- IN MeOH AT 0° *[18,50]*; (*i*) WITH Br^- IN gl. HOAc AND IN MeOH AT 100° *[1,27]*

Series	Leaving group	Rate constant k_2 ($l \cdot mole^{-1} \cdot sec^{-1}$)	Mobility rel. to		4′-substituent effect on mobility	$\Delta E^{\ddagger}$ ($kcal \cdot mole^{-1}$)	$\log_{10} B$ (or $\Delta S^{\ddagger}$ in e.u.)
			Cl = 1 (G.R.F.)	I = 1			
(*a*)	Cl	$7.47 \cdot 10^{-3}$	1	3.50	—	16.1	8.7_5
	I	$2.14 \cdot 10^{-3}$	0.287	1	—	17.4_5	9.1_5
	OMe	$1.13 \cdot 10^{-2}$	1.51	5.28	—	17.5_5	9.9_5
	OPh	$5.42 \cdot 10^{-3}$	0.727	2.53	—	17.7_5	9.7_5
(*b*)	F	1.74	890	5650	—	13.5	11.0_5
	Cl	$2.00 \cdot 10^{-3}$	1	6.50	—	17.4_5	11.2_5
	I	$3.05 \cdot 10^{-4}$	0.154	1	—	18.9_5	11.7
	OMe[a]	*ca.* $8 \cdot 10^{-5}$	*ca.* 0.04	*ca.* 0.4	—	19.5	11.5
	OPh	$4.57 \cdot 10^{-4}$	0.229	1.50	—	18.9_5	11.7
(*c*)[b]	OPh	$1.42 \cdot 10^{-3}$	—	—	1	18.8_5	12.2_5
	p-Nitrophenoxy	$2.36 \cdot 10^{-2}$	—	—	16.6	16.9_5	11.9_5
(*d*)[b]	OPh	$1.65 \cdot 10^{-3}$	—	—	1	18.7_5	12.2
	p-Iodophenoxy	$4.02 \cdot 10^{-3}$	—	—	2.44	17.7_5	11.8
	p-Phenoxyphenoxy	$3.07 \cdot 10^{-3}$	—	—	1.85	17.3	11.3_5
	p-Nitrothiophenoxy	$6.09 \cdot 10^{-4}$	—	—	0.369	19.7	12.4
(*e*)[c]	Cl[d]	$1.76_5 \cdot 10^{-2}$	1	—	—	—	—
	OPh	$4.73 \cdot 10^{-3}$	0.268	—	1	—	—
	p-Nitrophenoxy	$6.46 \cdot 10^{-2}$	3.66	—	13.6_5	—	—
	p-Chlorophenoxy	$1.07 \cdot 10^{-2}$	0.607	—	2.26	—	—
	β-Naphthoxy	$8.12 \cdot 10^{-3}$	0.460	—	1.72	—	—

	m-Cresoxy	5.07·10	0.209	[illegible]	0.781	—	—
	p-Cresoxy	$2.84 \cdot 10^{-3}$	0.162	—	0.601	—	—
	o-Cresoxy	$1.43 \cdot 10^{-3}$	0.0810	—	0.299	—	—
(*f*)	F	1.5	770	3280	—	—	—
	Cl	$1.95 \cdot 10^{-3}$	1	4.27	—	11.6	(−30.2)
	I	$4.57 \cdot 10^{-4}$	0.234	1	—	12.0	(−31.7)
	NO_2	$4.03 \cdot 10^{-1}$	207	882	—	—	—
	p-$CH_3C_6H_4SO_3$	$4.53 \cdot 10^{-2}$	24.2	99.0	—	—	—
	PhSO	$2.15 \cdot 10^{-3}$	1.10	4.70	—	10.8	(−33.3)
	$PhSO_2$	$1.43 \cdot 10^{-3}$	0.734	3.13	—	12.0	(−29.3)
	p-$NO_2C_6H_4O$	$1.35 \cdot 10^{-3}$	0.693	2.96	—	10.5	(−35.3)
(*g*)	F	$1.17 \cdot 10^{0}$	F/OMe (234)354[e]	—	—	16.9_5	11.5_5
	OMe	$4.99 \cdot 10^{-3}$	(1)1.51	—	—	20.5	11.5_5
	OEt	$2.83 \cdot 10^{-3}$	(0.568)0.858	—	—	19.9_5	10.9_5
(*h*)	NO_2, $PhSO_2$, PhSO, *ar.*-$C_5H_5N^+$	> 100	> 100	> 100	—	—	—
	F	129	33.2	25.0	—	10.7	(−12.5)
	Cl	3.89	1	0.753	—	10.3	(−20.0)
	I	5.17	1.33	1	—	10.7	(−17.9)
	p-$NO_2C_6H_4O$	1.56	0.400	0.302	—	12.1	(−15.2)
(*i*)[f]	*ar.*-$C_5H_5N^+$	$8.5 \cdot 10^{-3}$	—	*ca.* 10^3		12.5	3.6
	I	$4.48 \cdot 10^{-6}$	—	1		24.9_5	9.3

[a] Calculated by Miller's procedure *[2]*. [b] Lead to values of ρ (with 4′-substituents) of 0.96 and 1.21 respectively. [c] Lead to a value of ρ (with 4′-substituents) of 1.46. [d] The result for Cl is with OMe^-/MeOH, but any error in comparison is constant and likely to be extremely small *[92]*. [e] G.R.F. values obtained by assuming the F/Cl mobility ratio is the same with OH^- in H_2O at 50° as with OH^- in dioxane/water (60/40 v/v) at 50°. [f] Reaction with a pyridinium salt in HOAc, and with an iodo compound in MeOH. Data are not precisely comparable, because of differences in electrical character of the reactions and of the character and dielectric constants of solvent.

TABLE 39

LEAVING GROUP MOBILITY IN REACTIONS OF SOME 1-X-2,4,6-TRINITROBENZENES WITH OH^- IN H_2O *[45]*

Leaving group	Rate constant k_2 $(l \cdot mole^{-1} \cdot sec^{-1})$	Mobility rel. to Cl = 1 (G.R.F.)
F	$7 \cdot 10^2$	1400
Cl	$5 \cdot 10^{-1}$	1
OMe	$1._6 \cdot 10^0$	$3._2$
OPh	$1._5 \cdot 10^0$	3

The usual detailed rate data are available for the first three groups and estimates are made for the last. Thus Bevan and his co-workers *[91]* report that methoxide ion does not react with *p*-nitroso-*N*,*N*-dimethylaniline at 30° during 76 h, and from the data of Miller and his co-workers *[26,28,92]* the corresponding *p*-nitro compound is expected to be about fifty times less reactive still. In contrast, from kinetic data for *p*-chloronitrobenzene *[28]* it may be calculated that several percent reaction occurs in these conditions. Though no precise value is known therefore, the mobility of the dimethylamino group relative to chlorine must be very small. Its low mobility compared to neutral oxygen (methoxyl) and fluorine (Table 37) is due not only to the smaller electronegativity of nitrogen than either oxygen or fluorine, but also to an expected larger difference between the strengths of the Ar–X bond of the initial state and Alph–X bond of the rate-limiting transition state, which may be assumed to exceed values estimated for fluorine and methoxyl, *viz.* 4 and 9 $kcal \cdot mole^{-1}$ respectively *[2,93]*.

The azido group is more electronegative than the dimethylamino group but less electronegative than the nitro group and its mobility is expected to be intermediate between these, as is found. Miller *[2,18]* calculated that the $\Delta E^{\ddagger}$ value for the reaction of methoxide ion with *p*-nitrophenyl azide is approximately equal to that for the reaction of the chloro compound. This has been confirmed by Miller and Wong *[3]* and is in accord also with qualitative data *[94,95]*.

Information about the mobility of methoxyl *[96,97]* with alkoxides as reagent is mainly qualitative, but includes some kinetic data for methoxyl exchange *[88]* in the dinitro series. The data are in full accord with the mobility obtained by using hydroxide, and other reagents, in the dinitro series (Table 38).

The experimental data of Table 37 fully confirm the theoretical order $NMe_3^+ > NO_2 > NMe_2$; and $F > OMe > NMe_2$. Both the trimethylammonio and the nitro group are very mobile. The high mobility of the nitro group found here is also in accord with data from many earlier references quoted by Bunnett and Zahler *[11]*. The kinetic data of Table 37 suggest that the electronegativity of the nitro group, in terms of the α-substituent effect *[2]*, is equivalent to a reduction of 3 kcal·mole^{-1} in $\Delta E^{\ddagger}$.

The $\log_{10} B$ (and $\Delta S^{\ddagger}$) value for the reactions of the trimethylammonium compound with methoxide ion is about the same as that for the neutral 4-NO_2-1-X-compounds. This is in contrast with anion–cation aromatic S_N reactions, in which the cationic group is not displaced; a typical result there is a rise in $\log_{10} B$ of about 3 to 3.5, as compared with anion–neutral substrate reactions *[e.g. 33]*. The reaction of methoxide ion with the dimethylsulphonium compound, however, has an unusually large value of $\log_{10} B$, about 5 units higher than for anion–neutral substrate reactions, but the $\Delta E^{\ddagger}$ value is high also. These inconsistencies do not seem to stem from experimental error. While steric factors may be involved in these data, as suggested by Bolto and Miller *[33]*, a less speculative explanation must await corresponding measurements with neutral nucleophiles and larger reagents of both kinds.

The data of Tables 38 and 39 confirm and extend these patterns. The high mobility of cationic and dipolar nitrogenous groups is confirmed; and that of dipolar sulphur-containing groups (phenylsulphinyl and phenylsulphonyl) is added. An unusual feature of the latter two is that their mobility is substantially lower with piperidine than with thiophenoxide ion as reagent. This may involve a high rate of reversal of the original addition (k_{-1} as in Fig. 54) with piperidine. It may be recalled that with *N*-methylaniline a high k_{-1} value is observed in replacement of fluorine, but not the less electronegative heavier halogens. However even if this is the case, reasons for such a phenomenon with dipolar sulphur-containing but not dipolar nitrogenous groups are not obvious. There is substantial qualitative

References p. 176

evidence for the high mobility also of cationic and dipolar iodine *[98–102]*.

The displacement of the diazonium group seems to involve a special mechanism, except possibly in tetrazotised *p*-phenylene diamine (Chapter 2, p. 32). It seems safe therefore to assume intrinsically high mobility, lessened by the conjugation which would make the difference in bond energies of the aromatic bond of the initial state and aliphatic type bond in the transition state quite large, *cf.* alkoxyl and dialkylamino groups.

A fuller discussion of the mobility of alkoxyl groups is possible. The expected high mobility of a group such as methoxyl (a convenient standard), due to the high electronegativity of oxygen, is reduced by a larger difference in the strengths of the bonds to the group in the initial and transition state than is found with the halogens. Whereas the electronegativities of F, O, and Cl, as given by Pauling's indices *[103]*, are 4.0, 3.5, 3.0, the G.R.F. values (mobility relative to chlorine) of alkoxy and chlorine are close to equality, so that with many reagents mobility is between 10^{-2} and 10^{-3} times that of fluorine.

Changes in R of OR groups have also been investigated. A comparison of methoxyl and phenoxyl groups suggests a somewhat higher mobility for the latter since aromatic carbon is more electronegative than aliphatic carbon, and also there is an alternative conjugation path for the unshared electrons on oxygen to the ring in which substitution occurs, thus minimising the differences in the bond strengths in the initial and transition state. This view is supported by experimental confirmation of the order OPh > OMe with aniline, and the sodium salt of ethyl acetoacetate as reagent *[104]*. With hydroxide ion in dioxane/water *[44]* however, the order is OMe > OPh, but this probably reflects special features, noticeable also in the different stoichiometry of the reactions of the two ethers, *viz.*

$$\mathrm{ArOMe + OH^- \rightarrow ArO^- + MeOH} \tag{1}$$

$$\mathrm{ArOPh + 2OH^- \rightarrow ArO^- + H_2O + OPh^-} \tag{2}$$

In the displacement of methoxyl, the initially formed hydroxylic addition compound is deprotonated by the leaving methoxyl group; whereas in the other case the less basic leaving phenoxyl group does not fulfil this function, which is carried out instead by a second hydroxide ion. This suggests that there is a relative stabilisation of the rate-limiting transition state for the methoxy as compared with the phenoxy compound. This is illustrated in Fig. 55 in which dotted lines are used to represent hydrogen bonds and broken lines partial bonds.

Variation of substituents in a replaceable aryloxyl group demonstrates the effect of changed electronegativity, and this is presumably accompanied by changes in magnitude of the alternative conjugation: electron-releasing substituents increase mobility and electron-withdrawing substituents decrease it. The data of Table 38 can be used to estimate Hammett reaction constants (ρ) for such substituents in the 4′-position. Values so obtained are similar to ρ-values for 4′-substituents in attached rings as activating groups (Chapter 4). These effects of electron-withdrawing substituents are also illustrated by the

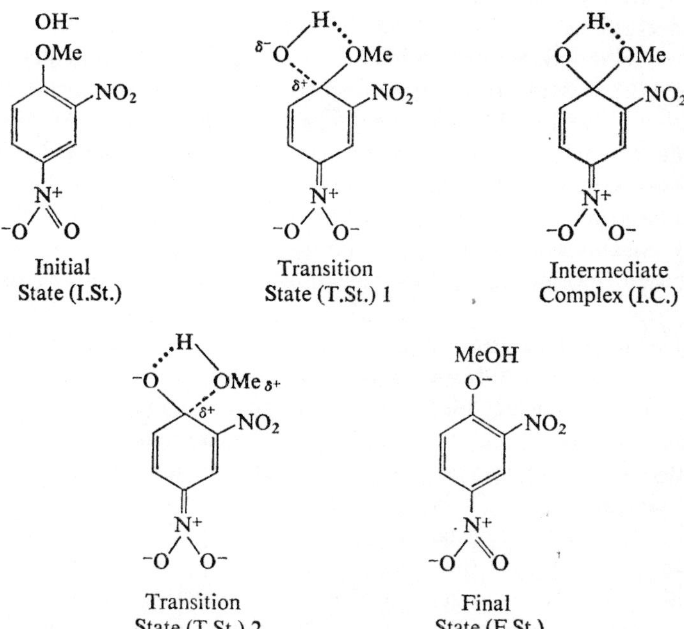

Fig. 55. Hydrolysis of 2,4-dinitroanisole, illustrating hydrogen-bonding in reaction intermediates and deprotonation of the added hydroxyl by the leaving methoxyl group.

References p. 176

higher mobility of the toluene-*p*-sulphonoxyl than either the methoxyl or the phenoxyl group. Among supporting qualitative data are those from the reactions of 2,4-dinitroanisole with a variety of reagents such as piperidine, aniline, hydroxylamine, alkoxide ions, and water *[81,89,104–109]*.

From the preceding discussion, whereas there is a large difference in mobility between fluorine and the heavier halogens, any difference in the chalcogens will be only a relatively small difference in mobility of oxygen-containing groups and those containing the heavier elements of the group. The application of Miller's calculations *[2,18]* suggests that the mobility of methoxyl is similar to that of thiomethoxyl. Curd and his co-workers *[110]* have reported on a number of mobile groups such as halogen, alkoxyl, aryloxyl and thioalkoxyl without differentiating their mobility. Gitis and his co-workers *[82]* however, report that the methoxyl group is more mobile than the thiomethoxyl group in transesterification. Barlotti and Cerniani *[111]* report the order OAr > SAr ~ SeAr with amine reagents. These data are qualitative, and kinetic data are lacking, except that results of Beckwith and Miller *[36b]* [Table 38 (*d*)] confirm the order OAr > SAr (Ar = *p*-nitrophenyl) with methoxide ion also. The mobility order OR > SR has been found also in heterocyclic systems *[112]*. The experimental results available all indicate that the difference in mobility between oxygen and sulphur groups is substantially less than is common between fluorine and chlorine, supporting the concepts discussed above.

The thiocyano group is presumably much more electronegative than the thiomethoxyl group, but this would not lead to a high mobility with first row nucleophiles unless the difference is enough to lead to an electronegativity greater than that of chlorine. Because of similarities between thiocyanogen and halogen systems, the former is classed as pseudo-halogenoid. The size factor has led to the common comparison with the heavier halogens, but there is no information on such factors as electron affinity and electronegativity. Unpublished results of Miller and his co-workers *[3c,22b]* show that the mobility of the thiocyano group is similar to that of fluorine in reactions with methoxide ion in methanol, and that its activating power as a 4-substituent in 1-halogeno-2-NO_2-4-X-benzenes is similar to that of the nitro and nitroso group. This suggests that the electro-negativity of the thiocyano group is large and probably more nearly comparable with that of fluorine than with the less electronegative heavy halogens.

The previous arguments, extrapolated to consideration of replaceable groups attached by Group V elements, suggest that the order $NR_2 < PR_2$ is likely for the first row as well as heavy nucleophiles, but experimental data are lacking.

While data are not available to permit precise calculations (as in ref. *2*) for displacement of such species as hydrogen and alkyl with their bonding electrons, it can be estimated that such reactions are very endothermic, and have very high values of the energy of T.St.2 relative to T.St.1 and the I.C. The formation of an intermediate complex by addition of a nucleophile may however be very facile. Thus even where there is also a readily displaceable group in a ring, complexes formed by addition at methine carbon may be formed as well as those leading to a normal substitution (see Chapter 1, pp. 8, 12).

Hydride ion displacements in strongly basic conditions are well-known in organic chemistry. The Chichibabin reaction is perhaps the best-known in aromatic S_N reactions and is discussed in Chapter 7 (p. 284). Hydride ion displacements in aromatic systems are also facilitated by oxidation (including oxidation by the substrate itself); or by an electron-redistribution which permits eventual displacement of the hydrogen as a proton (see Chapter 9, p. 373).

The formation of alizarin (1,2-dihydroxyanthraquinone) by oxidative alkali fusion of anthraquinone-2-sulphonic acid is an example of such a displacement, and, in view of the formation of resorcinol from benzene-*p*-disulphonic acid by alkali fusion, the alizarin synthesis may involve substitution with rearrangement as well as displacement of hydrogen.

Even groups with an overall negative charge may be displaced, but, from available data, only when dipolar (but not necessarily with a formal dipole), and attached to the ring by the positive end. Such groups may be assumed to be electron-withdrawing despite the overall charge. Sulphonate and arsonate groups are known examples, and their electron-withdrawing character is confirmed by their ability as substituent groups to *activate* displacement of other groups in aromatic S_N reactions (Chapter 4, p. 77).

References p. 176

The sulphonate and arsonate groups are quite readily displaced when attached to rings carrying substituents activating for aromatic S_N reactions, while substituent effects on their displacement correspond to those in typical activated substitutions. Thus 2,4-dinitrobenzene sulphonic acid reacts readily with sulphide, methoxide, and thiophenoxide ions, and ammonia *[113]*. Similarly a sulphonate group in the 2- or 4-position of a pyridine ring is mobile *[114]*. Sulphonic acids are strong acids, and the reagents used are bases as well as nucleophiles, so that displacement of the anionic form may be assumed. Though these data are qualitative it seems quite definite that the sulphonate group is less mobile than a sulphone group as would be expected. The extended theoretical order of mobility $X^+ > X^+\text{–}Y^- > X^0 > X^-$ is thus supported here.

Barber *[115]* has shown that the arsonate group is displaced from *o*-nitrophenylarsonic acid in concentrated sodium hydroxide solution. In 5-halogeno(Cl,Br,I)-2-nitrophenylarsonic acids the arsonate group *ortho* to the nitro group is displaced rather than a halogen *para* to it. Since with anionic reagents there is a small preference for activation *para* rather than *ortho* to a nitro group (Chapter 4, p. 101), it may be concluded that the arsonate group is more mobile than chlorine, bromine or iodine (*i.e.* its G.R.F. is greater than unity). Barber also commented on the arsonate group as an activating group. Bunnett *[116]* has referred to facile substitution of sulphonate and arsonate groups *inter alia*, by sodium amide in boiling piperidine (piperidide ion as reagent) indicating that these are apparently direct substitutions, *i.e.* by the addition–elimination and not the elimination–addition or benzyne S_N2 mechanism (Chapter 3).

5. Group Mobility in Aromatic S_N1 Reactions

There is very little information on this because the reactions known with any certainty involve only displacement of diazonium ions, discussed in Chapter 2. As regards reactions of unactivated compounds such as halogeno-benzenes and -toluenes, there may be uncatalysed S_N1 as well as parallel addition–elimination and benzyne S_N2 mechanisms, but there is no evidence of group mobility which can be ascribed to reactions proceeding by an uncatalysed S_N1 mechanism.

In some reactions catalysed by copper *[117–119]* the halogen reactivity order is I > Br > Cl > F. This is correlated with a key-step involving coordination by halogen to copper (see Chapter 8, p. 341). Details of the mechanism are uncertain, but it is convenient to mention these reactions in this section.

6. Group Mobility in Aromatic Elimination–Addition S_N2 Reactions

In these reactions we are concerned with halogen mobility. This has been discussed in Chapter 3. In summary, initial loss of hydrogen takes place more readily *ortho* to fluorine but the elimination of fluorine itself is slower. Overall therefore the loss of hydrogen halide in the elimination stage can never lead to a high mobility of fluorine compared with other halogens, and it is commonly less than that of chlorine. The subsequent stage does not involve the halogen, the intermediate being the same whichever halogen is eliminated. The points are clearly illustrated in Table 32 in which the reaction with methoxide is a normal S_N2 reaction, whereas piperidide and amide ion reactions are benzyne reactions.

The reactions with neutral piperidine would not be expected to proceed by the benzyne mechanism, and while the quoted values, if confirmed, could be regarded as offering some support for a one-stage mechanism in these unsubstituted compounds with this reagent, no supporting data were subsequently published. In their absence, and lacking the key results for the reaction with fluorobenzene, it seems unwise to attach much weight to them. In any case the order could arise from a solvation entropy effect, as occurs more mildly in reactions of neutral reagents with more activated substrates, in which the solvation of the species Hal–$\overset{-}{\text{B}}$zd–$\overset{+}{\text{P}}$ip is in the order F > Cl > Br > I, and thus $\Delta S^{\neq}$ values are in the order F < Cl < Br < I (*i.e.* F most negative).

REFERENCES

1 J. MILLER AND A. J. PARKER, *J. Am. Chem. Soc.*, 83 (1961) 117.

2*a* J. MILLER, *J. Am. Chem. Soc.*, 85 (1963) 1628; *b* D. L. HILL, K. C. HO AND J. MILLER, *J. Chem. Soc., B*, (1966) 299.

3 J. MILLER AND K. W. WONG, (*a*) *Austral. J. Chem.*, 18 (1965) 117; (*b*) *J. Chem. Soc.*, (1965) 5454; (*c*) unpublished work.

4 G. M. BENNETT AND I. H. VERNON, *J. Chem. Soc.*, (1938) 1783.

5 C. K. INGOLD, *Structure and Mechanism in Organic Chemistry*, Cornell Univ. Press, Ithaca, N.Y., 1953, (a) p. 338–345; (b) p. 357–360.

6 J. HINE, *Physical Organic Chemistry*, 2nd Ed., McGraw-Hill, New York, 1962, p. 182–185.

7 C. A. BUNTON, *Nucleophilic Substitution at a Saturated Carbon Atom*, Elsevier, Amsterdam, 1963, p. 72–74.

8 R. E. PARIER, in M. STACEY, J. C. TATLOW AND A. G. SHARPE (Eds.), *Advances in Fluorine Chemistry*, Vol. III, Butterworths, London, 1963, p. 71–75.

9 J. F. BUNNETT, in *Theoretical Organic Chemistry; Kekulé Symposium, Chem. Soc., London, 1958*, p. 150–154.

10 S. I. MILLER, C. E. ORZECH, C. A. WELCH, G. R. ZIEGLER AND J. I. DICKSTEIN, *J. Am. Chem. Soc.*, 84 (1962) 2020.

11 J. F. BUNNETT AND R. E. ZAHLER, *Chem. Rev.*, 49 (1951) 273.

12 J. MILLER, *Rev. Pure Appl. Chem.* (*Australia*), 1 (1951) 171.

13 J. F. BUNNETT, *Quart. Rev.*, 12 (1958) 1.

14 S. D. ROSS, in S. G. COHEN, A. STREITWIESER AND R. W. TAFT (Eds.), *Progress in Physical Organic Chemistry, Vol. 1*, Interscience, New York, 1963, p. 31.

15 D. N. GLEW AND E. A. MOELWYN-HUGHES, *Proc. Roy. Soc.* (*London*), 211A (1952) 254.

16 B. V. TRONOV AND E. A. KRUEGER, *J Russ. Phys.-Chem. Soc.*, 58 (1926) 1270.

17 C. G. SWAIN AND C. B. SCOTT, *J. Am. Chem. Soc.*, 75 (1953) 246.

18 K. C. HO, J. MILLER, AND K. W. WONG, *J. Chem. Soc., B* (1966) 310.

19 S. C. CHAN, K. Y. HUI, J. MILLER AND W. S. TSANG, *J. Chem. Soc.*, (1965) 3207.

20 B. A. BOLTO, M. LIVERIS AND J. MILLER, *J. Chem. Soc.*, (1956) 750.

21 C. W. L. BEVAN AND G. C. BYE, *J. Chem. Soc.*, (1954) 3091.

22*a* J. MILLER AND K. Y. WAN, *J. Chem. Soc.*, (1963) 3429; *b* J. MILLER AND K. Y. WAN, unpublished work; *c* M. LIVERIS AND J. MILLER, *Chem. and Ind.*, (1957) 954.

23 E. BERLINER, M. J. QUINN AND P. J. EDGERTON, *J. Am. Chem. Soc.* 72 (1950) 5305.

24 F. W. BERGSTROM, R. E. WRIGHT, C. CHANDLER AND W. A. GILKEY, *J. Org. Chem.*, 1 (1936) 170.

25*a* R. HUISGEN, in *Theoretical Organic Chemistry; Kekulé Symposium Chem. Soc., London, 1958*, p. 158; *b* R. HUISGEN AND J. SAUER, *Angew. Chem.*, 72 (1960) 91.

26 R. BOLTON, J. MILLER AND A. J. PARIER, *Chem. and Ind.*, (1963) 492.

27 R. BOLTON AND J. MILLER, unpublished work.

28 G. P. BRINER, M. LIVERIS, P. G. LUTZ AND J. MILLER, *J. Chem. Soc.*, (1954) 1265.

29 C. W. L. BEVAN, *J. Chem. Soc.*, (1951) 2340.

30 C. W. L. Bevan and J. Hirst, (a) *Chem. and Ind.*, (1954) 1422; (b) *J. Chem. Soc.*, (1956) 254.
31*a* N. B. Chapman and R. E. Parker, *Chem. and Ind.*, (1951) 248; *b* N. B. Chapman, R. E. Parker and P. W. Soanes, *J. Chem. Soc.*, (1954) 2109.
32 B. A. Bolto, J. Miller and V. A. Williams, *J. Chem. Soc.*, (1955) 2926.
33*a* B. A. Bolto and J. Miller, unpublished work; *b* B. A. Bolto and J. Miller, *Austral. J. Chem.*, 9 (1956) 74, 304.
34 J. Miller and V. A. Williams, unpublished work.
35 B. A. Bolto, J. Miller and A. J. Parker, *J. Am. Chem. Soc.*, 79 (1957) 93.
36*a* A. L. Beckwith, G. D. Leahy and J. Miller, *J. Chem. Soc.*, (1952) 3552; *b* A. L. Beckwith and J. Miller, unpublished work.
37*a* K. B. Lam and J. Miller, *Chem. Commun.* (1966) 642; *b* K. B. Lam and J. Miller, unpublished work.
38 J. D. Reinheimer, R. C. Taylor and P. E. Rohrbaugh, *J. Am. Chem. Soc.*, 83 (1961) 835.
39 N. B. Chapman and R. E. Parker, *J. Chem. Soc.*, (1951) 3301.
40 A. H. Rheinlander, *J. Chem. Soc.*, 123 (1923) 3099.
41 T. O. Bamkole, C. W. L. Bevan and J. Hirst, *Chem. and Ind.*, (1963) 119.
42 G. S. Hammond and L. R. Parks, *J. Am. Chem. Soc.*, 77 (1955) 340.
43 J. F. Bunnett and J. J. Randall, *J. Am. Chem. Soc.*, 80 (1958) 6020.
44 R. L. Heppolette and J. Miller, unpublished work.
45 M. A. Adeniran, C. W. L. Bevan and J. Hirst, *J. Chem. Soc.*, (1963) 5868.
46 A. J. Parker, in N. Kharasch, (Ed.), *Organic Sulfur Compounds*, Vol. 1, Pergamon Press, London, 1961, p. 105 *et seq.*
47 H. A. Talen, *Rec. Trav. Chim.*, 47 (1928) 782.
48 J. Miller and H. W. Yeung, unpublished work.
49 T. L. Chan, J. Miller and F. Stansfield, *J. Chem. Soc.*, (1964) 1213.
50 J. F. Bunnett and W. D. Merritt, *J. Am. Chem. Soc.*, 79 (1957) 5967.
51*a* J. Miller, *J. Chem. Soc.*, (1952) 3550; *b* M. Liveris, P. G. Lutz and J. Miller, *J. Am. Chem. Soc.*, 78 (1956) 3375; *c* J. Miller and R. Roper, unpublished work.
52 F. H. Kendall and J. Miller, *J. Chem. Soc., B*, (1967) 119.
53*a* J. F. Bunnett, E. W. Garbisch and K. M. Pruitt, *J. Am. Chem. Soc.*, 79 (1957) 385; *b* J. F. Bunnett and K. W. Pruitt, *J. Elisha Mitchell Sci. Soc.*, 73 (1957) 297.
54 J. Murto, *Acta Chem. Scand.*, 20 (1966) 310.
55 R. E. Parker and T. O. Read, (a) *J. Chem. Soc.*, (1962) 9; (b) (1962) 3149.
56 M. W. Fuller and J. Miller, unpublished work.
57 A. H. Fainberg and W. T. Miller, *J. Am. Chem. Soc.*, 79 (1957) 4164, 4170.
58*a* C. G. Swain, *J. Am. Chem. Soc.*, 70 (1948) 1119; *b* C. G. Swain and W. P. Langsdorf, *J. Am. Chem. Soc.*, 73 (1951) 2813; *c* C. G. Swain, R. B. Moseley and D. W. Bown, *J. Am. Chem. Soc.*, 77 (1955) 3731.
59 J. F. Bunnett and R. F. Snipes, *J. Am. Chem. Soc.*, 77 (1955) 5422.
60 R. L. Heppolette, J. Miller and V. A. Williams, *J. Am. Chem. Soc.*, 78 (1956) 1975.
61 J. F. Bunnett, *J. Am. Chem. Soc.*, 79 (1957) 5969.
62 J. F. Bunnett and J. D. Reinheimer, *J. Am. Chem. Soc.*, 81 (1959) 315.
63 O. L. Brady and F. R. Cropper, *J. Chem. Soc.*, (1950) 507.
64 I. R. Lantzke and J. Miller, unpublished work.
65 Z. J. Allan and J. Podstata, *Collection Czech. Chem. Commun.*, 29 (1964) 2264.

66 F. Pietra and A. Fava, *Tetrahedron Letters*, (1963) 1535.
67*a* C. Bernasconi and H. Zollinger, *Tetrahedron Letters*, (1965) 1083; *b* C. Bernasconi and H. Zollinger, *Helv. Chim. Acta*, 49 (1966) 113.
68 B. Capon and C. W. Rees, *Ann. Rep. Chem. Soc.*, 60 (1964) 278.
69 F. Sanger, *Biochem. J.*, 39 (1945) 507.
70 W. B. Whalley, *J. Chem. Soc.*, (1950) 2241.
71 H. B. Gottlieb, *J. Am. Chem. Soc.*, 58 (1936) 532.
72*a* G. C. Finger and C. W. Kruse, *J. Am. Chem. Soc.*, 78 (1956) 6034; *b* G. C. Finger and L. D. Starr, *Chem. and Ind.*, (1962) 1328.
73 H. G. Cook and B. C. Saunders, *Biochem. J.*, 41 (1947) 558.
74 W. M. Weaver and J. D. Hutchison, *J. Am. Chem. Soc.*, 86 (1964) 261.
75 J. V. Janovsky, *Ber.*, (a) 19 (1886) 2158; (b) 24 (1891) 971.
76 N. V. Sidgwick, *The Organic Chemistry of Nitrogen*, 2nd Ed., edited by T. W. Taylor and W. Baker, Clarendon Press, Oxford, 1942.
77 T. Canbäck, *Farm. Revy*, 48 (1949) 217, 234, 249; *Chem. Abstr.*, 43 (1949) 6175[a].
78 M. Ishidate and T. Sukaguchi, *J. Pharm. Soc. Japan*, 70 (1950) 444.
79 M. J. Newlands and F. Wild, *J. Chem. Soc.*, (1956) 3686.
80 M. Kimura and M. Thoma, *Yakugaku Zasshi*, 78 (1958) 1401; *Chem. Abstr.*, 53 (1959) 8056[d].
81 M. Akatsuka, *Yakugaku Zasshi*, 80 (1960) 375, 378; *Chem. Abstr.*, 54 (1960) 18407[b].
82 S. S. Gittis *et al.*, (a) *Zh. Obshch. Khim.*, 27 (1957) 1894; (b) 29 (1959) 2646, 2648; (c) 30 (1960) 3810; (d) *Dokl. Akad. Nauk, S.S.S.R.*, 144 (1962) 785.
83 N. N. Voroshtsov and G. G. Yakobson, *Zh. Obshch. Khim.*, (a) 31 (1961) 3705; (b) 28 (1958) 40.
84 C. A. Kingsbury, *J. Org. Chem.*, 29 (1964) 3262.
85 R. Mecke, *Disc. Faraday Soc.*, 9 (1950) 161.
86 F. M. Sacks and R. M. Fuoss, *J. Am. Chem. Soc.*, 75 (1953) 5172.
87 R. L. Heppolette, I. R. Lantzke and J. Miller, *Austral. J. Chem.* 9 (1956) 299.
88 R. W. G. Broadbank, A. H. E. Harhash and S. Kanchanalai, *Proc. Conf. Use Radioisotopes, Copenhagen, 1960*, 3 (1962) 179.
89*a* Y. Ogata and M. Okano, *J. Am. Chem. Soc.*, 71 (1949) 3212; *b* Y. Ogata and M. Okano, *J. Chem. Soc. Japan*, 69 (1948) 148.
90*a* J. Murto and E. Tommila, *Acta Chem. Scand.*, 16 (1962) 63; *b* J. Murto, *Acta Chem. Scand.*, 18 (1964) 1029.
91 C. W. L. Bevan, J. Hirst and A. J. Foley, *J. Chem. Soc.*, (1960) 4543.
92 J. Miller, *Austral. J. Chem.*, 9 (1956) 61.
93*a* P. Gray, *Trans. Faraday Soc.*, 52 (1956) 344; *b* P. Gray and A. Williams, *Trans. Faraday Soc.*, 55 (1959) 760.
94 M. O. Forster and H. E. Fierz, *J. Chem. Soc.*, 91 (1907) 1942.
95 A. Mangini and D. D. Casoni, *Boll. Sci. Facoltà Chim. Ind. Bologna*, 3 (1942) 173; *Chem. Abstr.*, 38 (1944) 4916[7].
96 J. J. Blanksma, *Chem. Weekblad.*, 6 (1909) 313.
97 L. Gattermann and A. Ritschke, *Ber.*, 23 (1890) 1738.
98 D. Vorländer, *Ber.*, 70 (1937) 146.
99 H. Lütgert, *Ber.*, 70 (1937) 151.
100 R. B. Sandin, R. G. Christiansen, R. L. C. Brown and S. Kirkwood, *J. Am. Chem. Soc.*, 69 (1947) 1550.

101 F. M. Beringer, A. Brierly, M. Drexler, E. M. Gindler and C. C. Lumpkin, *J. Am. Chem. Soc.*, 75 (1953) 2748.
102 E. S. Lewis and C. A. Stout, *J. Am. Chem. Soc.*, 76 (1954) 4619.
103 L. Pauling, *The Nature of the Chemical Bond*; 3rd Ed., Cornell Univ. Press, Ithaca, N.Y., 1960, p. 93.
104 W. Borsche, *Ber.*, 56 (1923) 1488, 1494.
105 H. Salkowski, *Ber.*, 5 (1872) 872.
106 L. Desvergnes, *Mon. Sci.*, 14 (1924) 249; *Chem. Abstr.*, 19 (1925) 1700.
107 R. J. W. LeFèvre, S. M. Saunders and E. E. Turner, *J. Chem. Soc.*, (1927) 1168.
108 R. S. Cahn, *J. Chem. Soc.*, (1931) 1121.
109 A. Oliverio, *Atti X⁰ Cong. Intern. Chim.*, 3 (1939) 258; *Chem. Abstr.*, 33 (1939) 9302[8].
110 F. H. S. Curd, C. G. Raison and F. L. Rose, *Brit. Pat.* 592,928 (1947); *Chem. Abstr.*, 42 (1948) 2291[a].
111 L. Bartolotti and A. Cerniani, *Boll. Sci. Facoltà Chim. Ind. Bologna*, 14 (1956) 33; *Chem. Abstr.*, 50 (1956) 15182[a].
112*a* D. J. Brown and J. M. Lyall, *Austral. J. Chem.*, 17 (1964) 794; *b* D. J. Brown and J. M. Lyall, *Austral. J. Chem.*, 18 (1965) 741, 1811; *c* D. J. Brown and R. V. Foster, *Austral. J. Chem.*, 19 (1966) 1487, 2321.
113 C. M. Suter, *The Organic Chemistry of Sulfur*, Wiley, New York, 1944, pp. 420–442.
114 A. Mangini and M. Colonna, *Gazz. Chim. Ital.*, 73 (1943) 313.
115 H. J. Barber, *J. Chem. Soc.*, (1929) 2333.
116 T. K. Brotherton and J. F. Bunnett, *Chem. and Ind.*, (1957) 80.
117*a* R. G. R. Bacon and H. A. O. Hill, *Quart. Rev.*, 19 (1965) 95; *b* R. G. R. Bacon and H. A. O. Hill, *Proc. Chem. Soc.*, (1962) 113; *J. Chem. Soc.*, (1964) 1097, 1108.
118. R. D. Stephen and C. E. Castro, *J. Org. Chem.*, 28 (1963) 3313.
119. H. Weingarten, *J. Org. Chem.*, 29 (1964) 977, 3624.

Chapter 6

THE NUCLEOPHILIC REAGENT

1. Introduction

The relative reactivity of nucleophiles in aromatic S_N reactions often depends on the nature of the substrate, more particularly of the leaving group. Thus Miller and his co-workers *[1]* have discussed the marked variation in the relative reactivity of azide, methoxide, thiomethoxide and thiophenoxide ions in their reactions with some fluoro- and iodo-nitrobenzenes, and shown for example, that with *p*-fluoronitrobenzene in methanol at 0°, the order of reactivity is $N_3^- < OMe^- < SPh^- < SMe^-$, with corresponding ratios: 0.000248:1:1.56:20.6; whereas with 1-iodo-2,4-dinitrobenzene the different order is $N_3^- < OMe^- < SMe^- < SPh^-$, with corresponding ratios: 0.0565:1:1.32:16800; the most striking change being that the SPh^-/OMe^- rate ratio increases from 1.56 to 16800.

In reversible reactions the role of reagent and replaced group is clearly interchangeable. While therefore it is impossible to discuss reagents without considering such factors as these, there are sufficient distinctive features to permit devotion of a separate chapter to the nucleophilic reagent.

2. Nucleophilicity and Basicity

A convenient definition of a nucleophilic reagent is one which supplies a pair of electrons to form a new bond between itself and an atom at the reactive centre in the (electrophilic) substrate *[2,3]*. Lewis *[4]* has given a similar definition, and these definitions include basic reagents as defined by Brønsted *[5]*. There is nevertheless some divergence of usage of the terms nucleophilicity and basicity. Many chemists [*e.g.* Hine in *[6]*], following Swain and Scott *[3]*, use the term nucleophilicity for rate phenomena, including the reactions of bases; and the term basicity for equilibria, including reactions of nucleophiles. One then speaks of carbon-basicity *[7]*, nucleophilicity towards hydrogen and so on.

However the terms nucleophile and electrophile and acid and base have generally agreed significance, and Miller and his co-workers *[8]* suggested that it is clearer and more consistent to speak of kinetic and thermodynamic nucleophilicity and basicity*, just as, for example, steric effects on rates and equilibria are distinguished as kinetic and thermodynamic *[9a]*. They have suggested the following abbreviations for a given species $\ddot{Y}$: KB_Y(H–X); TB_Y(H–X); KN_Y(A–X); TN_Y(A–X), kinetic and thermodynamic basicity of $\ddot{Y}$ towards an acid H–X; and kinetic and thermodynamic nucleophilicity of $\ddot{Y}$ towards an electrophile A–X. They are defined as the (standard) free energy of activation ($\Delta G^{\ddagger}$) or of reaction (ΔG^{0}), for kinetic or thermodynamic parameters respectively, in the general reaction: $\ddot{Y} + A\text{–}X \rightarrow A\text{–}Y + \ddot{X}$, where $A = H$ for basicity and any other element for nucleophilicity. A superscript zero, *e.g.* $KN_Y(C_{Ar}\text{–}X^0)$ is added when considering reactivity towards a standard substrate (see below).

3. Mechanism and Nucleophilicity

In a bimolecular nucleophilic substitution passing through a single transition state, kinetic nucleophilicity involves both formation of the bond by the nucleophile *and* rupture of the bond to the leaving group. Both are directly and markedly involved in the reactivity of the nucleophile. Saturated aliphatic S_N2 reactions are well-known examples of this type.

In a bimolecular reaction passing through an intermediate complex and two transition states, commonly accepted as the pattern of activated aromatic S_N2 reactions (as well as aromatic S_E and S_R reactions), where the formation of the first transition-state is rate-limiting, kinetic nucleophilicity involves bond formation by the nucleophile without rupture of the bond to the leaving group, and factors consequent on the latter have no influence; though as was discussed in

* Ingold in revising his text *[9]* uses the terms kinetic and thermodynamic nucleophilic strength and base strength.

References p. 230

Chapter 5, its electronegativity, if sufficiently high, has an effect lowering the activation energy of the bond-forming step *[8,10–12]*. Where the formation of the second transition-state is rate-limiting, however, factors resulting from bond rupture will be directly involved and the situation resembles that in reactions passing through a single transition state (see below). Miller and his co-workers *[13]* have suggested that this distinction between the one-stage and two-stage mechanisms is a very general one, commenting on it also in relation to inorganic systems. Nucleophilic substitution in unsaturated organic systems must also be considered in relation to this generalisation.

4. Quantitative Aspects of Nucleophilicity and Basicity (as defined in Section 2)

In his general review of nucleophilic reactivity *[2b]* Bunnett has listed seventeen factors which, from experiment or theory or both, he considers, influence nucleophilic reactivity, but recognised that some are better supported than others, and that some are interdependent.

Swain and his co-workers *[3]* have considered nucleophilic reactivity towards aliphatic and side-chain aromatic compounds in terms of a generalised "push-pull" mechanism, and correlated a substantial number of rates with the linear free energy equation: $\log k/k_0 = sn + s'e$, referring to a general reaction: $N + S + E \rightleftarrows$ [T.St.] $\rightarrow$ Products. In this N is the nucleophilic reagent, S an uncharged reactant, and E an electrophilic reagent. The overall rate constant k is compared with k_0 for which water is both standard nucleophile and electrophile; n is a measure of the nucleophilic reactivity (*i.e.* kinetic nucleophilicity as defined by Miller *[8]*) of a nucleophile N, equal to zero for water; and e is a measure of the electrophilic reactivity of E, equal to zero for water. They showed that the Brønsted catalysis law *[14]* is derivable from their equation, as is the Grunwald–Winstein correlation of solvolysis rates *[15]*. With water as the solvent, $e = 0$, so $s'e$ disappears, and the relationship simplifies to $\log k/k_0 = sn$. By definition MeBr is chosen as standard substrate for which s = unity and $\log k/k_0 = n$. From this, different n values are obtained and used to derive other s values and so on.

Tables 40 and 41 give some values of nucleophilic and substrate constants (*n* and *s*) from Swain's paper *[3b]*, from which iodide and thiosulphate, for example, show up as powerful reagents (high *n* values) and unsaturated substrates (such as acid halides and carboxylic esters), as being more susceptible than saturated substrates to differences in nucleophilic power of the reagent (high *s* values). This may be correlated with the greater importance of bond formation in the relevant reactions. Swain *et al.* were aware of limitations in their equation, and also that nucleophiles in which a heavy atom forms a bond to the reaction centre are often more reactive than one would expect from their basicities (this is, however, a thermodynamic quantity, and one should in any case not necessarily expect a close parallelism even between kinetic basicity and kinetic nucleophilicity), and suggested that the factor involved is polarisability. They also pointed out that a new *n* scale would be needed for displacements on atoms other than carbon. The relative reactivity of light and heavy nucleophiles, has also been commented on by Miller *[10c]*.

TABLE 40

NUCLEOPHILIC CONSTANTS *[3]*

Nucleophile	Nucleophilic constant
H_2O	0.00[a]
Picrate ion	1.9
SO_4^{2-}	2.5
$CH_3CO_2^-$	2.72
Cl^-	3.04
Pyridine	3.5
HCO_3^-, HPO_4^{2-}	3.8
Br^-	3.89
N_3^-	4.00
Thiourea	4.1
OH^-	4.2
$PhNH_2$	4.49
SCN^-	4.77
I^-	5.04
SH^-, SO_3^{2-}	5.1
$S_2O_3^{2-}$	6.36
$HPSO_3^{2-}$	6.6

[a] By definition.

References p. 230

TABLE 41

SUBSTRATE (SUSCEPTIBILITY) CONSTANTS *[3]*

Substrate	Substrate constant
Ethyl tosylate	0.66
β-Propiolactone	0.77
Benzyl chloride	0.87
Epichlorhydrin	0.93
Mustard cation	0.95
Glycidol	1.00
Methyl bromide	1.00[a]
Benzene sulphonyl chloride	1.25
Benzoyl chloride	1.43
Acetyl fluoride	1.78
Benzoyl fluoride	1.85
Ethyl acetate	2.48

[a] By definition.

Edwards *[16]* set up a similar equation to that of Swain and Scott with separate parameters for basicity and polarisability as components of nucleophilicity, *viz.*, $\log k/k_0 = \alpha E_n + \beta H$ where values of k are rate or equilibrium values relative to those for water (k_0); E_n is a nucleophilic constant characteristic of an electron donor; H the relative basicity of the donor towards protons; and α and β are corresponding substrate constants. For the H scale he used the normal pK_a of the conjugate acids in water plus 1.74, this being the correction for the pK_a of H_3O^+, *i.e.* $H = pK_a + 1.74$.

For E_n he used the electrode potentials corresponding to $2Y^- \rightleftarrows Y_2 + 2e$, where Y = the nucleophile, for which Foss *[17]* had pointed out a relationship with nucleophilic character, and showed also that this was correlated with Swain and Scott's *[3b]* n values. Since values are relative to water, he calculated the potential for $2H_2O \rightleftarrows H_4O_2^{2+} + 2e$ as –2.60 V, and then his E_n values are the oxidation potentials (E_0) of the electron donors plus 2.60, *i.e.* $E_n = E_0 + 2.60$. His initial correlation of E_n with electrode potentials as a fundamental set of data is akin to the use of electron affinity or ionisation energy, though less clearly related to the reactions of nucleophilic reagents.

The use of H (basicity) is really a weighting device, since basicity is nucleophilicity towards hydrogen. Its success as such is because the strengths of bonds formed by nucleophiles were not considered, and bonds formed with hydrogen are commonly stronger than those with other elements. However some negative as well as positive values of β modifying H, had to be used. Nevertheless the approach is also a valuable one, but was not applied to aromatic S_N reactions. Tables 42 and 43, from Edwards' paper *[16a]* give values of nucleophilic constants (E_n) and basicity (H), and corresponding substrate (susceptibility) constants, α (for E_n) and β (for H).

Davies *[18]* also considers an oxibase scale relating oxidation electrode potential and basicity of a nucleophile to its reactivity.

TABLE 42

NUCLEOPHILIC AND BASICITY CONSTANTS *[16a]*

Nucleophile	Nucleophilic constant (E_n)	Basicity constant (H)
NO_3^-	0.29	0.40
SO_4^{2-}	0.59	3.74
$ClCH_2CO_2^-$	0.79	4.54
$CH_3CO_2^-$	0.95	6.46
C_5H_5N	1.20	7.04
Cl^-	1.24	−3
OPh^-	1.46	11.74
Br^-	1.51	−6
N_3^-	1.58	6.46
OH^-	1.65	17.48
NO_2^-	1.73	5.09
$PhNH_2$	1.78	6.28
SCN^-	1.83	1
NH_3	1.84	11.22
$(OMe)_2POS^-$	2.04	4
$EtSO_2S^-$	2.06	−5
I^-	2.06	−9
$(OEt)_2POS^-$	2.07	4
$CH_3C_6H_4SO_2S^-$	2.11	−6
$SC(NH_2)_2$	2.18	0.80
$S_2O_3^{2-}$	2.52	3.60
SO_3^{2-}	2.57	9.00
CN^-	2.79	10.88
S^{2-}	3.08	14.66

References p. 230

TABLE 43

SUBSTRATE (SUSCEPTIBILITY) CONSTANTS *[16b]*

Substrate	Substrate constants	
	α (for E_n)[a]	β (for H)[a]
Ethyl tosylate	1.68	0.014
Benzyl chloride	3.53	−0.128
β-Propiolactone	2.00	0.069
Diazoacetone	2.37	0.191
Mustard cation	2.45	0.074
Epichlorohydrin	2.46	0.036
Methyl bromide	2.50	0.006
Glycidol	2.52	0.000
Iodoacetate ion	2.59	−0.052
Benzoyl chloride	3.56	0.008

[a] See Table 42.

Edwards has also suggested a possible correlation of his E_n values with polarisability *[16b]*.

There is unlikely to be disagreement with the view that there is a relationship between nucleophilic displacements on hydrogen (basicity) and on other elements (nucleophilicity); and that there is a relationship between values of polarisability and of other parameters such as strength of bonds formed, and electron affinity which are considered by Miller and his co-workers *[1,8,10c]*.

Edwards and Pearson *[19]*, considering the general reaction: N + S–X → N–S + X (N = nucleophile and S–X = electrophilic substrate), have added an additional factor which they call the α-effect, *viz.* an enhancement of reactivity when the atom α to the nucleophilic (bond-forming) atom has a pair of unshared electrons. This is tentatively correlated with the greater stability of cations in systems with unshared electrons on the α-atom: the analogy being that the nucleophilic atom, in forming the transition state, is losing a share of electrons and becoming less negative. This should not be confused with the α-substituent (electronegativity) effect discussed by Miller *[8]* (see below). They excluded consideration of solvation effects, steric factors, effects of hydrogen bonding, the consequences of forming cyclic transition states, and more generally have excluded entropy effects. Ingold *[9b]* agreeing that the α-effect is associated with the presence of unshared electrons on the α-atom interprets it in a somewhat different way. In the anomalous nucleophiles the highest occupied molecular orbital, centred largely on the nucleophilic atom, is antibonding with a node normal to the bond to the α-atom. These electrons provide a special factor of inhomogeneous polarisability towards any substrate able to make large use of the polarisability of the nucleophiles, *viz.* the empty or emptying orbital of an electrophilic substrate.

Pearson *[20]* has elaborated the observations of Chatt and his co-workers *[21]*, and classified generalised acids and bases as Class (*a*) or hard (low polarisability), and Class (*b*) or soft (high polarisability), and commented on stronger bonding between hard and soft pairs as compared with hard–soft combinations. He recognised, however, that polarisability itself might not be the factor primarily responsible, but others which have a relationship to it, *e.g.* ionisation energy (potential), and electronegativity. He considered several explanations for differences in behaviour such as the varying degree of ionic or covalent bonding; π-bonding; electron correlation theory; and solvation. According to the first of these *[22–24]* strong ionic bonding occurs when both acid and base are of Class (*a*). When both are of Class (*b*) the repulsive part of the potential energy curve rises less steeply than in the former combination leading to closer approach, better overlaps, and this also leads to stronger covalent bonding.

On the π-bonding theory *[21,25–27]* the important feature of Class (*b*) acids is the presence of outer d-orbitals which can form π-bonds with suitable ligands, *viz.*, those of Class (*b*), which have empty d-orbitals in the atoms which form bonds with the acid, *e.g.* P, As, S, I. Conversely, while Class (*a*) acids would have tightly held outer electrons, they would have empty orbitals available and Class (*a*) bases, *e.g.* which have O and F as the atoms forming bonds with the acid, could then form π-bonds in the opposite sense.

From the electron correlation theory *[28]*, London forces between atoms or groups in the same molecule may lead to stabilisation of the molecule and such forces depend on the product of the polarisabilities and could thus account for the stability of bonding of Class (*b*) type. An alternative explanation *[29]* is that orbital rehybridisation leads

References p. 230

to stronger bonds. These theories have not been applied to explain strong Class (*a*) bonding.

Solvation theory *[7,8,10c,30–31]* is applied to explain how solvents reduce the Class (*a*) character, and enhance the Class (*b*) character of basic anions. The change from a protic to an aprotic solvent enhances the reactivity of these, most for those belonging to Class (*a*), which are most solvated in protic solvents.

Working independently, Hudson *[32]* and Miller *[1,8,10c,30]* have focused attention on solvation energy of ionic nucleophiles, their electron affinities, and the strength of the bonds they form to the electrophilic centre. Hudson applied his ideas more generally but in less detail than Miller who concentrated almost exclusively on nucleophilicity in activated aromatic S_N2 reactions, and showed how it is possible to calculate and compare kinetic and thermodynamic nucleophilicity in these reactions with thermodynamic basicity. He pointed out that the principles used could be extended to reactions other than those between anions and neutral substrates, and also to exchange reactions on atoms other than carbon and hydrogen. While the methods of Hudson, and of Miller, are not very closely related to those of Swain and Scott, and of Edwards and Pearson, the ability of all these to correlate a wide variety of reactions suggests the interdependence of the factors considered. Since the procedure of Miller has been directly and successfully applied to aromatic S_N reactions, it is now considered in detail.

5. Reactivity of Anionic Reagents

(*a*) *General*

Miller's procedure is based on the two-stage addition–elimination mechanism. Potential-energy profiles are calculated as described below, leading to values of $\Delta E^{\ddagger}$, approximately equal to $\Delta H^{\ddagger}$ for these solution reactions, and ΔH^0. Entropy factors are then considered, in order to estimate values of $\Delta S^{\ddagger}$ and ΔS^0. From the usual equations: $\Delta G^{\ddagger} = \Delta H^{\ddagger} - T\Delta S^{\ddagger}$ and $\Delta G^0 = \Delta H^0 - T\Delta S^0$, the free energies of activation and reaction which measure kinetic and thermodynamic nucleophilicity in aromatic S_N reactions, are then calculated. They may be compared with experimental values of reaction rates and heats of reaction. A similar calculation of ΔG^0 for the exchange (acid–base) reactions on hydrogen measures thermodynamic basicity.

(*b*) *Factors affecting reactivity*

The first part of the calculation uses mainly bond and solvation energies, and electron affinities to give energy differences between the Intermediate (σ or benzenide) Complex (I.C.) and Initial and Final States (I.St. and F.St.). In all, a considerable number of terms is used, many of them constant for a reaction series. These values are not all known with a high degree of precision, and the use of an empirical correction term had been expected, though not in fact required.

By use of a semi-empirical curve, relating percentage bond dissociation energy to thermicity, based on Hammond's postulate *[33]*, *viz.* that for highly exothermic reactions the transition state (T.St.) resembles the initial species, and for a highly endothermic reaction resembles the final species, an estimate is then made of the energy of each of the two transition states (T.St.1, and T.St.2) relative to the I.C. This gives the complete potential energy profile, provided additional effects are also considered, *viz.* the presence of a highly electronegative group attached at the reaction centre which reduces the energy of the bond-forming transition state—this has been called the α-substituent effect *[1,8,10c]*; differences in bond strength of bonds at the reaction centre being aromatic-type in initial and final states, and aliphatic-type in transition states and the intermediate complex.

The transition states occur at reaction coordinates such that the magnitude of the bond dissociation energy term (proceeding from I.C. to I.St. or F.St.) is balanced by the sum of electron affinity and solvation energy terms. Thus expressing the energy as a percentage of bond dissociation energy alone necessarily underestimates the percentage of actual bond dissociation. Bearing this in mind, it is clear nevertheless from the calculations on the reactions of anions with neutral substrates that in forming the T.St.'s from the I.C. the amount of bond dissociation is not large, so that they resemble structurally the fully bonded benzenide intermediate complexes. This is also strongly supported by the large substituent effects, illustrated by large values of the Hammett reaction constants (ρ).

References p. 230

Adverse non-bonding interactions presumably occur in the reaction intermediates, between the nucleophile and the group present at the point of attachment which are approximately at the tetrahedral angle to each other. Often this factor will have little influence in differentiating reactivity and is not considered as a general factor, though reference has been made previously (and see below) to a heavy nucleophile interaction. Conceivably the non-bonding interactions are still less or even absent for attachment of a nucleophile to a ring carbon bearing a hydrogen atom. This would result in especially facile addition, as distinct from overall substitution, and such cases have been reported by Servis in his NMR study of Meisenheimer complexes (see Chapter 1, p. 12).

Entropy factors in kinetic nucleophilicity affecting values of $\Delta S^{\ddagger}$ are allowed for by generalisation of experimental data. In the absence of major steric factors, such as may arise with *ortho*-substituents, the values of $\Delta S^{\ddagger}$ follow a simple pattern, *e.g.*, with anions in methanol values of $\Delta S^{\ddagger}$ are almost always in the range −5 to −14 e.u. when either or both the reagent and leaving group are attached to the ring by a first row atom, and −15 to −24 e.u. when both are attached by heavy atoms *[1,8,10c]* (heavy nucleophile interaction). In contrast, values of ΔS^0 in thermodynamic nucleophilicity and basicity are calculated from values of ionic solvation entropies *[8]*.

Even in a two-stage reaction with formation of T.St.1 rate-limiting, the nucleophilic power of a reagent may be affected by the nature of the group displaced (see above), and it is necessary for general discussion to have a standard substrate. For this Miller *[8]* has selected *p*-iodonitrobenzene for several reasons, but especially that with this substrate the formation of T.St.1 (bond-formation by the nucleophile) is rate-limiting; that there is no α-substituent (electronegativity) effect of the iodine; and despite the size of the replaced group (iodine) steric effects of substituents, at least when the reagent is not *also* sterically demanding, are unimportant. Typically with nucleophiles in which the bond to the electrophilic centre (the ring carbon to which the iodine is attached) is formed by a heavy atom, the value of $\Delta S^{\ddagger}$ is low, and this is possibly a steric effect. However it is not a major factor and the magnitude of the effect is known (see above). A minor disadvantage is that *p*-iodonitrobenzene is but moderately reactive, and kinetic nucleophilicity values (in terms of $\Delta G^{\ddagger}$) are thus rather high even for quite powerful reagents. Correspondingly Miller *[8]* has chosen hydriodic acid as the standard acid substrate for consideration of thermodynamic basicity.

Tables 44 and 45 give full details of calculations for some reactive species, confirmed whenever tested by experimental measurements of rates and heats of reaction *[8]*, and Tables 46 to 48 summarise the results of the corresponding calculations *[8]*, giving values of kinetic and thermodynamic nucleophilicity with the standard aromatic substrate (*p*-iodonitrobenzene) and of thermodynamic basicity with the standard acid (hydriodic acid). Some of these too have been tested by experiment and results are also in satisfactory agreement. Examples of these calculations are illustrated by potential energy–reaction coordinate curves shown as Figs. 40 to 53 in Chapter 5 (p. 149). Some additional curves are shown as Figs. 56 to 60 of this Chapter.

(*c*) *Quantitative estimates of reactivity*

A number of these are now considered in detail in the reactions of the standard aromatic substrate, *p*-iodonitrobenzene.

Methoxide. This first row nucleophile is conveniently considered as a reference standard for comparison with other nucleophiles, being both a reactive and generally well-behaved nucleophile.

The kinetic nucleophilicity towards the standard aromatic substrate or KN($C_{Ar}X^0$) is given, as explained, in terms of the free energy of activation ($\Delta G^{\ddagger}$). The calculated value is 27.4 kcal·mole^{-1} and this is also the experimental value. From Table 46 (see also Fig. 42) the value of $\Delta H^{\ddagger} = 25.0$ and of $T\Delta S^{\ddagger} = -2.4$ kcal·mole^{-1}. The strength of this reagent is associated with the strong C–OMe bond, and but moderate values of electron affinity (E_{aff}) and heat of solvation (H_{solv}) *[8,10c]*. The corresponding thermodynamic nucleophilicity or TN($C_{Ar}X^0$) value is that of the standard free energy of reaction (ΔG^0) and this is –24.5 kcal·mole^{-1}. From Table 47 (see also Fig. 42 of Chapter 5) the value of $\Delta H^0 = -20.5$ and of $T\Delta S^0 = +4$ kcal·mole^{-1}. The thermodynamic basicity or TB(HX^0) is also that of a standard free energy of reaction and equals –25 kcal·mole^{-1}. From Table 48 the value of $\Delta H^0 = -21$ and of $T\Delta S^0 = +4$ kcal·mole^{-1}.

[*Text continued on p. 196*]

TABLE 44

CALCULATED RELATIVE ENERGY LEVELS, WITH CALCULATED AND EXPERIMENTAL KINETIC DATA FOR REACTION OF SOME ANIONIC NUCLEOPHILES (Y^-) WITH 1-X-2,4-DINITROBENZENES (IN MeOH AT 25°). (VALUES IN kcal·mole^{-1} *[8b]*)

Exchange reaction number	Reagent $(Na^+)Y^-$	Substrate X	Calculated energy levels relative to I.C. = 0[a]					Kinetic data[b] (*i*) for forward reaction					(*ii*) for reverse reaction		
			I.St.	T.St.1	I.C.	T.St.2	F.St.	$\Delta E^{\ddagger}$ calcd.	$\simeq\Delta H^{\ddagger}$ exptl.	$T\Delta S^{\ddagger}$ exptl. or estd.[c]	$\Delta G^{\ddagger}$ calcd.	$\Delta G^{\ddagger}$ exptl.	$\Delta E^{\ddagger}$ $(\simeq\Delta H^{\ddagger})$ calcd.	$T\Delta S^{\ddagger}$ estd.	$\Delta G^{\ddagger}$ calcd.
(1)	OMe^-	F	−2	+13	0	+11.5	−26.5	15	13.5	−2.8	17.8	16.3	39.5	−3	42.5
(2)	OMe^-	I	−2	+17	0	+4.5	−22.5	19	19.0	−2.0	21.0	21.0	39.5	−4	43.5
(3)	OMe^-	N_3	−4	+14.5	0	+13	−9	18.5	18.1	−1.4	19.9	19.5	23.5	−3	26.5
(4)	N_3^-	F	−4	+13.5	0	+13	−23.5	17.5	16.6	−2.2	19.7	18.8	37.5	−3	40.5
(5)	SMe^-	F	+5	+15.5	0	+15.5	−21.5	10.5	10.5	−3.6	14.1	14.1	37	−4	41
(6)	SMe^-	OMe	0	+14	0	+17	−2	17	—	(−4)	21	(21)	19	−4	23
(7)	SMe^-	Cl	+5	+18.5	0	+11	−19	13.5	14.1	−5.9	19.4	20.0	37.5	−6	43.5
(8)	SMe^-	I	+5	+19.5	0	+8.5	−17.5	14.5	14.3	−7.0	21.5	21.3	37	−7	44
(9)	SMe^-	N_3	+3	+16.5	0	+18	−4	15	14.5	−3.3	18.3	17.8	22	−3	25
(10)	SPh^-[d]	F	+5	+12.5	0	+15.5	−21.5	10.5	10.7	−3.7	14.2	14.4	37	−3	40
(11)	SPh^-[d]	I	+5	+16.5	0	+8.5	−17.5	11.5	10.7	−5.3	16.8	16.0	34	−7	41

[a] I.St. and F.St. = Initial and Final state; T.St.1 and T.St.2 = Transition state 1 and 2; I.C. = Intermediate (benzenide) complex.
[b] Approx. expt. error ±0.4 kcal·mole^{-1} in $\Delta H^{\ddagger}$ and 2 e.u. in $\Delta S^{\ddagger}$. [c] By consideration of experimental values, this may be estimated generally as about −3 for first row nucleophiles; about −4 for heavy nucleophiles, provided the replaced group is attached to the ring by a first row element; values to −7 kcal·mole^{-1} for heavy nucleophiles when replaced group is attached to the ring by a heavy element. [d] Calculated data use additional assumptions discussed in ref. *1*.

TABLE 45

CALCULATED AND EXPERIMENTAL THERMODYNAMIC DATA FOR REACTIONS OF SOME ANIONIC REAGENTS (Y^-) WITH 1-X-2,4-DINITROBENZENES AND WITH ACIDS H–X (IN MeOH AT 25°) (VALUES IN kcal·mole^{-1} *[8b]*)

Exchange reaction number	Reagent $(Na^+)Y^-$	Substrate X	For reaction $Y^- + ArX \rightarrow Ar{-}Y + X^-$ ΔH^0 calcd.	ΔH^0 exptl.[a]	$T\Delta S^0$ calcd.	ΔG^0 calcd.	ΔG^0 exptl.	For reaction $Y^- + HX \rightarrow HY + X^-$ ΔH^0 calcd.	$T\Delta S^0$ calcd.	ΔG^0 calcd.
(1)	OMe^-	F	−24.5	−26.2	−4	−20.5	−22.2	−15.5	−4	−11.5
(2)	OMe^-	I	−20.5	—	+4	−24.5	—	−21	+4	−25
(3)	OMe^-	N_3	−5	−2.4	+0.5	−5.5	−2.9	−11	+0.5	−11.5
(4)	N_3^-	F	−19.5	−17.8	−4.5	−15	−13.3	−4.5	−4.5	0
(5)	SMe^-	F	−26.5	−28.6	−7	−19.5	−21.6	−14.5	−7	−7.5
(6)	SMe^-	OMe	−2	*ca.* 0	−3	+1	+3	+1	−3	+4
(7)	SMe^-	Cl	−24	−25.0	−2.5	−21.5	−22.5	−17	−2.5	−14.5
(8)	SMe^-	I	−22.5	—	+1	−23.5	—	−20	+1	−21
(9)	SMe^-	N_3	−7	−7.1	−2.5	−4.5	−4.6	−10	−2.5	−7.5
(10)	SPh^-	F	−26.5[b]	−28.1	−9	−17.5	−19.1	—	—	—
(11)	SPh^-	I	−22.5[b]	−25.2	−1	−21.5	−24.2	—	—	—

[a] Individual run results: (1) −25.9, −26.5; (2) –; (3) −2.2, −2.6; (4) −17.2, −17.3, −17.8, −18.2, −18.4; (5) −28.3, −28.9; (6) *ca.* 0, *ca.* 0; (7) −24.2, −25.8; (8) –; (9) −6.8, −7.4; (10) −28.0, −28.2; (11) −24.5, −25.8. [b] Assuming same as with SMe^- as reagent *[1]*. The experimental values −28.1 and −28.6 for reaction of SPh^- and SMe^- with the fluoro compound are regarded as equal within experimental error.

TABLE 46

CALCULATED AND EXPERIMENTAL ENTHALPY AND FREE ENERGY BASED VALUES OF KINETIC NUCLEOPHILICITY TOWARDS A STANDARD AROMATIC SUBSTRATE, *viz. p*-IODONITROBENZENE, KN_Y–($C_{Ar}X^0$) AT 25° IN MeOH (VALUES IN kcal·mole^{-1} *[1,8b]*)

Nucleophile		SPh^-	SMe^-	OMe^-	N_3^-	SCN^-	I^-	Br^-	Cl^-	F^-
$\Delta H^{\ddagger}$	calcd.	17.5	19.5	25	27.5	30.5	33.5	37	38.5	44
	exptl.	19.4	19.7	25.0	27.8	—	—	—	—	—
$T\Delta S^{\ddagger}$	estd.	—	—	—	—	− 7	− 7	− 7	− 7	− 3
	exptl.	− 4.3	− 6.7	− 2.4	− 2.9	—	—	—	—	—
$\Delta G^{\ddagger}$	calcd.	21.8	26.2	27.4	30.4	37.5	40.5	44	45.5	47
	exptl.	23.7	26.4	27.4	30.7	—	—	—	—	—

TABLE 47

CALCULATED ENTHALPY AND FREE ENERGY BASED VALUES OF THERMODYNAMIC NUCLEOPHILICITY TOWARDS A STANDARD AROMATIC SUBSTRATE, *viz*, *p*-IODONITROBENZENE TN_Y–($C_{Ar}X^0$) AT 25° IN MeOH (VALUES IN kcal·mole^{-1} *[1,8b]*)

Nucleophile	SPh^-	SMe^-	OMe^-	N_3^-	SCN^-	I^-	Br^-	Cl^-	F^-
ΔH^0	− 22.5	− 22.5	− 20.5	− 15.5	− 7.5	0	+ 2	+ 1.5	+ 4
$T\Delta S^0$	− 1	+ 1	+ 4	+ 3.5	+ 1.5	0	+ 1.5	+ 3.5	+ 8
ΔG^0	− 21.5	− 23.5	− 24.5	− 19	− 9	0	+ 0.5	− 2	− 4

TABLE 48

CALCULATED ENTHALPY AND FREE ENERGY BASED VALUES OF THERMODYNAMIC BASICITY TOWARDS A STANDARD ACID, *viz.* HYDRIODIC ACID TB_{Y}–(HX^0) AT 25° IN MeOH (VALUES IN kcal·mole^{-1} [8b])

Base	SPh^-	SMe^-	OMe^-	N_3^-	SCN^-	I^-	Br^-	Cl^-	F^-
ΔH^0	—	− 20	− 21	− 10	—	0	0	− 3	− 5.5
$T\Delta S^0$	—	+ 1	+ 4	+ 3.5	—	0	+ 1.5	+ 3.5	+ 8
ΔG^0	—	− 21	− 25	− 13.5	—	0	− 1.5	− 6.5	− 13.5

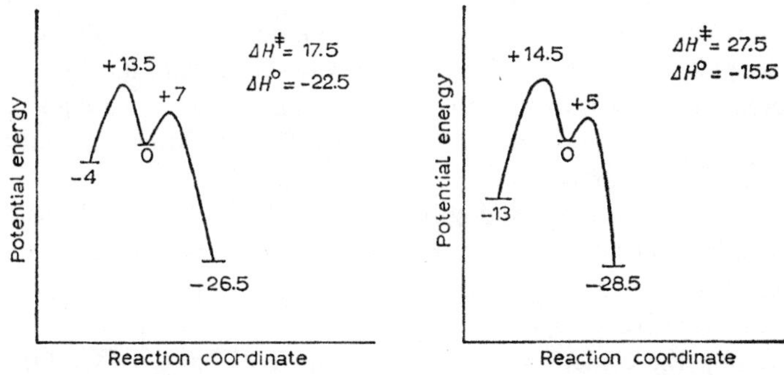

Fig. 56. Potential energy/reaction coordinate profile of SPh^-/*p*-$NO_2C_6H_4I$.
Fig. 57. Potential energy/reaction coordinate profile of N_3^-/*p*-$NO_2C_6H_4I$.

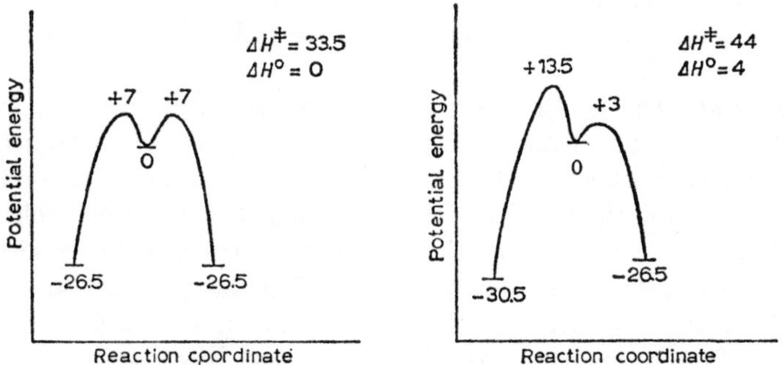

Fig. 58. Potential energy/reaction coordinate profile of I^-/*p*-$NO_2C_6H_4I$.
Fig. 59. Potential energy/reaction coordinate profile of F^-/*p*-$NO_2C_6H_4I$.

References p. 230

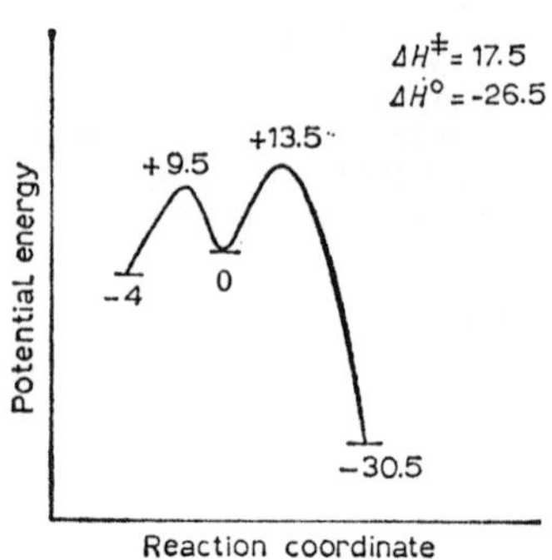

Fig. 60. Potential energy/reaction coordinate profile of $SPh^-/p\text{-}NO_2C_6H_4F$.

Methoxide ion is seen to be powerful as a kinetic and thermodynamic nucleophile and as a thermodynamic base, enhancing its value as a reference standard. Results for other exchange reactions are given in Tables 44 and 45 and have been discussed fully by Miller *[8]* but are not considered further here.

In considering these and subsequent nucleophilicity and basicity values, it should be noted that the smaller numerically the positive value of KN, the more powerful is a reagent; while the larger numerically the negative value of TN and TB, the more powerful the reagent.

Among the chalcogenide ions XR^- those of oxygen ($X = O$) are the most solvated in protic solvents, and therefore gain most in strength when reactions are carried out in aprotic instead of protic solvents. This is discussed further below, but only qualitatively since there are insufficient fundamental solvation data referring to aprotic solvents.

Fluoride. This is also a first row nucleophile but from Group VII instead of Group VI, and in protic solvents is well known experimentally to be a much weaker kinetic nucleophile than methoxide. Corresponding to this the $KN(C_{Ar}X^0)$ value is calculated as 47 kcal·mole^{-1}, 19.6 kcal·mole^{-1} higher than the value for methoxide ion. Its weakness is seen to be due to the very high value of the heat of solvation and high value of the electron affinity swamping the overall favourable effect of the strong bond formed by the reagent. It is a weaker kinetic nucleophile in protic solvents than any of the other halide ions.

It is important to comment further and generally here on the relationship between nucleophilicity and strength of the bond formed by the nucleophile. A weak bond between nucleophile and substrate tends to contribute to a low value of the energy of the I.St. relative to that of the I.C. and T.St.'s, and thus to numerically high positive values of $\Delta H^{\ddagger}$ and $\Delta G^{\ddagger}$, and to less negative values of ΔH^0 and ΔG^0. For kinetic but *not* thermodynamic nucleophilicity there is, however, a *partial* compensation in that the T.St. levels relative to the I.C. and thus to the I.St. stem from use of a percentage bond dissociation energy (B.D.E.)—thermicity curve for which a low value of the B.D.E. leads to less positive values for the T.St.'s, and thus partly counteracts the more negative value of the I.St. Therefore the bond strength factor contributes more to thermodynamic nucleophilicity (and basicity too) than to kinetic nucleophilicity. It is however important in both, and thus the nucleophiles which form weak bonds, *viz.* the heavy nucleophiles (they are also the most polarisable) show up better in relation to the first row nucleophiles in reactions such as the saturated aliphatic S_N2 reactions, in which in formation of the rate-limiting T.St. there is less bond formation by the nucleophile than in these aromatic S_N2 reactions. This distinction has been pointed out by Miller *[10c]*, and discussed further above.

The $TN(C_{Ar}X^0)$ value for fluoride ion is only –4 kcal·mole^{-1}, thus it is also a very weak thermodynamic nucleophile, though in contrast with the kinetic nucleophilicity pattern, stronger than the other halide ions. In contrast also the $TB(HX^0)$ value is –13.5 kcal·mole^{-1}, and while this is a value for a weak reagent (base) fluoride ion is much stronger as a thermodynamic base than as either a thermodynamic or kinetic nucleophile, and also much stronger as a base than any other halide ions. These distinctions are largely correlated with the outstandingly strong H–F bond.

Small anions such as fluoride and hydroxide ions have outstandingly high heat of solvation in protic solvents, so that their basicity and nucleophilicity are particularly enhanced when they react instead in aprotic solvents in which solvation of anions is reduced *[7,10c,30,31]*. This factor is the basis of the general method of Finger *et al.* *[34]* for making aryl fluorides by treating aryl halides (Cl, Br, I compounds) with potassium fluoride in solvents such as nitrobenzene, dimethyl sulphoxide or dimethylformamide. Miller and his co-workers *[35a]* have further investigated the high reactivity of fluoride ion in such solvents.

References p. 230

Iodide. This is a specially interesting case since it is both a halide ion and a heavy nucleophile. The $KN(C_{Ar}X^0)$ value is 40.5 kcal·mole^{-1}, *i.e.* it is substantially weaker than methoxide ($\Delta KN = +13.1$), but substantially stronger than fluoride ion ($\Delta KN = -6.5$). Further the heavy nucleophile interaction previously discussed reduces the $KN(C_{Ar}X^0)$ value in comparison with methoxide and fluoride. Considering $\Delta H^{\ddagger}$ values alone, the $\Delta\Delta H^{\ddagger}$ value is +8.5 kcal·mole^{-1} relative to methoxide and –10.5 kcal·mole^{-1} relative to fluoride ion. The values of kinetic nucleophilicity in protic solvents originally used by Miller *[10c]* in terms of $\Delta E^{\ddagger}$ ($\approx \Delta H^{\ddagger}$) are thus a good guide to intrinsic nucleophilicity towards aromatic carbon. Iodide ion may thus be regarded as a kinetic nucleophile of moderate strength, which is stronger in protic but not aprotic solvents than the lighter halide ions. In contrast the $TN(C_{Ar}X^0)$ and $TB(HX^0)$ values are zero, iodide ion being very weak as a thermodynamic nucleophile and thermodynamic base, and this was also implicit in the choice of standard substrates.

This difference in kinetic and thermodynamic behaviour is striking. Weakness is essentially a consequence of weak bonds formed, together with relatively high electron affinity. Only for kinetic nucleophilicity, as discussed above, is a substantial compensatory effect of the weakness of the bond possible. It should not be forgotten, however, that in reactions of iodide ion the nature of the displaced group has an especially important effect on its nucleophilic strength. When the displaced group is attached to the reaction centre by a first row element in particular, the energy level of the T.St. involving the rupture of this bond (T.St.2) is very high compared to that involving formation of a bond between iodine and the reaction centre (T.St.1), and the formation of T.St.2 is then rate-limiting by a wide margin, so that iodide ion then behaves as a very weak kinetic nucleophile also. An example of this is the very low reactivity of iodide ion with 1-fluoro-2,4-dinitrobenzene *[30,36]*, commonly a highly reactive substrate.

Also relevant here as well as more generally, is a factor affecting group mobility discussed in Chapter 5 (p. 168), *viz.* that a few groups such as alkoxy and dialkylamino groups, have a relatively large difference between the strengths of aromatic bonds in I.St.'s and F.St.'s and the aliphatic-type bonds in I.C.'s and T.St.'s. Thus an I.St. of the type I^-/ArOMe will have lower energy than most I.St.'s of the type I^-/ArX, and thus lead to an especially high value of $\Delta H^{\ddagger}$, since with such an I.St. the formation of T.St.2 would be rate-limiting and both factors lead to a high value of $\Delta H^{\ddagger}$. From Miller's calculations *[8,10c]* the adverse effects of these two factors on the kinetic nucleophilicity of iodide ion are illustrated by $\Delta H^{\ddagger}$ and $\Delta G^{\ddagger}$ values for its reaction with *p*-iodonitrobenzene (iodine exchange), *p*-fluoronitrobenzene, and *p*-nitroanisole, *viz.* 33.5 and 40.5; 39.5 and 43.5; and 45.5 and 49.5 kcal·mole^{-1} respectively (see also Figs. 58 and 59, and Fig. 42 of Chapter 5).

Azide. This is also of interest, for it is a first row nucleophile but is very different from amide ion and in some respects is regarded as resembling chloride ion. Its considerable nucleophilic power is well known and in accord with this the KN($C_{Ar}X^0$) value is calculated as 30.4 kcal·mole^{-1}, confirmed by the experimental value of 30.7 kcal·mole^{-1}. It is thus somewhat weaker than methoxide ion and the reasons for this emerge as the higher electron affinity more than counteracting slightly higher bond strength and lower heat of solvation, both of which favour azide ion. The TN($C_{Ar}X^0$) value is −19 kcal·mole^{-1} showing that it is a fairly strong thermodynamic nucleophile, though somewhat weaker than methoxide ion. It is, however, a much stronger kinetic and thermodynamic nucleophile than any of the halide ions (see above and below).

In contrast the TB(HX^0) value is −13.5 kcal·mole^{-1}, so that azide is substantially weaker as a base than as a nucleophile. This is due to the relatively weak H–N_3 bond, which is weaker than the H–OMe bond whereas the C–N_3 bond is stronger than the C–OMe bond.

Thiomethoxide. This has particular interest since it is generally regarded as a stronger nucleophile but weaker base than methoxide ion. The KN($C_{Ar}X^0$) value is 26.2 kcal·mole^{-1}, in good agreement with experiment (26.4 kcal·mole^{-1}), as compared with 27.4 for methoxide ion (ΔKN = 1.2 kcal·mole^{-1}). It is also noteworthy that the difference favouring thiomethoxide ion would be greater if it were not for the heavy nucleophile interaction causing $\Delta S^{\ddagger}$ to be more negative. The $\Delta H^{\ddagger}$ values are 19.5 and 25 kcal·mole^{-1} respectively, favouring thiomethoxide ion by 5.5 kcal·mole^{-1}. The higher reactivity of thiomethoxide than methoxide ion is seen to be due to the lower heat of solvation which more than counteracts bond energy and

References p. 230

electron affinity differences, which mildly favour methoxide. As a corollary, in the chalcogenides as in the halides, a reversal of the order of reactivity of heavy and light nucleophiles is to be expected in changing from a protic to an aprotic solvent.

The TN($C_{Ar}X^0$) value is –23.5 kcal·mole^{-1}. This shows that thiomethoxide though a strong thermodynamic nucleophile is a little weaker than methoxide; whereas it is stronger as a kinetic nucleophile. The difference stems from stabilisation of the final state in the methoxide ion reaction.

The TB(HX^0) value is –21 kcal·mole^{-1}. This is a value for a strong base but noticeably weaker than methoxide, for which the value is –25 kcal·mole^{-1}. The ΔTB value of 4 kcal·mole^{-1} corresponds well with known differences in acidity of alcohols and corresponding thiols *[37,38]*.

Bromide and chloride. Having considered both fluoride and iodide ions, it is worth including the two intermediate halide ions. Their KN($C_{Ar}X^0$) values are 44 and 45.5 kcal·mole^{-1}. These are values expected for very weak kinetic nucleophiles and lead to the order of reactivity in protic solvents $I^- > Br^- > Cl^- > F^-$. Their TN($C_{Ar}X^0$) values are +0.5 and –2 kcal·mole^{-1}, *i.e.* bromide is very slightly weaker and chloride slightly stronger than iodide. The order of the halide ions is markedly affected by entropy differences, for as regards the ΔH^0 contribution alone, the values are (I^- to F^-) 0, +2, +1.5, +4 kcal·mole^{-1}, *i.e.* $I^- > Cl^- > Br^- > F^-$; whereas with $T\Delta S^0$ values at 25° in the order (I to F) equal to 0, +1.5, +3.5, +8 kcal·mole^{-1}, the ΔG^0(TN) values are 0, +0.5, –2, –4 kcal·mole^{-1} *i.e.* $F^- > Cl^- > I^- > Br^-$.

Values of TB(HX^0) for Cl^- and Br^- are –6.5 and –1.5 kcal·mole^{-1}. These correspond to very weak thermodynamic bases, and with earlier results lead to the overall basicity order $F^- > Cl^- > Br^- > I^-$ *i.e.* the reverse of the kinetic nucleophilicity order though roughly paralleling that of thermodynamic nucleophilicity. However the basicity differences are substantially larger, with fluoride a base of not inconsiderable strength. This is only partly due to entropy differences, for the ΔH^0 components (I^- to F^-) are 0, 0, –3, –5.5 kcal·mole^{-1}, *i.e.* $I^- = Br^- < Cl^- < F^-$. Incorporation of the $T\Delta S^0$ values at 25° leads to the ΔG^0(TB) values 0, –1.5, –6.5, –13.5 kcal·mole^{-1}, *i.e.*

$I^- < Br^- < Cl^- < F^-$. The basicity order is a consequence of larger differences in strength between H–Hal than C–Hal bonds.

Thiocyanate. In some respects thiocyanate ion is regarded as resembling the heavier halide ions and it is therefore of interest to consider the results of calculations of its reactivity as a nucleophile even though the fundamental data used are not as reliable as for other reagents considered.

The $KN(C_{Ar}X^0)$ value is 37.5 kcal·mole^{-1} *i.e.* it is of moderate strength, greater than any of the halide ions, but weaker than azide ion. The differences are almost entirely $\Delta H^{\ddagger}$ dependent. The level of reactivity is largely a consequence of a rather high electron affinity (estimated) and but a moderate value for bond strength.

The $TN(C_{Ar}X^0)$ value is –9 kcal·mole^{-1} *i.e.* it is a weak thermodynamic nucleophile but stronger than any of the halide ions, and weaker than azide ion. Regrettably there are insufficient data to permit a calculation of a $TB(HX^0)$ value.

Hydroxide and hydroperoxide. A reliable calculation for hydroxide ion is impracticable because of the complication that the I.C., possessing an aliphatic type hydroxyl, interacts partly and reversibly but to an unknown extent with another molecule of hydroxide ion, whereas the final product reacts more or less completely with hydroxide ion to form a phenoxide ion.

Qualitatively, however, it appears that the reactivity of hydroxide is markedly reduced in comparison with methoxide by a much higher heat of solvation, which is substantially but not entirely counterbalanced by the stronger bond formed by hydroxide ion and because the phenol initially formed has a substantial heat of solvation compared with that of anisole, which is small enough to be neglected in comparison with ionic heats of solvation.

Bunnett and Davis *[39a]* have estimated that the OMe^-/OH^- rate ratio for reaction with 1-chloro-2,4-dinitrobenzene is 33/1 in dioxane/water (60/40), by use of phenoxide ion as a link to gauge relative reactivities. Murto *[40a]* has discussed the rate constants for reactions of hydroxide and methoxide ions in methanol/water mixtures with 1-fluoro-2,4-dinitrobenzene, and obtained the approximate ratio 50/1 (OMe^-/OH^-). With *o*- and *p*-dinitrobenzene he obtained a somewhat

References p. 230

higher ratio *[40b,c]*. All these results are in accord with the explanation now offered.

An assessment of the comparative reactivity of hydroxide and hydroperoxide ions is difficult because of lack of data, but the evidence available *[41,42]* suggests that bond-dissociation energy and electron-affinity factors have relatively small differentiating effect whereas an expected substantial difference in solvation energy in protic solvents ($OH^- > O_2H^-$) should lead to the reactivity order $O_2H^- > OH^-$. Translation of the views of Edwards and Pearson *[19]*, and of Ingold *[9b]* on the α-effect, applicable to hydroperoxide ion, in terms of these energy quantities is uncertain: values of both bond-dissociation energy and electron affinity terms may be involved.

Hydride ion. From the discussion already given in Chapter 5 (p. 173) concerning displacement of hydride, one may correspondingly conclude qualitatively that the low electron affinity of hydrogen leads to very high values of energy of initial states, H^-/ArX, compared to those of intermediate complexes, $\left[\begin{matrix} H \\ X \end{matrix} > \mathrm{Bzd}\right]^-$. Thus values of $\Delta H^{\ddagger}$ are small, the reagent extremely reactive, and there is less bond-formation in transition-state 1 than is usual in aromatic S_N reactions.

Amide ions. There are insufficient data to make satisfactory calculations for amide and substituted amide ions. Nevertheless some useful discussion is possible. The bonds formed by such nucleophiles to carbon are of similar strength to those formed by corresponding oxygen nucleophiles; and heats of solvation are of similar magnitude. The main difference from corresponding oxygen nucleophiles is then in the electron affinity term.

There are no very reliable data here either, but Ogg *[43]* has suggested a value of 20–30 $kcal \cdot mole^{-1}$ for the NH_2 group. This may be compared with 20 $kcal \cdot mole^{-1}$ reported by Baughan *et al.* *[44]* for the CH_3 group, and the considerably higher values for methoxyl and fluoro groups *[45,46]*. A value of 30–40 $kcal \cdot mole^{-1}$, below that of corresponding oxygen groups but higher than that of the methyl group, is an acceptable estimate. With the other information available this leads to the conclusion that amide ions are considerably more reactive than corresponding oxygen nucleophiles. This is in accord with what we know of such species, and now explained as essentially a consequence of their low electron affinity. A complicating factor is the concurrent very high basicity which is conspicuous in the benzyne mechanism (Chapter 3). As with hydroxide ion this leads to acid–base equilibria involving first-formed products, when the product formed by reaction with the amide ion bears an ionizable proton, as in reactions of NH_2^- and NHR^-.

Carbanions. Similar considerations to those described for amide ions apply. Their generally very high reactivity is well known and now explained as essentially a consequence of their low electron affinity, for C–C bonds while strong are not too different in this respect from C–N and C–O bonds.

It should be recalled that acid–base reactions with the great majority of protic solvents prevent us from obtaining values of nucleophilicity in these solvents to compare with values for weaker reagents.

Other anionic reagents. It is well known that reagents such as sulphite and thiosulphate are quite powerful nucleophiles. In terms of the preceding discussion this probably stems from relatively low values of electron affinity involving these doubly negatively charged species. Differences in electron affinity are probably also responsible for the higher reactivity of sulphites than sulphates, sulphinates than sulphonates, phosphites than phosphates, and in other similar pairs or sets. A further contribution to the reactivity of these named species is that their relatively large size minimises the adverse effect of high solvation energy due to the double charge. Data are lacking for calculations with these species, but the qualitative discussion above is confirmed experimentally, for example, by the results of Bevan and his co-workers *[47]* (see also Chapter 5, Table 34B) which show that the reactivity of sulphite ion is high and comparable with that of alkoxides.

In conclusion the calculations of this section readily predict and explain the orders of reactivity (kinetic nucleophilicity) in protic solvents of anionic reagents in which the nucleophilic atom varies according to its position along a horizontal row or down a vertical column of the Periodic Table, *e.g.* $C^- > N^- > O^- > F^-$; $O^- < S^-$; $F^- < Cl^- < Br^- < I^-$; and though only qualitatively, since the fundamental solvation data are lacking, the reversal of the reactivity order

References p. 230

down a vertical column, in changing from a protic to an aprotic solvent.

The procedure is applicable also to reagent behaviour as bases as well as nucleophiles. The main limitation to its full use is the lack of data, which in particular largely prevents its general extension for the time being to substitution on other elements except hydrogen, and to aprotic solvents.

6. Reactivity of Neutral Reagents

(*a*) *General*

It should be possible to apply the same general principles to reactions of neutral nucleophiles as to those of anionic nucleophiles *[1,8,10c]*, taking account of some additional complexities. The lack of some of the fundamental data precludes fully predictive calculations without some semi-empirical estimates, but acceptable and useful results are obtainable even from qualitative discussions. To facilitate this, and for comparison with the calculations attempted, a reasonable selection of quantitative data for reactions of neutral nucleophiles is required and this is given in Tables 49 to 51, (data from refs. *35b*, *39b*, *40*, *48* to *56*). Reference should be made also to relevant sections of Tables 32 to 35 of Chapter 5 (p. 140).

It is necessary to recall that complexes formed by the initial co-ordination of neutral nucleophiles $\dot{\text{Y}}$ with typical neutral substrates are species $\overset{+}{\text{Y}}$–$\overset{-}{\text{Bzd}}$–X, and since cationic groups have high mobility (Chapter 5, p. 140) the reverse reaction is expected to be relatively facile. The positive charge remains on the nucleophilic atom unless it is lost by transfer of an ionisable proton attached to Y^+ to another molecule of reagent acting as a base, or the charge is delocalised as occurs in a few special cases, *e.g.* with thiourea as reagent. Proton transfer leaves intermediates corresponding to those formed by anionic nucleophiles Y^-, with consequent reduction in facility of the reverse reaction.

Reactions of neutral nucleophiles are nevertheless generally second order (first order in reagent and substrate), and though complicating factors have been recorded *[57,58,59]* these are not in conflict with the conclusion which follows; they are not subject to general base catalysis; and have the replaced group mobility order F > Cl, Br, I. It is inferred that the energy levels of I.St. and T.St.1 for reactions of these nucleophiles are sufficiently high for T.St.1 formation to be rate-limiting even with fluoro compounds, for which T.St.2 levels are relatively high (Chapter 5).

There are some other features of these reactions worthy of preliminary note. First, despite the above general statement, some reactions have been shown to be subject to general base catalysis, *e.g.* the reaction of *N*-methylaniline with 1-fluoro-2,4-dinitrobenzene *[50,52,58]* and of piperidine with 1-phenoxy-2,4-dinitrobenzene *[58]*. The former significantly shows a fluorine/chlorine mobility ratio of unity instead of a hundred or more, as is common, and this mobility change is enhanced by carrying out the reaction in a dipolar aprotic solvent *[50]*. It is also significant that this phenomenon is characteristic of displacement of groups attached by first row elements, in which bonds between the group and the ring are relatively strong.

Only in the DABCO (1,4-diaza-[2.2.2]-bicyclooctane) catalysed reaction of 1-chloro-2,4-dinitrobenzene with *p*-anisidine *[59a]* is there an unequivocal example in a protic solvent of a base-catalysed aminolysis in displacement of a group (chlorine) not attached by a first-row element. The carbon–chlorine bond is also strong, and the result would be explicable if the transition-state for formation of a bond by *p*-anisidine is sufficiently lower in energy than that for other amines considered.

Secondly, the most common and most investigated neutral nucleophiles are amines and their reactivity pattern *viz.*, $NH_3 < RNH_2$, $R_2NH > R_3N$; and $RNH_2 > ArNH_2$, (this includes the comparison of aliphatic and aromatic heterocyclic species such as piperidine and pyridine) is not a simple one. This is illustrated by values from Table 49 of rate constants, with ratios in parentheses, for reactions of 1-chloro-2,4-dinitrobenzene with (*a*) ammonia, methylamine, dimethylamine, piperidine, and trimethylamine in methanol or ethanol at 25° (the relatively minor effect of the difference between methanol and ethanol as a solvent may be neglected for this comparison) *viz.* $4 \cdot 10^{-6}$ (1), $3.16 \cdot 10^{-3}$ (790), $3.55 \cdot 10^{-2}$ (8900), $1.53 \cdot 10^{-2}$ (3800), $1.1 \cdot 10^{-6}$ (0.27); (*b*) of methylamine and aniline in methanol at 25°, *viz.* $3.16 \cdot 10^{-3}$ (1); $6.20 \cdot 10^{-5}$ (0.0196); (*c*) piperidine and pyridine in methanol at 50°, *viz.* $6.41 \cdot 10^{-2}$ (1), $5.53 \cdot 10^{-6}$ (0.0386). A third very

TABLE 49

REACTIVITY OF SOME NEUTRAL NUCLEOPHILES WITH 1-CHLORO-2,4-DINITROBENZENE

Reagent	Solvent	Temperature (°C)	Rate constant k_2 ($l \cdot mole^{-1} \cdot sec^{-1}$)	Kinetic data $\Delta E^{\ddagger}$ ($kcal \cdot mole^{-1}$)	$\log_{10} B$ (or $\Delta S^{\ddagger}$, e.u.)	Rate ratios for Solvent	Rate ratios for Reagent
NH_3	EtOH	25	$4 \cdot 10^{-6}$	—	—	1	1
$MeNH_2$	EtOH	25	$3.16 \cdot 10^{-3}$	10.7	5.3_5	—	760
Me_2NH	EtOH	25	$3.55 \cdot 10^{-2}$	9.3	5.3_5	—	9000
[piperidine ring, NH]	EtOH	0 25	$3.10 \cdot 10^{-3}$ $1.53 \cdot 10^{-2}$	10.2	5.6_5	—	3700
$EtNH_2$	EtOH	25	$9.2 \cdot 10^{-4}$	11.7	5.5_5	—	240
Et_2NH	EtOH	25	$1.9 \cdot 10^{-4}$	12.7	5.6	—	48
iso-$PrNH_2$	EtOH	25	$1.0 \cdot 10^{-4}$	15.9	7.6_5	—	25
iso-PrEtNH	EtOH	25	*ca.* 10^{-5}	—	—	—	$\geqslant 1$
iso-Pr_2NH	EtOH	25	*ca.* 10^{-6}	—	—	—	$\geqslant 0.1$
iso-$BuNH_2$	EtOH	25	$6.8 \cdot 10^{-4}$	—	—	—	170
iso-Bu_2NH	EtOH	25	$5.8 \cdot 10^{-5}$	—	—	—	15
tert-$BuNH_2$	EtOH	25	$3.8 \cdot 10^{-6}$	—	—	—	0.85
NH_2NH_2	EtOH	25	$5.98 \cdot 10^{-3}$	—	—	—	750[a]
[3-CF_3-piperidine ring, NH]	EtOH	25	$9.65 \cdot 10^{-4}$	—	—	—	240
[4-CF_3-piperidine ring, NH]	EtOH	25	$2.83 \cdot 10^{-3}$	—	—	—	710

O NH	EtOH	25	$4.17 \cdot 10^{-3}$	—	—	—	1000
HN NH	EtOH	25	$2.74 \cdot 10^{-2}$	—	—	—	3400[a]
NH	EtOH	25	$4.6 \cdot 10^{-2}$	—	—	—	1200
$PhNH_2$	EtOH	25 50 67.2 100 131.5	$6.20 \cdot 10^{-5}$ $2.69 \cdot 10^{-4}$ $6.46 \cdot 10^{-4}$ $2.76 \cdot 10^{-3}$ $8.94 \cdot 10^{-3}$	11.2 (10.4)[b]	4.0 (3.5)[b]	1	16 (1)[c]
m-$CH_3C_6H_4NH_2$	EtOH	50	$3.91 \cdot 10^{-4}$	11.6	4.4	—	(1.45)[c]
o-$ClC_6H_4NH_2$	EtOH	50	$6.32 \cdot 10^{-5}$	12.3	4.1	—	(0.235)[c]
o-$BrC_6H_4NH_2$	EtOH	50	$5.44 \cdot 10^{-5}$	14.6	5.6	—	(0.202)[c]
m-$ClC_6H_4NH_2$	EtOH	50	$1.97 \cdot 10^{-5}$	13.0	4.1	—	(0.073)[c]
m-$BrC_6H_4NH_2$	EtOH	50	$2.19 \cdot 10^{-5}$	15.1	5.5	—	(0.082)[c]
PhNHMe	EtOH	67.2 131.5	$7.3 \cdot 10^{-5}$ $4.8 \cdot 10^{-4}$	—	—	1	(0.11)[c] (0.055)[c]
NH_3	MeOH	0 25 50 67.2 131.5	$1.55 \cdot 10^{-7}$ $2.02 \cdot 10^{-6}$ $1.74 \cdot 10^{-5}$ $6.61 \cdot 10^{-5}$ $3.24 \cdot 10^{-3}$	16.6	6.45 (− 31.1)	(25°) 0.5	1
Me_3N	MeOH	25 131.5	*ca.* 10^{-6} *ca.* 10^{-3}	*ca.* 14	*ca.* 4.5	—	0.5
$PhNH_2$	MeOH	25 50 100 131.5	$4.86 \cdot 10^{-5}$ $1.90 \cdot 10^{-4}$ $2.28 \cdot 10^{-3}$ $7.53 \cdot 10^{-3}$	11.4	4.0	(25°) 0.77 (131.5°) 0.84	(25°) 23.6 (1)[c] (131.5°) 2.32

continued on p. 208

TABLE 49 (*continued*)

Reagent	Solvent	Temperature (°C)	Rate constant k_2 ($l \cdot mole^{-1} \cdot sec^{-1}$)	Kinetic data $\Delta E^{\ddagger}$ ($kcal \cdot mole^{-1}$)	$\log_{10} B$ (or $\Delta S^{\ddagger}$, e.u.)	Rate ratios for: Solvent	Rate ratios for: Reagent
N (pyridine ring)	MeOH	25	$6.04 \cdot 10^{-7}$	16.9_5	6.2	—	0.3
$PhNMe_2$	MeOH	131.5	$3 \cdot 10^{-5}$	—	—	—	0.01 (0.004)[c]
NH (piperidine ring)	MeOH	0 25	$1.61 \cdot 10^{-3}$ $1.00 \cdot 10^{-2}$	11.8_5	6.7	(25°) 0.65	(25°) 4950
Me_2NH	DMF	25	$5.10 \cdot 10^{-3}$	8.4_5	3.9	0.144	—
NH (piperidine ring)	$MeOH/C_6H_6$ (1:1 v/v)	25	$6.45 \cdot 10^{-3}$	—	—	0.421 (0.645)[d]	— —
NH (piperidine ring)	$Dioxane/H_2O$ (6:4 v/v)	0	$1.38 \cdot 10^{-2}$	—	—	4.45 (8.58)[d]	—
NH (piperidine ring)	C_6H_6	25	$8.64 \cdot 10^{-2}$	6.5	3.7	5.65	—
N_2H_4	$Dioxane/H_2O$ (6:4 v/v)	25.2	$3.87 \cdot 10^{-3}$	—	—	0.63	—
$PhNH_2$	$MeOH/C_6H_6$ (1:1 v/v)	100	$8.89 \cdot 10^{-4}$	—	—	0.32 (0.39)[d]	—
$PhNH_2$	Me_2CO	50	$1.75 \cdot 10^{-5}$	—	—	0.065 (0.09)[d]	—
PhNHMe	$PhNO_2$	131.5	$4.4 \cdot 10^{-4}$	—	—	0.9	—

TABLE 50

REACTIVITY OF SOME NEUTRAL NUCLEOPHILES WITH PICRYL CHLORIDE

Reagent	Solvent	Temperature (°C)	Rate constant k_2 ($l \cdot mole^{-1} \cdot sec^{-1}$)	Kinetic data $\Delta E^{\ddagger}$ ($kcal \cdot mole^{-1}$)	$\log_{10} B$	Rate ratio
MeOH	MeOH	25	$7.55 \cdot 10^{-9}$	18.1	6.6_5	1
$PhNH_2$	MeOH	25	$6.77 \cdot 10^{0}$	7.8_5	5.6	$8.97 \cdot 10^{8}$

important feature is that the values of $\Delta S^{\ddagger}$ for reactions of neutral nucleophiles with neutral substrates are much more negative than those for reactions of anionic nucleophiles, *e.g.* from Table 49, for reactions with 1-chloro-2,4-dinitrobenzene in methanol, $\Delta S^{\ddagger}$ (50°) values are –30.2 for piperidine, and –42.4 e.u. for aniline, whereas with methoxide ion the value is –9.20 e.u. *[10a]*. From Table 50, for the reaction with picryl chloride in methanol, the value of $\Delta S^{\ddagger}$ is –30.2 for methanol (as reagent), and –35.0 e.u. for aniline, whereas with *p*-nitrophenoxide ion it is –13.4 e.u. *[60]*. Miller and his co-workers *[35b]* have also compared azide ion and dimethylamine in their reactions with *p*-fluoronitrobenzene in an aprotic solvent (dimethylformamide) and the values of $\Delta S^{\ddagger}$ are –10.6 and –42.9 e.u. respectively. Also in Table 49 there is a result for the reaction of piperidine with 1-chloro-2,4-dinitrobenzene from which the $\Delta S^{\ddagger}$ value is –44 e.u. It therefore appears that this is a characteristic phenomenon in the reactions of neutral as compared with anionic reagents in aprotic as well as protic solvents, though Suhr *[61]* has reported results for a number of amines in reactions with *p*-fluoronitrobenzene in dimethyl sulphoxide and quotes $\Delta S^{\ddagger}$ values, which in some cases are little more negative than values common for anionic reagents.

(*b*) *Factors affecting reactivity*

(*i*) *The bond strength factor*

The strength of the bond formed by the reagent is a key factor in assessing its strength as a nucleophile, as has been seen. A neutral nucleophile, $\dot{Y}$, in forming a bond with an electrophilic centre, here

TABLE 51

REACTIVITY OF SOME NEUTRAL NUCLEOPHILES WITH 1-FLUORO-2,4-DINITROBENZENE

Reagent	Solvent	Temperature (°C)	Rate constant k_2 ($l \cdot mole^{-1} \cdot sec^{-1}$)	Kinetic data $\Delta E^{\ddagger}$ $kcal \cdot mole^{-1}$	$\log_{10} B$	Rate ratio	
						Solvent	Reagent
H_2O	H_2O	40	$3.65 \cdot 10^{-9}$	—	—	—	$8.05 \cdot 10^{-8}$[a]
NH_3	MeOH	0	$2.10 \cdot 10^{-3}$				1
		40	$4.55 \cdot 10^{-2}$				1
		50	$5.89 \cdot 10^{-2}$	12.1	6.9_5		1
		67.2	$1.51 \cdot 10^{-1}$				1
Piperidine	MeOH	0	$1.5 \cdot 10^{0}$	—	—		715
$PhNH_2$	EtOH	0	$2.70 \cdot 10^{-3}$			1	1.29[b]
		50	$1.68 \cdot 10^{-2}$	6.4	2.6	1	0.203[b]
		67.2	$2.78 \cdot 10^{-2}$			1	0.184[b]
$PhNH_2$	Me_2CO	50	$2.20 \cdot 10^{-4}$	—	—	0.013	—
PhNHMe	EtOH	67.2	$7.30 \cdot 10^{-5}$	—	—	1	$4.80 \cdot 10^{-4}$[b,c]
PhNHMe	$PhNO_2$	67.2	$2.3 \cdot 10^{-6}$	—	—	0.031	—

[a] Includes a relatively minor component due to the difference in solvent effects of water and methanol. [b] Includes a small and unimportant component due to the difference in solvent effects of ethanol and methanol. [c] Reagent ratio $PhNHMe/PhNH_2$ in ethanol at 67.2° is $2.6 \cdot 10^{-3}$.

a suitable C_{Ar}–X, becomes positively charged, whereas an anionic nucleophile becomes neutral.

In aromatic S_N reactions we have therefore to compare $\overset{+}{Y}$–B$\overset{-}{z}$d– with Y–B$\overset{-}{z}$d– bonds. Fundamental data required are scarce, but Price *et al.*, for example *[62]*, quote bond dissociation energy (B.D.E.) values indicating the approximate equality of strengths of H–NH_2 and H–NH_3^+ bonds; and near equality of those of H–OH and H–OH_2^+. While there is not too precise a parallelism between strengths of H–X and C–X bonds, their strengths do not differ markedly, and it is thought reasonable to infer that strengths of C–X and corresponding C–X^+ bonds (X = N or O) differ very little and that differences are small generally. It is concluded that the bond strength factor, though very important in the absolute sense, is unlikely to cause important differences between reactivity of neutral and corresponding anionic nucleophiles.

(*ii*) *The solvation factor*

The picture here is quite different, for this factor is not only very important in itself but also in its differential impact on reactions of anionic and neutral nucleophiles.

In reactions of neutral reagents with neutral substrates the solvation energies of I.St.'s may be neglected by comparison with those of I.St.'s in which reagents are anions, and with charged T.St.'s, though no doubt some small allowance should be made for solvation of species such as water and lower alcohols, ammonia and lower amines in protic solvents, and aromatic substrates in aromatic and many dipolar aprotic solvents (see below). The T.St.'s and I.C.'s formed by reactions of anionic reagents with neutral substrates are themselves anions but larger and less strongly solvated than I.St.'s. In contrast, the reactions of neutral reagents with neutral substrates form T.St.'s and I.C.'s having the character of large dipolar molecules of the type $\overset{+}{Y}$–B$\overset{-}{z}$d–X with well separated centres of positive and negative charge, and such species should be highly solvated with an energy of solvation likely to exceed the sum of the solvation energy of two ions, one being of the type ArY^+ and the other ArX^-; though less than the sum of solvation energies of ions Y^+ and X^-. The reaction products consist of an anion and a neutral molecule from the reaction of an anion and neutral molecule, and an anion and a cation from a pair of neutral reactants (the incursion of a second mole of reagent acting as a base in the latter case will not substantially affect this picture). However the difference in solvation energy from intermediates Y–$\overset{-}{\text{Bzd}}$–X to ArY plus X^- is likely to be of the same order as that of intermediates $\overset{+}{\text{Y}}$–$\overset{-}{\text{Bzd}}$–X proceeding to ArY^+ plus X^-. The main differential effect of the solvation energy is thus in the change from I.St.'s to T.St.'s.

The solvation entropy component of $\Delta S^{\ddagger}$ also markedly differentiates anionic and neutral reagents in their reactions with neutral substrates. The very large increase in solvation energy in forming the T.St. in the latter class is clearly paralleled by a large decrease in $\Delta S^{\ddagger}$ values which should be much more negative than in the anion reactions where there is a small *decrease* in solvation in forming the T.St.'s. This has been mentioned above. Further as was pointed out by Miller and his co-workers *[13]*, in relation to aliphatic S_N2 reactions, $\Delta S^{\ddagger}$ values for the reactions of C–Hal compounds are in the order –F < –Cl < –Br < –I *i.e.* fluorine most negative. If there is a similar effect in these reactions it will result in (*i*) a reduction in the reactivity of fluorine relative to the other halogens and (*ii*) the probable upset of the mobility order of Cl, Br, and I for which $\Delta H^{\ddagger}$ values are similar, and more particularly of chlorine and bromine, which not only differ very little, but also are less likely to be differentially affected by adverse steric contributions to $\Delta S^{\ddagger}$ than are reactions of iodo compounds. There is clear support for both of these suggestions in Tables 49 to 51, and Tables 32 to 35 of Chapter 5.

(*iii*) *Ionization energy*

This is also a key term in assessing nucleophilic reactivity and has been discussed at length in relation to anionic nucleophiles. It plays an important part in differentiating the activity of different nucleophiles. The magnitude of this term is much higher for neutral than anionic nucleophiles, but since many reactions of the former do proceed readily, despite also the adverse effect of very negative values of $\Delta S^{\ddagger}$, the view that the T.St.'s and I.C.'s formed by reactions of neutral nucleophiles with neutral substrates are very strongly solvated is well supported. Indeed it is difficult to see how such reactions can otherwise be energetically possible.

(iv) The α-substituent effect

This is likely to be important in these reactions of neutral nucleophiles, for this electronegativity effect *e.g.* of groups of the type $-\overset{+}{N}\Leftarrow$ is probably of the same order as that of fluorine; and is so assumed in calculations (see below).

The timing of proton transfer in the case of nucleophiles of the type $\ddot{Y}$–H is not easy to decide except that the general kinetic form of the reactions indicate it is subsequent to formation of the rate-limiting T.St. The high α-substituent effect of $-\overset{+}{N}\Leftarrow$ leading to a low value of the energy of T.St.2, favours elimination of replaced groups before proton transfer, though this is counteracted at least in part by the enhanced stability of F.St.'s of the type Ar–$\ddot{O}$– and Ar–$\dot{N}\langle$, and consequent reduction of the level of T.St.2 if the proton is lost first. While proton transfer from H–$\overset{+}{Y}$–B$\overset{-}{z}$d–X, where the C–Y bond is aliphatic in type, to $\ddot{Y}$–H is expected to be energetically much less favoured than from Ar–$\overset{+}{Y}$–H to $\ddot{Y}$–H it may nevertheless be exothermic, and the consequent entropy changes may also be favourable. While the full answer to the question is in doubt, the answer to the most important question is known from experimental evidence, *viz.* that proton loss is generally subsequent to formation of the rate-limiting T.St. Species H–$\overset{+}{Y}$–B$\overset{-}{z}$d–X where X = O or S are more acid than when X = N, but the basicity of the species $\ddot{Y}$–H is also less so that in this respect there may be little difference between different classes of neutral nucleophiles, though the larger α-substituent effect of $-\overset{+}{O}\langle$ may cause a significant difference.

(c) *Comparison of neutral nucleophiles differing in the nucleophilic atom*

It is first instructive to make a comparison in the same horizontal row of the Periodic Table, *e.g.* of ammonia and water (and hydrogen fluoride), and of alcohols and amines.

The bond energy term favours the reactivity order HF > H_2O > NH_3 and ROH > RNH_2, though differences between the nitrogen and oxygen species are small.

References p. 230

The solvation term is less important in distinguishing different neutral nucleophiles of similar size, as in the same horizontal row of the Periodic Table, than in the absolute sense, but would also favour the order $HF > H_2O > NH_3$ and $ROH > RNH_2$.

The most important item in this comparison is, however, the ionization energy term, differences in which are very large. Values in kcal · mole^{-1} are NH_3, 234 *[63]*; H_2O, 290 *[64]*; HF, 401 *[65]*; $MeNH_2$, 207 *[63]*; and MeOH, 249 *[64]*. The much higher reactivity of ammonia and amines than of water and alcohols is thus easily predictable, as is the lack of nucleophilic character of hydrogen fluoride. The difference in reactivity between amino and hydroxy compounds is large, and there are very few comparative kinetic data relating to aromatic S_N reactions. However, Murto *[40a]* reports that the bimolecular rate constant (k_2) at 40° for hydrolysis (water as solvent and reagent) of 1-fluoro-2,4-dinitrobenzene is $3.65 \cdot 10^{-9}$. From the data of Reinheimer *et al.* *[53]* the rate constant (k_2) for the reaction with ammonia in methanol at 40° is $4.55 \cdot 10^{-2}$. The ratio of these rate constants, which does not take into account the effect of the solvent difference, is $1.25 \cdot 10^7$, and since these neutral reagent and substrate reactions are expected to be faster in water, a larger reagent rate ratio, probably of the order of 10^9 is a reasonable estimate. Similarly Bevan and his co-workers *[51]* reported results (Table 50) by which may be compared methanol and aniline reacting with picryl chloride in methanol as solvent at 25°. The rate ratio is $8.97 \cdot 10^8$, and for comparison of methylamine and methanol, a rate ratio of the order of 10^{10} is probably a fair estimate. The relevant data in Table 50, suggest that a $\Delta\Delta E^{\neq}$ value of about 10 kcal · mole^{-1} is mainly responsible for the large difference in reactivity.

It is also of great interest to compare nucleophiles in the same column of the Periodic Table *e.g.* amines and phosphines, alcohols and thiols. There is but a small difference in the strengths of C–P and C–N bonds, *e.g.* the Me–PMe_2 bond is 4 kcal · mole^{-1} weaker than the Me–NMe_2 bond *[66,67]*, the difference thus favouring nitrogen.

Intermediates of the type $\text{H–}\overset{+}{\text{P}}\text{–}\overset{-}{\text{Bzd}}\text{–X}$ may be assumed to have less solvation energy, though in the absence of steric effects, higher entropy, than the corresponding $\text{H–}\overset{+}{\text{N}}\text{–}\overset{-}{\text{Bzd}}\text{–X}$ types. This factor is thus rather indeterminate, and most likely to differentiate markedly only when steric effects diminish reactivity, probably in the $\Delta S^{\ddagger}$ term *e.g.* reduction of reactivity of tertiary amines relative to tertiary phosphines.

The ionisation energies of ammonia and phosphine are about equal *[68]* so that this factor is unlikely to differentiate the reactivity of nitrogen and phosphorus nucleophiles to an important extent. In summary, the factor in general terms most likely to differentiate nucleophilic reactivity of nitrogen and phosphorus nucleophiles is that of bond strength. In the case of reactivity towards carbon the difference is relatively small, though favouring nitrogen, and could be offset by steric factors.

There is little experimental evidence but the comments of Ahrland *et al. [21]* in cases where the relative reactivity is $N > P$, and in other cases where it is $P > N$, appear to be correlated with bond strengths. At the same time a comparison of tri-n-butylphosphine with triethylamine in a saturated aliphatic S_N2 reaction (in reaction with ethyl iodide in acetone) *[69]* indicates a slight inferiority of the amine because of adverse steric factors. There are no suitable data for aromatic S_N reactions, but overall the few available data support the above discussion. With the emphasis on the bond strength factor the low nucleophilicity of nucleophiles derived from higher members of Group V can be predicted.

There is a small difference in the strength of C–S and C–O bonds *e.g.* the Me–SMe bond is 4 kcal·mole^{-1} weaker than the Me–OMe bond *[70,71]* thus favouring oxygen. The solvation factor, similarly to the nitrogen–phosphorus comparison, would only mildly favour oxygen. The difference in ionisation energy is, however, large and should be enough to counteract this difference, *e.g.* values for ethanol and ethanethiol are 242 and 214, and of water and hydrogen sulphide 290 and 242 kcal·mole^{-1} *[64,72,73]*. The evidence thus favours higher reactivity of sulphur nucleophiles. These compounds are, however, weak acids, particularly the thiols, and the alcohol–thiol difference is of the order of 3 pK_a units *[37,38]*, so that thiol reactions are characteristically via the anionic forms and information available refers to conditions where anion production is favoured. As a result, there is no satisfactory kinetic comparison of alcohols and thiols in the aromatic series to confirm the above discussion.

Qualitative comparison of Group V and Group VI neutral nucleo-

References p. 230

philes can include that of neutral nitrogen and sulphur reagents. Bond-dissociation energy factors do not differentiate but the ionisation energy term moderately favours nitrogen, and the solvation energy term is also expected to favour nitrogen. The difference in nucleophilic strength of nitrogen and sulphur is, however, less than that between nitrogen and oxygen nucleophiles, and is likely to be obscured by reaction of thiols via anionic forms.

(*d*) *Reactivity of amines*

The general procedure previously described and applied qualitatively readily leads to reasonable conclusions about reactivity of neutral nucleophiles having different nucleophilic atoms. There remains the problem of the great majority of investigated reactions of neutral nucleophiles which are those of amines.

Since Brady and Cropper *[48]* (see Table 49) have shown that steric effects can be large in reactions of bulky amines, the intrinsic differences in amine reactivity are most clearly seen by comparing the members of the series NR_3 where R = H or Me, for which we have quoted reactivity ratios which are of the order of NH_3 (1), $MeNH_2$ (10^3), Me_2NH (10^4); Me_3N (10^{-1}). This is not an obvious pattern. With the same nucleophilic atom, the bond energy factor is a relatively minor one, as is seen from the bond dissociation energies Me–NH_2, 79; Me–NHMe 73; Me–NMe_2, 69 *[67]*; and Me–$\overset{+}{N}Me_3$ assumed also to be 69 kcal·mole^{-1} (see above). This leads to the reactivity order $NH_3 > MeNH_2 > Me_2NH = Me_3N$. The order of solvation energy of species of the type $-\overset{+}{N}\text{–Bzd–}\overset{-}{X}$ is: H–$\overset{+}{N}H_2$– > H–$\overset{+}{N}$HMe– > H–$\overset{+}{N}Me_2$– > Me–$\overset{+}{N}Me_2$–. Amine reactivity is then in the same order and these differences are large. Another large factor is, however, in the opposite direction, *viz.* the ionisation energy term. Values are: NH_3, 234 *[63,74]*; $MeNH_2$, 207 *[63]*; Me_2NH, 190 *[75]*; and Me_3N, 180 kcal·mole^{-1} *[63]*, and this favours the order $Me_3N > Me_2NH > MeNH_2 > NH_3$. While therefore the order $NH_3 < MeNH_2 < Me_2NH > Me_3N$ is readily understandable as a resultant of the terms discussed above, it could not have been predicted with any certainty. We may similarly consider the pair $MeNH_2$ and $PhNH_2$. Bond dissociation energies Me–NHMe and Me–NHPh of 73 and 60 kcal·mole^{-1} *[67,76]* are relevant and similar values may be assumed for $\mathrm{Me{-}\overset{+}{N}H_2Me}$ and $\mathrm{Me{-}\overset{+}{N}H_2Ph}$ bonds. Values are in general accord with a difference of 10 kcal·mole^{-1} between the C–OMe and the C–OPh bond *[77]*. The species $\mathrm{Ph\overset{+}{N}H_2{-}\overset{-}{B}zd{-}X}$ are less solvated than $\mathrm{Me\overset{+}{N}H_2{-}\overset{-}{B}zd{-}X}$, and this leads to the reactivity order $\mathrm{MeNH_2} >$ $\mathrm{PhNH_2}$. While solvation entropy differences may minimise this, steric contributions favour $\mathrm{MeNH_2}$. In contrast the ionisation energy of aniline is substantially less than that of methylamine, values being 178 and 207 kcal·mole^{-1} *[63,74,75]* and this favours aniline.

The experimental order $\mathrm{PhNH_2} < \mathrm{MeNH_2}$ is thus also clearly understandable as a resultant of these terms, but again could not have been predicted with any certainty.

(*e*) *Quantitative estimates of reactivity*

Despite the lack of some fundamental data an attempt to carry out quantitative calculations by using semi-empirical estimates, is clearly very desirable. To be really valuable there must however be internal consistency in arbitrary estimates of fundamental data which are lacking, and these estimates should have some theoretical basis.

The deficiency is in the values to be allotted to solvation energies of species $\mathrm{\overset{+}{Y}{-}\overset{-}{B}zd{-}X}$, particularly needed when Y = N. From previous discussions values of the order of 100 kcal·mole^{-1} are reasonable, though clearly quite markedly dependent on the number of R groups attached to nitrogen. Ten such calculations are now reported, beginning with a pair of calculations using trimethylamine as reagent, for which the general question of timing and energetics of proton transfer from intermediates $\mathrm{H{-}\overset{+}{Y}{-}\overset{-}{B}zd{-}X}$ to $\mathrm{\ddot{Y}{-}H}$ does not arise, though this is further considered later. The terms and procedure are as described by Miller *[8,10c]* and discussed earlier for anionic reagents.

The essence of the problem is that an arbitrary value should be chosen first for the heat of solvation of species $\mathrm{\overset{+}{N}Me_3{-}\overset{-}{B}zd{-}X}$, though this choice is not entirely free. From the previous discussion it was concluded that the heat of solvation of a species $\mathrm{Y^+{-}\overset{-}{B}zd{-}X}$ should

References p. 230

exceed the sum of the heats of solvation of $[Ar–Y]^+$ and $[Ar–X]^-$, but be less than the sum of heats of solvation of species Y^+ and X^-. For similarly sized X and Y the values of heats of solvation are also similar. For $[Ar–X]^-$ which corresponds to a singly negatively charged benzenide complex, a value of 54 kcal·mole^{-1} has already been successfully used in calculations *[8,10c]*. The value chosen for the species $Y^+–\overset{-}{Bzd}–X$ should therefore be approximately $\geqslant$100 kcal·mole^{-1}. Further the values selected for different species $Y^+–\overset{-}{Bzd}–X$ must bear a relationship to each based essentially on size, and all must lead consistently to conclusions compatible with experiment. It must be emphasised that the argument leads to the correct order of magnitude for the values of heats of solvation, but the actual values chosen are arbitrary within the framework specified.

Ten calculations are shown in Table 52 (the first pair for $Y^+ = NMe_3^+$) and the values of heats of solvation in kcal·mole^{-1} selected in this way are $–Y^+ = –NMe_3^+$, 103; $–NHMe_2^+$, 117; $–NH_2Me^+$, 131; $–NH_3^+$ 145; $–NHMePh^+$, 98; $–NH_2Ph^+$, 112. The first value is arbitrary except that it is of the order of 100 kcal·mole^{-1} and in view of the size of the NMe_3^+ group less than double the value for a singly-charged benzenide complex. The stepwise removal of methyl groups from $–NMe_3^+$ to $–NH_3^+$ is arbitrarily assumed to give the same increase of 14 kcal·mole^{-1} in solvation energy for each step. The small solvation energy of neutral amines is not considered, being absorbed in the ion solvation energy terms: even for NH_3 it is only about 6 kcal·mole^{-1} *[78]*. Whereas the difference in heat of solvation between $–NH_2Me^+$ and $–NHMe_2^+$ is assumed to be 14 kcal·mole^{-1}, that between $–NH_2Me^+$ and $–NHMePh^+$, which must be substantially larger, is assumed to be 33 kcal·mole^{-1}, so that the solvation energy of $–NH^+MePh$ is 98 and $–NH_2Ph^+$ then 112 kcal·mole^{-1}.

These values are arbitrary but reasoned and consistent. All other data used in calculations are known fundamental data.

Figures 61 to 70, illustrate the results of the ten calculations shown in Table 52, for which additional data are contained in references *8,10c,76,79,80*.

The conclusions are of great interest, for the characteristic low values of $\Delta E^{\ddagger}$ ($\simeq \Delta H^{\ddagger}$ in the solution reactions) are reproduced, and though $\Delta S^{\ddagger}$ values are only predicted qualitatively the very marked

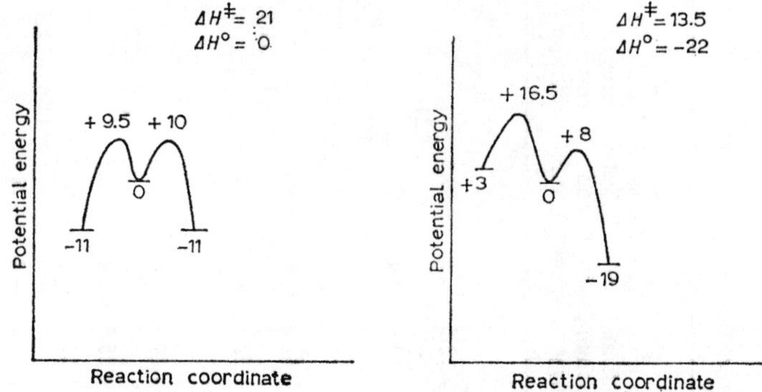

Fig. 61. Potential energy/reaction coordinate profile of $Me_3N/p\text{-}NO_2C_6H_4OMe$.
Fig. 62. Potential energy/reaction coordinate profile of $Me_3N/2,4\text{-}(NO_2)_2C_6H_4Cl$.

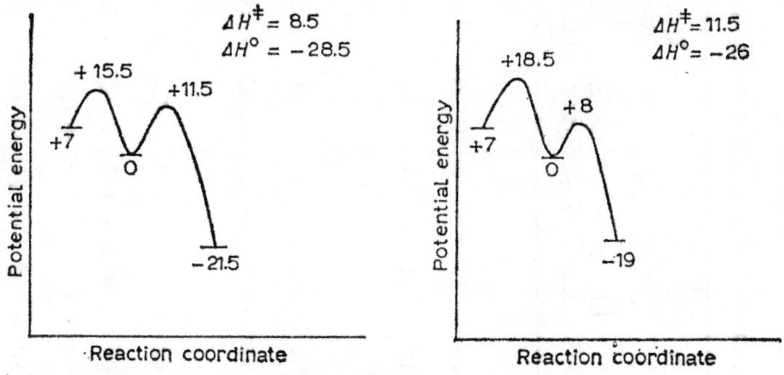

Fig. 63. Potential energy/reaction coordinate profile of $Me_2NH/2,4\text{-}(NO_2)_2C_6H_3F$.
Fig. 64. Potential energy/reaction coordinate profile of $Me_2NH/2,4\text{-}(NO_2)_2C_6H_3Cl$.

solvation of intermediates as compared with initial states must lead to very large negative values of $\Delta S^{\ddagger}$. Thus $\Delta G^{\ddagger}$ values are high despite low values of $\Delta H^{\ddagger}$, and rates of reaction relatively low. For example with 1-chloro-2,4-dinitrobenzene in methanol the experimental $\Delta S^{\ddagger}$ (25°) values are of the order of −30 to −40 e.u. leading to $\Delta G^{\ddagger}$ (25°) values in the range 20–25 as compared with 20.2 kcal·mole^{-1} for methoxide ion *[10a]*.

Individual reactions are now considered and comparisons made. In the reaction of trimethylamine with *p*-nitroanisole (Fig. 61) it should

TABLE 52

CALCULATION OF ACTIVATION ENERGIES FOR REACTIONS OF SOME AMINES (VALUES IN $kcal \cdot mole^{-1}$)

Reaction and I.C.	Fission	B.D.E.	I.E. or E_{aff}	$H_{solv.}$	Series term	Energy levels relative to I.C. = 0	% B.D.E. and corresponding energy	Less α-subst. effect	Energy levels of T.St.1 and T.St.2	$\Delta E^{\ddagger}$ ($\simeq \Delta H^{\ddagger}$) forward	$\Delta E^{\ddagger}$ ($\simeq \Delta H^{\ddagger}$) reverse reaction
$^{-}O_2N^{+}$=C₆H₃(NO₂)(C(Cl)$\overset{+}{N}Me_3$) (D.N.)[a]	C–$\overset{+}{N}Me_3$	+ 69	− 180	0	+ 100	− 11	18, + 12.5	− 3	+ 9.5	21	21
	C–OMe	+ 77	− 61	− 83	+ 56	− 11	18, + 14	− 4	+ 10		
$^{-}O_2\overset{+}{N}$=C₆H₄(C(OMe)$\overset{+}{N}Me_3$) (M.N.)[a]	C–$\overset{+}{N}Me_3$	+ 69	− 180	0	+ 114	+ 3	25.5, + 17.5	− 1	+ 16.5	13.5	35.5
	C–Cl	+ 80	− 85	− 79	+ 65	− 19	15, + 12	− 4	+ 8		
D.N.(F)($\overset{+}{N}HMe_2$)	C–$\overset{+}{N}HMe_2$	+ 69	− 190	0	+ 128	+ 7	28, + 19.5	− 4	+ 15.5	8.5	37
	C–F	+ 107	− 80.5	− 113	+ 65	− 21.5	14.5, + 15.5	− 4	+ 11.5		
D.N.(Cl)($\overset{+}{N}HMe_2$)	C–$\overset{+}{N}HMe_2$	+ 69	− 190	0	+ 128	+ 7	28, + 19.5	− 1	+ 18.5	11.5	37.5
	C–Cl	+ 80	− 85	− 79	+ 65	− 19	15, + 12	− 4	+ 8		

D.N.($\overset{+}{N}H_2Me$)(Cl)	$C\text{–}\overset{+}{N}H_2Me$	+ 73	− 207	0	+ 142	+ 8	28.5, + 21	− 1	+ 20	12	39
	C–Cl	+ 80	− 85	− 79	+ 65	− 19	15, + 12	− 4	+ 8		
D.N.($\overset{+}{N}H_3$)(Cl)	$C\text{–}\overset{+}{N}H_3$	+ 79	− 236	0	+ 156	− 1	23, + 18	− 1	+ 17	18	36
	C–Cl	+ 80	− 85	− 79	+ 65	− 19	15, + 12	− 4	+ 8		
D.N.($\overset{+}{N}HMePh$)(F)	$C\text{–}\overset{+}{N}HMePh$	+ 57	− 169	0	+ 109	− 3	21.5, + 12	− 4	+ 8	14.5	33
	C–F	+ 107	− 80.5	− 113	− 65	− 21.5	14.5, + 15.5	− 4	+ 11.5		
D.N.($\overset{+}{N}HMePh$)(Cl)	$C\text{–}\overset{+}{N}HMePh$	+ 57	− 169	0	+ 109	− 3	21.5, + 12	− 1	+ 11	14	30
	C–Cl	+ 80	− 85	− 79	+ 65	− 19	15, + 12	− 4	+ 8		
D.N.($\overset{+}{N}H_2Ph$)(F)	$C\text{–}\overset{+}{N}H_2Ph$	+ 60	− 178	0	+ 123	+ 5	26.5, + 16	− 4	+ 12	7	33.5
	C–F	+ 107	− 80.5	− 113	+ 65	− 21.5	14.5, + 15.5	− 4	+ 11.5		
D.N.($\overset{+}{N}H_2Ph$)(Cl)	$C\text{–}\overset{+}{N}H_2Ph$	+ 60	− 178	0	+ 123	+ 5	26.5, + 16	− 1	+ 15	10	34
	C–Cl	+ 80	− 85	− 79	+ 65	− 19	15, + 12	− 4	+ 8		

M.N. = mononitro-; D.N. = dinitro-.

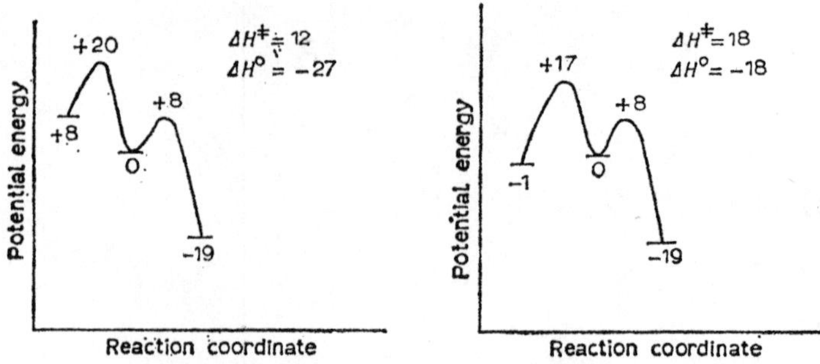

Fig. 65. Potential energy/reaction coordinate profile of $MeNH_2/2,4-(NO_2)_2C_6H_3Cl$.
Fig. 66. Potential energy/reaction coordinate profile of $NH_3/2,4-(NO_2)_2C_6H_3Cl$.

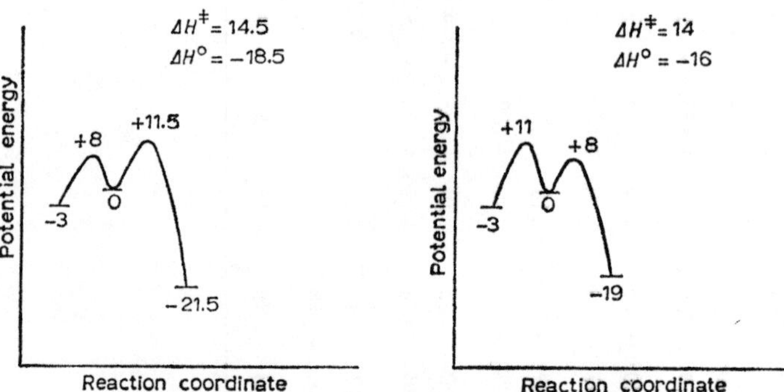

Fig. 67. Potential energy/reaction coordinate profile of $PhNHMe/2,4-(NO_2)_2C_6H_3F$.
Fig. 68. Potential energy/reaction coordinate profile of $PhNHMe/2,4-(NO_2)_2C_6H_3Cl$.

be noted that the numerical value of the series term for breaking the $C–NMe_3^+$ bond of the intermediate complex allows for a specially stable I.St., thus involving a large difference between the C_{Ar}–OMe bond of the initial state and C_{Alph}–OMe type bond of intermediates. The value of $\Delta H^{\ddagger}$ in both forward and reverse directions is calculated as 21 kcal·mole^{-1}, but since the $\Delta S^{\ddagger}$ value for the forward reaction should be of the order of –40 e.u. [$T\Delta S^{\ddagger}$ (25°) = –12 kcal·mole^{-1}] whereas that of the reverse reaction should be of the order of –10 e.u. [$T\Delta S^{\ddagger}$ (25°) = –3 kcal·mole^{-1}] or still less negative, the $\Delta G^{\ddagger}$ value

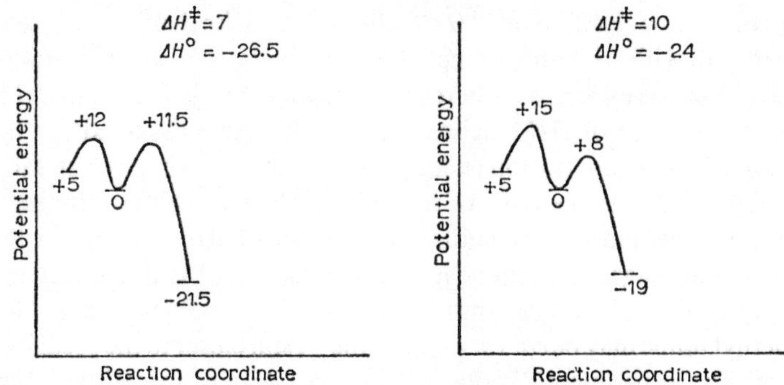

Fig. 69. Potential energy/reaction coordinate profile of $PhNH_2/2,4\text{-}(NO_2)_2C_6H_3F$.
Fig. 70. Potential energy/reaction coordinate profile of $PhNH_2/2,4\text{-}(NO_2)_2C_6H_3Cl$.

for the reverse reaction is predicted to be about 10 kcal·mole^{-1} lower than for the forward reaction. Thus reaction should effectively proceed completely in the reverse direction. This agrees with the work of Bolto and Miller *[10b]* who showed further that the experimental value of kinetic parameters in kcal·mole^{-1} for the reaction, which in fact does proceed in the reverse direction, are: $\Delta H^{\ddagger}$ 20.0; $T\Delta S^{\ddagger}$ (25°) −1.95; $\Delta G^{\ddagger}$ (25°) 21.95, in good agreement with predictions.

The reaction of trimethylamine with 1-chloro-2,4-dinitrobenzene in methanol, though also investigated, is not a satisfactory one for kinetic studies. The reactivity of the amine is low because of weaker bonds to highly substituted nitrogen and relatively low solvation energy of the highly substituted T.St., and these more than counteract the favourable effect of a relatively low ionisation energy. Since trimethylamine is a base there is also some competition from the lyate ion. Further, side reactions of the ammonium ion product are facile, and steric factors also reduce reactivity. Indeed Ross *[58]* has reported that triethylamine which is still more sterically demanding does not react. Lantzke and Miller *[56]*, however, were able to obtain usable if somewhat imprecise results with triethylamine and found $\Delta H^{\ddagger} \simeq$ 14 kcal·mole^{-1} as compared with the predicted value of 13.5. The experimental value of $\Delta S^{\ddagger}$ (25°) is about −40 e.u., a typical value, so that $\Delta G^{\ddagger}$ (25°) is 25.7, in good agreement with the predictions.

Calculations for the reaction of 1-chloro-2,4-dinitrobenzene with

References p. 230

dimethylamine lead to a $\Delta H^{\ddagger}$ value of 11.5 kcal·mole^{-1}, 2 kcal lower than with trimethylamine, and less adverse steric contributions to $\Delta S^{\ddagger}$ may be expected to increase the favourable difference in $\Delta G^{\ddagger}$. The lower value of $\Delta H^{\ddagger}$ arises because the value assumed for the necessarily higher solvation energy of the T.St. more than counteracts the higher ionisation energy of the amine. There are no experimental values for methanol, but in ethanol the value of $\Delta H^{\ddagger} = 9.3$. For the very similar reagent piperidine it is known that $\Delta H^{\ddagger}$ is 10.2 in ethanol and 11.6 in methanol, so that in methanol the value of $\Delta H^{\ddagger}$ for dimethylamine may be *ca.* 10.5 kcal·mole^{-1}, still closer to the predicted value. The experimental value of $\Delta S^{\ddagger}$ is –36 e.u. (piperidine data suggest that $\Delta S^{\ddagger}$ would be less negative also for dimethylamine in methanol). The value is in any case 4 e.u. less negative than for trimethylamine, corresponding to a $T\Delta S^{\ddagger}$ difference of 1.2 kcal·mole^{-1}, a $\Delta\Delta G^{\ddagger}$ value of 2.3 kcal·mole^{-1} and thus a substantially higher rate for the dimethylamine reaction.

For the reaction with 1-fluoro-2,4-dinitrobenzene, a value of $\Delta H^{\ddagger}$ which is 3 kcal·mole^{-1} lower than with the corresponding chloro compound is predicted, so that the G.R.F. of fluorine (fluorine/chlorine mobility ratio) is high, though somewhat reduced by the entropy factor discussed above (p. 212). For the very similar reagent piperidine, Tables 49 and 51 demonstrate such a high value of the G.R.F.

For reaction of methylamine with 1-chloro-2,4-dinitrobenzene the changes in bond strength and solvation compared with dimethylamine approximately counteract that due to the higher ionisation energy of methylamine and the predicted value of $\Delta H^{\ddagger}$ is 12 kcal·mole^{-1}. Experimentally in ethanol it is 10.7 kcal·mole^{-1} so that there is good agreement here also. The results predict methylamine to be slightly less reactive than dimethylamine, as is found experimentally.

The reaction of ammonia with 1-chloro-2,4-dinitrobenzene is predicted to be considerably less facile than that of methylamine or dimethylamine since the further increase in ionisation energy more than counterbalances favourable changes in bond strength and solvation. The predicted value of $\Delta H^{\ddagger}$ is 18 kcal·mole^{-1} and the experimental value is 16.6, in good agreement. The experimental value of $\Delta S^{\ddagger}$ is –31 e.u. and this still less negative value is in accord with qualitative predictions, based on absence of adverse steric contributions such as are expected for the more substituted amines.

The results so far thus satisfactorily predict $\Delta H^{\ddagger}$ values for ammonia and the three types of aliphatic amine as reagents, and though adverse steric contributions to $\Delta S^{\ddagger}$ are not calculated, but considered qualitatively, the resultant depression of rates, largest for trimethylamine, then bring all into the correct free energy order, *i.e.* the $\Delta H^{\ddagger}$ order $NH_3 < NMe_3 < NH_2Me < NHMe_2$ is modified to $NMe_3 < NH_3 < NH_2Me < NHMe_2$, by transposing NH_3 and NMe_3.

The reaction of *N*-methylaniline with 1-chloro-2,4-dinitrobenzene may be compared most conveniently with that of dimethylamine. Apart from a more adverse steric contribution to $\Delta S^{\ddagger}$ which will substantially reduce the reactivity of the former, it is seen that the favourable lower value of the ionisation energy is more than counterbalanced by the weakness of the bond formed, and the low value of the solvation energy of the T.St. The reported kinetic data (Table 49) are rather imprecise so that one can only say that there is a reasonable agreement with these conclusions, though the value of $\Delta H^{\ddagger}$ obtained here seems somewhat high. It is of special interest that the calculation for reaction of *N*-methylaniline with 1-fluoro-2,4-dinitrobenzene indicates that the formation of T.St.2 is rate-limiting, being higher by 3.5 kcal·mole^{-1}, so that rates are about the same for the fluoro and chloro compounds, and this is confirmed by experiment (Tables 49 and 51). Further the facile formation and reversal of the bond-forming step is shown by calculation and this would be affected by added base which makes reverse reaction less facile and may make the elimination stage more facile. This is in full accord with the results of Bunnett and of Hammond and their co-workers *[50,52]*. It is predicted that aniline is more reactive than *N*-methylaniline with 1-chloro-2,4-dinitrobenzene, and this is confirmed experimentally, though the difference is less marked than predicted $\Delta H^{\ddagger}$ values suggest. With 1-fluoro-2,4-dinitrobenzene the formation of T.St.1 is calculated as rate-limiting for reaction with aniline, in contrast with the situation with *N*-methylaniline, so that the fluorine mobility and halogen mobility pattern resembles that with the aliphatic amines.

The lower reactivity of aniline than methylamine is $\Delta S^{\ddagger}$ and not $\Delta H^{\ddagger}$ dependent and may be assumed to be due to more adverse steric factors with the aromatic amine.

The only important group of amines not considered is that comprising the aromatic heterocycles *e.g.* pyridine, and this is only in part

References p. 230

because of the need to estimate heats of solvation for species such as $\overset{+}{\mathrm{Py}}$–$\overset{-}{\mathrm{Bzd}}$–X. The main problem is that the ionisation energy of 227 kcal·mole^{-1} is believed to refer to loss of a ring electron and not one from nitrogen *[81]*. Inasmuch however as this implies a still *higher* value for the ionisation energy from nitrogen, the low reactivity of pyridine due to a high value of $\Delta H^{\ddagger}$ is a corollary, and the experimental pattern is predictable.

The calculations, with illustrative curves, show that in most cases the formation of T.St.1 is rate-limiting. It is therefore unimportant as far as rates are concerned whether the fast step of transference of a proton from H–$\overset{+}{\mathrm{Y}}$–$\overset{-}{\mathrm{Bzd}}$–X to $\ddot{\mathrm{Y}}$–H occurs before completion of reaction as long as it is subsequent to formation of T.St.1, or whether the proton is transferred from the final product Ar$\overset{+}{\mathrm{Y}}$–H to $\ddot{\mathrm{Y}}$–H. If the former, it is likely that it is exothermic in $\Delta H^{\ddagger}$ as well as $\Delta G^{\ddagger}$, for $\Delta S^{\ddagger}$ is probably more negative in forming a smaller and more solvated cation. If not then the elimination of X as X^{-} must take precedence, and proton loss is a final step. It is known *[67]* that H–N bond strengths in H–$\overset{+}{\mathrm{N}}$-aliphatic species are higher than in H–$\overset{+}{\mathrm{N}}$-aromatic species by about 10 kcal·mole^{-1}, so that as a final step it is more exothermic than transfer from the aliphatic type H–$\overset{+}{\mathrm{N}}$–$\overset{-}{\mathrm{Bzd}}$–X. However, even the latter may be more than enough to overcome the loss of the α-substituent effect of N^{+}, which is also partly offset by a more stable final state, due to $\ddot{\mathrm{N}}$–Ar conjugation. Proton loss from cationic intermediate complexes would give new intermediate complexes corresponding to those formed by reaction of conjugate bases of the amines, and while data are not available for precise calculations, enough is known to say that this will have the effect of raising both I.St. and T.St.1, thus hindering reversal of reaction, and this would be conspicuous if it were competitive, as has been predicted *e.g.* for the reaction of *N*-methylaniline with 1-fluoro-2,4-dinitrobenzene, and reported by Bunnett and Randall *[52]*.

A pair of curves based on the above analysis and qualitatively comparing the results of a mildly but sufficiently exothermic proton transfer from the intermediate complex to another molecule of amine, with the situation in which this is delayed until the final state is formed, are shown as Fig. 71. In this the full curve represents the latter case

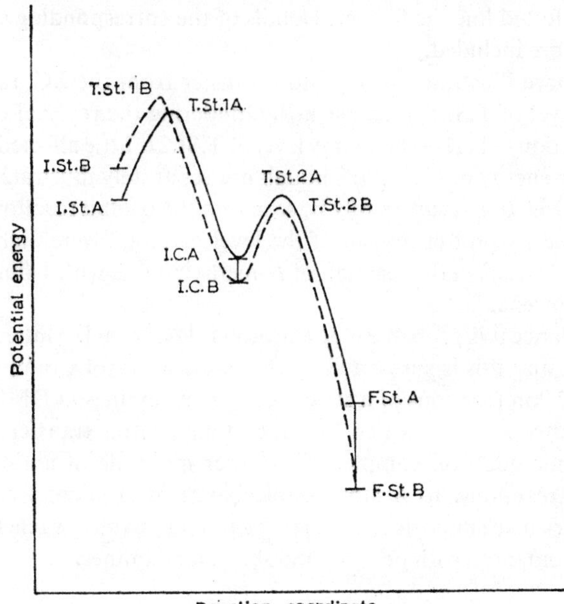

I.St.A RNH_2 Hal–$C_6H_3(NO_2)$–$\overset{+}{N}(=O)O^-$ RNH^- Hal–$C_6H_3(NO_2)$–$\overset{+}{N}(=O)O^-$ I.St.B

↓ via TSt1A ↓ via TSt1B

I.C.A $\overset{+}{R}NH_2$, Hal (σ-complex, NO_2) $=\overset{+}{N}(O^-)O^-$ $\xrightarrow{RNH_2}$ RNH, Hal (σ-complex, NO_2) $=\overset{+}{N}(O^-)O^-$ $(\overset{+}{R}NH_3)$ I.C.B.

↓ via TSt2A ↓ via TSt2B

F.St.A $\overset{+}{R}NH_2$–$C_6H_3(NO_2)$–$\overset{+}{N}(=O)O^-$ Hal^- $\xrightarrow{RNH_2}$ RNH–$C_6H_3(NO_2)$–$\overset{+}{N}(=O)O^-$ Hal^- $(\overset{+}{R}NH_3)$ F.St.B

Fig. 71. Potential energy/reaction coordinate profiles of $RNH_2/2,4\text{-}(NO_2)_2C_6H_3Hal$ and $RNH^-/2,4\text{-}(NO_2)_2C_6H_3Hal$.

and the dotted line the former. Details of the corresponding stages of reaction are included.

The curve illustrates how proton transfer from the I.C. raises the energy level of T.St.1, thus markedly hindering the reversal of initial coordination, whereas the energy level of T.St.2 is little affected despite the lower energy of the F.St. Though not separately illustrated (however, see Fig. 67), it can readily be seen that the overall reaction would be affected by proton transfer if the level of T.St.2 were higher than that of T.St.1, as in the reaction of *N*-methylaniline with 1-fluoro-2,4-dinitrobenzene.

On balance it is probable that a proton is lost from the intermediate complex, and this is supported by the recent work of Crampton and Gold *[82]* on reactions of a series of aliphatic amines in DMSO with 1,3,5-trinitrobenzene, which indicates that proton transfer from a zwitterionic addition complex to another molecule of amine leads, in those reactions, to a more stable benzenide (anionic) complex. Their work also supports the earlier discussion on the low reactivity of tertiary compared with primary and secondary amines.

7. Reactivity in S_N1 and Benzyne Mechanisms

In Chapter 2 reasons were advanced for the non-occurrence of aromatic S_N1 reactions proceeding via heterolysis to form an aryl cation (Ar^+). There is evidence, however, for a special S_N1 mechanism for diazonium ions, and for some S_N1-like reactions.

In respect of the reactions of diazonium ions, Lewis and his co-workers *[83]* have shown that reactions proceed via two intermediates, possibly in parallel and not sequential; one highly reactive and unselective and the other less reactive and more selective (Chapter 2, Fig. 35). The reactive species is regarded as an excited state of the diazonium ion containing most of the energy of the C_{Ar}–N bond and thus resembles a phenyl cation in behaviour. With this unselective species the nucleophilic reactivity of thiocyanate ion relative to that of water is only 2.8. The less reactive species has been shown to have its two nitrogen atoms equivalent and a spiro-benzenium ion structure (A) was suggested. With this species the nucleophilic reactivity of thiocyanate ion relative to that of water is 470.

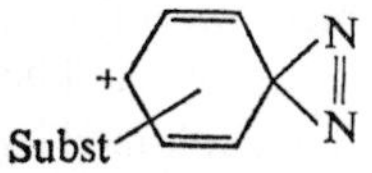

(A)

Crossley *et al.* *[84]* have shown that while decomposition of benzene diazonium chloride in aqueous solutions follows the first-order kinetic form in both dilute and concentrated solutions, the proportion of phenol formed drops from 95.2% in very dilute ($M/1280$) solution to 24.3% in concentrated ($M/2.36$) solution; and that in the latter the main by-product is chlorobenzene. They showed also that while rates are almost the same in 0.1 as $12M$ hydrochloric acid, the yield of chlorobenzene is much higher in the latter, reaching some 60%. It is well known that in Sandmeyer reactions *[85]*, copper (preferably cuprous) catalysts are required to obtain good yields of chloro or bromo compounds, whereas iodo compounds are readily obtained by use of potassium iodide without catalyst. These and similar facts suggest that in the S_N reactions of diazonium ions the products formed depend on the kinetic nucleophilicity of the reagent towards intermediates of the type discussed above.

With compounds such as chlorobenzene and similar simple benzene derivatives, despite early reports *[86,87]* which suggest the possibility of an S_N1 mechanism, more recent mechanistic evidence *[88]* indicates that these nucleophilic substitutions occur by concurrent benzyne and activated S_N2 mechanisms; depending on such factors as the basicity of the reagent and temperature at which reaction is carried out. With halogenotoluenes, for example, hydrolysis with the weakly basic aqueous sodium acetate proceeds at 340° by the normal (activated type) S_N2 mechanism. With the more strongly basic aqueous sodium hydroxide, hydrolysis proceeds at 250°, giving largely unrearranged products of the normal S_N2 mechanism, whereas at 340° the products are those of hydrolysis with rearrangement by the benzyne mechanism. The only rearranged products are those obtainable by this mechanism, and further, the same products are obtained whatever the halogen displaced. The effect of temperature indicates that the activation energy of the hydrolysis by the benzyne mechanism is higher than that of the normal S_N2 reactions of these compounds. Reaction by an S_N1 mechanism appears to be ruled out by the non-

References p. 230

formation of cresols by reaction of *p*-iodotoluene with aqueous sodium iodide or chloride at 340°. The known relatively facile hydrolysis of *o*-chlorophenol is explained by Bottini and Roberts *[88]* as the result of an internal S_N reaction by phenoxide oxygen forming initially a benzyne oxide.

In reactions proceeding by the benzyne mechanism (Chapter 3) a reagent acts in two ways: initially as a base removing a proton rendered acid by a group X, which is eliminated following proton removal to form a benzyne; and subsequently as a nucleophile adding to the benzyne triple bond. This is well illustrated by the results of Bergstrom and his co-workers *[89]*, who showed that while the triphenylmethyl anion has almost no action on chlorobenzene in liquid ammonia unless some amide ion is added, the major product is nevertheless tetraphenylmethane (Chapter 3, Fig. 24, p. 46). Unlike amide ion the triphenylmethyl anion is thus insufficiently basic to deprotonate the chlorobenzene, but in excess it can compete successfully with amide ion as a nucleophile adding to the benzyne, which is highly reactive and thus unselective.

Precise data are scarce, but the data of Table 4B (p. 48), which refer to the nucleophilic addition stage to a relatively unreactive and thus selective benzyne (9,10-phenanthryne), are of relative reactivities commensurate with kinetic nucleophilicity in reactions in which bond-formation to carbon is the rate-limiting step. The high reactivity of thiophenoxide relative to phenoxide ion *[1,8b]* is particularly significant in this respect, though its high reactivity relative to amide and carbanions is surprising. It is likely that this is the result both of using lithium, instead of, for example, potassium salts, and of a poorly ionising solvent, ether. In these conditions the salts exist largely as unreactive intimate ion-pairs, in which kinetic availability of second row nucleophiles is greater.

REFERENCES

1 K. C. Ho, J. Miller and K. W. Wong, *J. Chem. Soc., B*, (1966) 310.

2*a* J. F. Bunnett and R. E. Zahler, *Chem. Rev.*, 49 (1951) 273; *b* J. F. Bunnett, *Ann. Rev. Phys. Chem.*, 14 (1963) 271.

3*a* C. G. Swain, *J. Am. Chem. Soc.*, 70 (1948) 1119; *b* G. C. Swain and C. B.

SCOTT, *J. Am. Chem. Soc.*, 75 (1953) 141; *c* C. G. SWAIN AND W. P. LANGSDORF, *J. Am. Chem. Soc.*, 73 (1951) 2813; *d* C. G. SWAIN, R. B. MOSELEY AND D.W. BOWN, *J. Am. Chem. Soc.*, 77 (1955) 3731.
4 G. N. LEWIS, *J. Franklin Inst.*, 42 (1923) 718.
5 J. N. BRØNSTED, *Rec. Trav. Chim.*, 42 (1923) 718.
6 J. HINE, *Physical Organic Chemistry*, McGraw Hill, New York, 1956, p. 88.
7 A. J. PARKER, *Proc. Chem. Soc.*, (1961) 371.
8*a* J. MILLER in *Organic Reaction Mechanisms*, Chemical Society Special Publication No. 19, (1965) 193; *b* D. L. HILL, K. C. HO AND J. MILLER, *J. Chem. Soc.*, *B*, (1966) 299.
9*a* C. K. INGOLD, *Structure and Mechanism in Organic Chemistry*, Cornell Univ. Press, Ithaca, N.Y., 1953, p. 400; *b* C. K. INGOLD, *James Flack Norris Award Address*, 1965.
10*a* A. L. BECKWITH, G. D. LEAHY AND J. MILLER, *J. Chem. Soc.*, (1952) 3552; *b* B. A. BOLTO AND J. MILLER, *Australian J. Chem.*, 9 (1956) 74 and 304; *c* J. MILLER, *J. Am. Chem. Soc.*, 85 (1963) 1628.
11 J. F. BUNNETT, E. W. GARBISCH AND K. M. PRUITT, *J. Am. Chem. Soc.*, 79 (1957) 385.
12 R. E. PARKER AND T. O. READ, *J. Chem. Soc.*, (1962) 9.
13 S. C. CHAN, K. Y. HUI, J. MILLER AND W. S. TSANG, *J. Chem. Soc.*, (1965) 3207.
14*a* J. N. BRØNSTED AND K. J. PEDERSEN, *Z. Physik. Chem.*, 108 (1924) 185; *b* J. N. BRØNSTED, *Chem. Revs.*, 5 (1928) 320.
15 E. GRUNWALD AND S. WINSTEIN, *J. Am. Chem. Soc.*, 70 (1948) 846.
16 J. O. EDWARDS, *J. Am. Chem. Soc.*, (a) 76 (1954) 1540; (b) 78 (1956) 1819.
17*a* O. FOSS, *Kgl. Norske Vid. Selsk. Skrifter*, 2 (1945); *b* O. FOSS, *Acta Chem. Scand.*, 1 (1947) 8, 307; 3 (1949) 1385.
18 R. E. DAVIES, private communication.
19 J. O. EDWARDS AND R. G. PEARSON, *J. Am. Chem. Soc.*, 84 (1962) 16.
20*a* R. G. PEARSON, *J. Am. Chem. Soc.*, 85 (1963) 3533; *b* R. G. PEARSON, *Chemistry in Britain*, (1967) 103; *c* R. G. PEARSON AND J. SONGSTAD, *J. Am. Chem. Soc.*, 89 (1967) 1827.
21 S. AHRLAND, J. CHATT AND N. R. DAVIES, *Quart. Rev.*, 12 (1958) 265.
22 A. A. GRINBERG, *An Introduction to the Chemistry of Complex Compounds*, Pergamon Press, London, 1962, Chapter 7.
23 G. SCHWARZENBACH, *Advan. Inorg. Chem., Radiochem.*, (1961) 3.
24 R. J. P. WILLIAMS, *Proc. Chem. Soc.*, (1960) 20.
25 G. PORTER AND J. A. SMITH, *Proc. Roy. Soc.*, A261 (1961) 28.
26 M. J. CHRISTIE, *J. Am. Chem. Soc.*, 84 (1962) 4066.
27*a* J. CHATT, *Nature*, 165 (1950) 859; 177 (1956) 852; *b* J. CHATT, L. A. DUNCANSON AND L. M. VENANZI, *J. Chem. Soc.*, (1955) 4456.
28*a* K. S. PITZER, *J. Chem. Phys.*, 22 (1955) 1735; *b* K. S. PITZER AND E. CATALANO, *J. Am. Chem. Soc.*, 78 (1956) 4844.
29 R. S. MULLIKEN, *J. Am. Chem. Soc.*, 77 (1955) 884.
30 J. MILLER AND A. J. PARKER, *J. Am. Chem. Soc.*, 83 (1961) 117.
31*a* A. J. PARKER, *J. Chem. Soc.*, (1961) 1328; *b* A. J. PARKER, *Quart. Rev.*, 14 (1962) 163.
32*a* R. F. HUDSON, *Chimia*, 16 (1962) 173; *b* R. F. HUDSON AND M. GREEN, *J. Chem. Soc.*, (1962) 1055.
33 G. S. HAMMOND, *J. Am. Chem. Soc.*, 77 (1955) 334.

34*a* G. C. FINGER AND C. W. KRUSE, *J. Am. Chem. Soc.*, 78 (1956) 6034; *b* G. C. FINGER AND L. D. STARR, *J. Am. Chem. Soc.*, 81 (1959) 2674.
35*a* R. BOLTON, J. MILLER AND A. J. PARKER, unpublished work; *b* R. BOLTON, J. MILLER AND A. J. PARKER, *Chem. and Ind.*, (1963) 492.
36 P. J. C. FIERENS AND A. HALLEUX, *Bull. Soc. Chim. Belges*, 64 (1955) 717.
37 J. MAURIN AND R. A. PÂRIS, *Compt. Rend.*, 232 (1951) 2428.
38 J. F. BUNNETT, C. F. HAUSER AND K. V. NAHABEDIAN, *Proc. Chem. Soc.*, (1961) 305.
39 J. F. BUNNETT AND G. T. DAVIS, *J. Am. Chem. Soc.*, (a) 76 (1954) 3021; (b) 80 (1958) 4337.
40*a* J. MURTO, *Acta Chem. Scand.*, 18 (1964) 1024; *b* E. TOMMILA AND J. MURTO, *Acta Chem. Scand.*, 16 (1962) 53; *c* J. MURTO, *Ann. Acad. Sci. Fennicae*, AII (1962) No. 117.
41 K. B. YATSIMIRSKII, *Izvest. Vysshikh Uchebn. Zavednii Khim. i Khim. Tekhnol.*, 2 (1959) 480; *Chem. Abstr.*, 54 (1960) 6289.
42 J. H. S. GREEN, *Quart. Rev.*, 15 (1960) 125.
43 R. A. OGG, *J. Chem. Phys.*, 14 (1946) 399.
44 E. C. BAUGHAN, M. G. EVANS AND M. POLANYI, *Trans. Faraday Soc.*, 37 (1941) 377.
45 F. M. PAGE AND T. M. SUGDEN, *Trans. Faraday Soc.*, 53 (1957) 1092.
46 D. D. CUBICIOTTI, *J. Chem. Phys.*, 31 (1959) 1646.
47 M. A. ADENIRAN, C. W. L. BEVAN AND J. HIRST, *J. Chem. Soc.*, (1963) 5868.
48 O. L. BRADY AND F. R. CROPPER, *J. Chem. Soc.*, (1950) 507.
49 N. B. CHAPMAN AND R. E. PARKER, *J. Chem. Soc.*, (1951) 3301.
50 G. S. HAMMOND AND L. R. PARKS, *J. Am. Chem. Soc.*, 77 (1955) 340.
51*a* C. W. L. BEVAN AND J. HIRST, *Chem. and Ind.*, (1954) 1422; *b* C. W. L. BEVAN AND J. HIRST, *J. Chem. Soc.*, (1956) 254.
52 J. F. BUNNETT AND J. J. RANDALL, *J. Am. Chem. Soc.*, 80 (1958) 6020.
53 J. D. REINHEIMER, R. C. TAYLOR AND P. E. ROHRBAUGH, *J. Am. Chem. Soc.*, 83 (1961) 835.
54 W. GREIZERSTEIN, R. A. BONELLI AND J. A. BRIEUX, *J. Am. Chem. Soc.*, 84 (1962) 1026.
55 H. K. HALL, *J. Org. Chem.*, 29 (1964) 3539.
56 I. LANTZKE AND J. MILLER, unpublished work.
57*a* S. D. ROSS AND M. FINKELSTEIN, *J. Am. Chem. Soc.*, 79 (1957) 6547; *b* S. D. ROSS, *J. Am. Chem. Soc.*, 81 (1959) 2113; *c* S. D. ROSS, M. FINKELSTEIN AND R. C. PETERSEN, *J. Am. Chem. Soc.*, 81 (1959) 5336.
58 J. F. BUNNETT AND R. H. GARST *J. Am. Chem. Soc.*, (a) 87 (1965) 3875; (b) 87 (1965) 3879.
59*a* C. F. BERNASCONI AND H. ZOLLINGER, *Helv. Chim. Acta*, 49 (1966) 2563, 2570; *b* C. F. BERNASCONI AND H. ZOLLINGER, *Helv. Chim. Acta*, 50 (1967) 3; *c* G. BECKER, C. F. BERNASCONI AND H. ZOLLINGER, *Helv. Chim. Acta*, 50 (1967) 10.
60 G. D. LEAHY, M. LIVERIS, J. MILLER AND A. J. PARKER, *Australian J. Chem.*, 9 (1956) 382.
61 H. SUHR, *Ann.*, 687 (1965) 175.
62 W. C. PRICE, T. R. PASSMORE AND D. M. ROESSLER, *Disc. Faraday Soc.*, 35 (1963) 201.
63 A. JULG AND M. BONNET, *J. Chim. Phys.*, 60 (1963) 742.
64 H. SJOGREN AND E. LINDHOLM, *Phys. Letters*, 4 (1963) 85.
65 R. S. NEALE, *J. Phys. Chem.*, 68 (1964) 143.

66 C. T. Mortimer, *Pure Appl. Chem.*, 2 (1961) 71.
67 J. A. Kerr, R. C. Sekhar and A. F. Trotman-Dickenson, *J. Chem. Soc.*, (1963) 3217.
68 Y. Wada and R. W. Kiser, *Inorg. Chem.*, 3 (1964) 174.
69 W. A. Henderson and S. A. Buckler, *J. Am. Chem. Soc.*, 82 (1960) 5794.
70 T. L. Cottrell, *The Strengths of Chemical Bonds*, 2nd Edn., Butterworths, London, 1958, Table 11.5.1.
71*a* P. Gray, *Trans. Faraday Soc.*, 52 (1956) 344; *b* P. Gray and A. Williams, *Trans. Faraday Soc.*, 55 (1959) 760.
72 B. G. Hobrock and R. W. Kiser, *J. Phys. Chem.*, 67 (1963) 1283.
73 W. C. Price, *J. Chem. Phys.*, 16 (1948) 551.
74 A. D. Walsh and P. A. Warsip, *Trans. Faraday Soc.*, 57 (1961) 345.
75 K. Watanabe and J. R. Mottl, *J. Chem. Phys.*, 26 (1957) 1773.
76 G. L. Esteban, J. A. Kerr and A. F. Trotman-Dickenson, *J. Chem. Soc.*, (1963) 3879.
77 J. D. Cox, *Tetrahedron*, 18 (1962) 1337.
78 J. R. Morton, *J. Chem. Eng. Data*, 4 (1959) 251.
79 A. Terenin and F. I. Vilesov, *Advan. Photochem.*, 2 (1964) 385.
80 J. A. Kerr, A. F. Trotman-Dickenson and M. Wolter, *J. Chem. Soc.*, (1964) 3584.
81 M. R. Basila and D. J. Clancy, *J. Phys. Chem.*, 67 (1963) 1551.
82 M. R. Crampton and V. Gold, *Chem. Commun.*, (1965) 549.
83 E. S. Lewis and J. E. Cooper, *J. Am. Chem. Soc.*, 84 (1962) 3847.
84 M. L. Crossley, R. H. Kienle and C. H. Benbrook, *J. Am. Chem. Soc.*, 62 (1940) 1400.
85 K. H. Saunders, *The Aromatic Diazo Compounds*, 2nd Ed., Longmans, Green and Co., London, 1949, p. 277, 290.
86 W. J. Hale and E. C. Britton, *Ind. Eng. Chem.*, 20 (1928) 114.
87 C. F. Boehringer und Söhne, *German Patents* 269544 (1914); 284533 and 286266 (1915), quoted in ref. *85*.
88 A. T. Bottini and J. D. Roberts, *J. Am. Chem. Soc.*, 79 (1957) 1458.
89*a* R. E. Wright and F. W. Bergstrom, *J. Org. Chem.*, 1 (1936) 179; *b* R. A. Seibert and F. W. Bergstrom, *J. Org. Chem.*, 10 (1945) 544.

Chapter 7

NUCLEOPHILIC SUBSTITUTION IN NON-BENZENOID AROMATIC SYSTEMS

1. Introduction

The systems which most closely parallel the homocyclic fully aromatic systems, and for which the fullest information is available are the aza- and the polyaza-benzenes and -naphthalenes (azines), but nucleophilic substitution has been studied also in tropylium compounds, for example, and in inorganic systems such as the phosphonitrilic halides.

The scope of heteroaromatic substitution, even if confined to nucleophilic substitutions, is vast, and has been referred to in a recent review *[1]*. The more restricted subject of reactivity of azines towards nucleophiles, which is the major area of knowledge and activity in the field, has just been reviewed at length *[2]*.

A discussion of the essential features of nucleophilic substitution in non-benzenoid aromatic systems, related to the discussion already given of benzenoid systems, is a necessary task in a study of aromatic S_N reactions. In the former as in the latter there is no evidence for S_N1 reactions except the special type found in replacement of the diazonium group (Chapter 2, p. 31), whereas there is ample evidence that these nucleophilic substitutions utilise the S_N2 mechanisms, distributed similarly between the normal addition–elimination (activated) mechanism, and the elimination–addition or aryne mechanism.

2. Comparison of Homo- and Hetero-cyclic Organic Systems

The great majority of comparisons of homo- and hetero-cyclic organic systems are comparisons of systems in which one or more carbon atoms in a ring is replaced by a more electronegative atom (nitrogen being typical). This results in a greater electron-density on that atom, with a corresponding reduction on the remaining ring carbon atoms. It is illustrated, for example, by pyridine which has a dipole moment of 2.23D *[3]*. The intra-annular redistribution of electrons is seen clearly in aromatic systems, with which we are now concerned.

The resulting electron deficiency on ring carbon atoms facilitates the approach of a nucleophilic reagent, for the aromatic π-electron system is intrinsically unfavourable to reaction with nucleophiles as compared with electrophiles, because of repulsion between the nucleophile and the π-electrons *[4,5]*. It is also more particularly related to the ability of a hetero-atom to stabilise the arenide system of the transition states and intermediate complex in which the originally aromatic π-electron system (with any attached conjugated groups) is redistributed, with the π-electrons delocalised on all ring atoms except that at which substitution takes place, this being tetrahedrally hybridised, and bonded with both the incoming (nucleophilic) and replaceable (leaving) group. This is illustrated in Fig. 72 by intermediate complexes in examples of the addition–elimination S_N2 mechanism.

Electron-densities may be calculated by the methods of quantum mechanics or inferred from theories of electron displacement *[6a]*, but since the relationship of this to reactivity is not a direct one, they can be used only to predict qualitatively activation or deactivation compared to corresponding benzenoid systems, and orders of positional reactivity. It is easy to predict for example that nucleophilic substitution of a suitable leaving group at the α- or γ-position in the pyridine ring is more facile than at the β-position, and that all are more

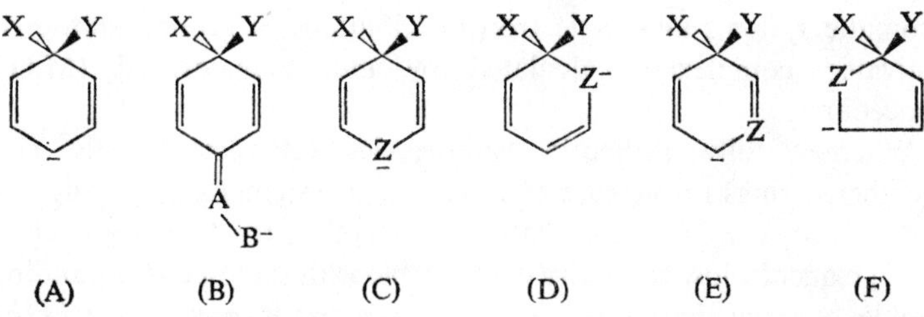

Fig. 72. Arenide systems. Intermediate complexes for some addition–elimination S_N2 reactions of benzenoid (A, B) and heteroaromatic (C, D, E, F) substrates.

facile than in the benzene ring. For substitution α- or γ- to the hetero-atom, the negative charge always associated with the delocalised π-electrons of an arenide system, can be located substantially on the hetero-atom, whereas β- to the hetero atom or in benzene it must be on carbon. The more facile reaction β- to the hetero atom, as compared with that in a benzene derivative, occurs because the negative charge can be placed on a carbon atom rendered more electronegative than in benzene by the neighbouring hetero-atom. These points are illustrated by Fig. 72C,D,E (Z = N), in which, for simplicity of representation, intermediate-complexes are shown rather than the actual transition states, which however closely resemble them *[7]*. In a similar way one can predict qualitatively differences in reactivity depending on the hetero-atom itself, *e.g.* in a monocyclic system (Fig. 72C), in the order: $Z = O^+ > R\text{–}\overset{+}{N}$ (R = alkyl or hydrogen) $> \overset{+}{N}\text{–}O^- > N >$ C–H (Fig. 72A).

Despite some early references *[e.g. 8]*, such as that to the formation of 2-aminopyridine from pyridine by reaction with amide ion, little attention was paid to S_N reactions in heteroaromatic substrates until relatively recent work, notably that initiated by Chapman and Amstutz and their co-workers *[9,10]*. The field is now an extremely active one.

While Bunnett and Zahler, and Miller, in their 1951 reviews *[5,11]* referred briefly to S_N reactions of these systems, an important early review was that of Chapman *[12]* in which correlations of reactivity and approximately calculated charge localisation energies of heterocyclic compounds were made. This followed theoretical papers by Wheland and Pauling *[13,14]* and Longuet-Higgins and Coulson *[15,16]*, in which were correlated qualitatively reactivity in aromatic systems and π-electron distributions in ground states, or activated complexes, calculated by molecular orbital (M.O.) procedures.

Whereas fully aromatic heterocycles (characteristically six-membered rings in the case of monocyclic systems) are classified as π-deficient *[17a]*, and are relatively susceptible to attack by nucleophilic reagents, low reactivity comparable with that of corresponding simple benzene derivatives is to be expected in π-excessive *[17b]* quasi-aromatic systems (five-membered rings with one hetero-atom in the case of monocyclic systems) such as furan, thiophen and pyrrole;

and still lower reactivity in pyrrole when, as is common, the reagent is also basic and the conjugate base of pyrrole is formed. Their π-excessive character is indicated by the requirement for a normally unshared pair of electrons of the hetero-atom to form part of the aromatic sextet, and this is indicated also by theoretical calculations *[18–20]*. More directly, it is readily seen that in transition states for S_N reactions at any position in the ring of such systems, the negative charge of the arenide system must always be on a carbon atom and not on the hereto atom (Fig. 72F). Also the difference in delocalisation energy of parent arene and arenide intermediates may be adverse in these quasi-aromatic π-excessive systems, as compared with benzene and the π-deficient hetero-aromatic systems. In support of the predicted low level of reactivity, Amstutz and Manly *[21]* have shown that in solvolysis by piperidine, 2-chloro-, 2-bromo-, and 2-iodo-furan are only slightly more reactive than the corresponding halogenobenzenes, the minor enhancement being presumably due to the σ-inductive effect of the hetero-atom increasing the electronegativity of neighbouring carbon atoms on which the negative charge of the arenide system may be placed (Fig. 72F).

3. Mechanisms

The majority of nucleophilic substitution reactions of non-benzenoid aromatic compounds have been shown to follow the pattern of the characteristic addition–elimination S_N2 mechanism in that, for example, the reactions are bimolecular, with the entering group taking the place of the leaving group *[e.g. 22]*; are activated by electron-withdrawing substituents; have high values of the Hammett reaction constant (ρ) *[23–25]*; give the characteristic mobility pattern, F > other halogens, with first-row nucleophiles in protic solvents *[26–31]* (a switch in mobility patterns with heavy nucleophiles seems not to have been investigated); and are characterised by there being but small differences in the mobility of many groups which are attached by bonds of markedly different strength. There are nevertheless well-documented cases, with the less reactive substrates in reactions with powerfully basic nucleophiles, where the elimination–addition aryne S_N2 mechanism occurs instead or in parallel. Such

References p. 306

reactions in the case of the heteroaromatic compounds are discussed in a recent review *[32]* and have been discussed generally in Chapter 3 (p. 41). They are characterised by formation of products in which the entering group takes a position neighbouring that to which the leaving group was attached (vicinal *cine*-substitution).

The data for mobility of fluorine relative to the other halogens are especially characteristic of mechanism *[33, cf. 34]*. Evidence from the relative mobility of chlorine, bromine and iodine alone is less conclusive, but data quoted by Illuminati *[1]*, and Shepherd and Fedrick *[2]* demonstrate typical aromatic patterns such as Cl > Br > I with alkoxides; and Cl < Br > I with amines, it being recalled that the order of mobility of the heavy halogens is more closely spaced than in saturated aliphatic S_N reactions. This is the case even with 2-halogenofurans of low reactivity *[21]* in which relative reactivity is in the order Cl < Br < I. It is significant too that in the more reactive halogenonitrothiophens the reactivity order is Cl > Br > I, even towards piperidine *[35]*.

Substituent effects are large, correlated in general terms with electron-withdrawing power, and more specifically the ability to stabilise or destabilise the arenide transition state. This especially involves the possibility of placing the negative charge on a heteroatom. For attached substituents, this requires conjugative withdrawing power. Such substituent effects are characteristically similar for a wide variety of reagents.

Evidence for a hetaryne mechanism *[32,36–42]* is provided not only by the occurrence of vicinal *cine*-substitution, but also by trapping of intermediates by cycloaddition reactions *[42]* and by comparisons of reactions where a change in position of a substituent, or of the halogen displaced, leads in some cases to the normal and in others to the aryne S_N2 mechanism *[37–40]*. Reasons have been given (*e.g.* Chapter 1, p. 19) for rejecting the view that some aromatic S_N reactions occur by typical aliphatic S_N1 or synchronous S_N2 mechanisms, and there is no evidence of a stronger kind in non-benzenoid aromatic S_N reactions to support a different conclusion there, though there are of course many reactions insufficiently studied to permit the drawing of definite conclusions as to mechanism. Reactivity towards nucleophiles in non-benzenoid aromatic systems is therefore considered in the following sections by classes of compounds, and in terms of the S_N2 mechanisms, but it is first necessary to consider some special features apposite to the largest of the groups, *viz.* heteroaromatic compounds, especially the azines.

4. Some Special Features of Nucleophilic Substitution of Heteroaromatic Compounds

Many of the common heterocyclic systems are bases so that in some reactions their conjugate acids or hydrogen-bonded forms are present. Since the conjugate acids are much more reactive (see below), and hydrogen-bonded forms also more reactive, the resultant effect can be large, even when only a small proportion of a heterocyclic species is protonated.

This simple effect may be complex in practice, for many nucleophiles are also bases, and competition for acidic species occurs, which may change during the course of a reaction and be affected also by solvent changes. Again, the effects of a substituent on the reactivity of the parent heterocyclic system, and on the basicity of a hetero-atom in it are likely to have opposite kinetic effects when reactions take place in conditions of acid catalysis.

Acid catalysis and acidic auto-catalysis were reported by Banks *[43]* and Chapman and his co-workers *[44]* in reactions of a number of azines with neutral reagents of the type $\ddot{Y}$–H. With the weakly basic aniline as reagent, addition of a little acid caused a marked increase in rate, since this could activate the heterocyclic substrate by protonation without protonating all the aniline.

When the reagent is a neutral nucleophile of the type $\ddot{Y}$–H, the reaction mixture contains acidic species such as X–Ar–$\overset{+}{Y}$–H intermediates, and Ar–$\overset{+}{Y}$–H and H–$\overset{+}{Y}$–H products. Transfer of a proton to azine nitrogen is facilitated by high basicity of the azine nitrogen and low basicity of $\ddot{Y}$–H. The resulting activation causes autocatalysis, and is recorded for example in reactions of chloropyrimidines and their methyl derivatives with aniline in ethanol *[44]*, but not with the more basic reagents morpholine and piperidine. The chloroquinolines are more strongly basic than the chloropyrimidines and so better able to compete with more basic reagents for acids produced during reaction. Thus 2-chloroquinoline exhibits autocatalysis in its reaction with morpholine, while the still more basic 4-chloroquinoline also exhibits autocatalysis in its reaction with piperidine. Another interesting variation of the autocatalytic effect is the demonstration by Amstutz and his co-workers *[45]* that the reaction of 5-chloroacridine with piperidine is autocatalytic in ethanol but not in toluene, in which the acid formed is precipitated as piperidinium chloride.

In summary, the reactions of nucleophiles with the azines, whether solvolytic or otherwise, are subject to acid catalysis unless the nucleophile and/or solvent are sufficiently basic by comparison with the substrate so that the latter cannot compete for acid added to the reaction mixture, or in the case of nucleophiles of the type $\ddot{Y}$–H, produced during reaction (autocatalysis).

A detailed study by Horrobin *[46]* of the stepwise hydrolysis of 2,4,6-trichloro-, 2-arylamino-4,6-dichloro-, and 2,4-bis(arylamino)-6-chloro-1,3,5-triazine in acidic, neutral, and alkaline conditions has demonstrated both acid and base catalysis, but only the former is specifically related to the heteroaromatic system, involving formation of conjugate acids of the substrate. The base catalysis is concerned with the formation of species more nucleophilic than water. Zollinger *[47,48]* has however demonstrated base catalysis involving the substrate directly, in reactions of the 2,4,6-trichloro compound (cyanuric chloride) with aniline in benzene. In this solvent, removal of the proton from the intermediate complex by a basic catalyst is kinetically effective (Fig. 73A), and it is relevant that Brieux and his co-workers *[49]* have shown that the reaction of *p*-chloronitrobenzene with piperidine is subject to base catalysis in benzene but not in ethanol. Zollinger also demonstrated a form of bifunctional catalysis (Fig. 73B) by some amphoteric compounds, which can be represented by the general structure H–Acid–X–$\ddot{\text{B}}$ase. In this, protonation of ring nitrogen and deprotonation of the intermediate complex, both kinetically effective, are suggested. Since there are necessarily overriding steric requirements, support for this suggestion is given by the ability of α-pyridone, but not γ-pyridone, to act as catalyst.

Somewhat similar to this is Shepherd and Fedrick's *[2]* suggestion that a reagent such as ethanolamine, in which the nitrogen is the nucleophilic element, may have enhanced reactivity by using its hydroxyl proton to form a hydrogen bond with azine nitrogen. Its

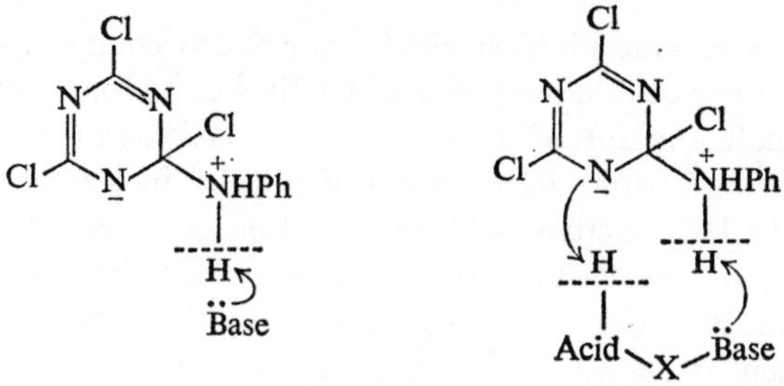

Fig. 73. A, base catalysis; B, bifunctional catalysis, in reactions of aniline with cyanuric chloride.

very high reactivity with *o*-chloronitrobenzene has been similarly explained *[50]*. It is worth recalling the related intramolecular hydrogen bond and accompanying general electrostatic interaction (Chapter 4, p. 101), which results in the rate-limiting transition states formed by 1-halogeno-2-nitrobenzenes, in their reactions with neutral reagents of the type $\ddot{Y}$–H, being more stable than those formed by 1-halogeno-4-nitrobenzenes. This is demonstrated by the higher reactivity of the former with primary and secondary amines, whereas with tertiary amines and alkoxides the latter are more reactive *[44a,51–54]*. The role of intramolecular hydrogen-bond acceptor, as might be expected, has also been assigned to ring nitrogens of heteroaromatic systems, *e.g.* in comparisons of reactivity of 2- and 4-chloroquinoline *[55]* and of 2- and 4-chloropyridine-*N*-oxide *[56.]* Many other examples of the general activating effect of coordination of the unshared electrons of ring nitrogen are known *[2]*.

5. Reactivity of Heterocyclic Aromatic Systems *via* the Addition–Elimination Mechanism

(*a*) *Aza-substituted benzenes*

(*i*) *Pyridine derivatives*

The S_N reactions of these well-studied systems are facile. Except for less reactive species, and then only with strongly basic reagents, the

References p. 306

evidence favours the addition–elimination S_N2 mechanism. Thus with the strongly basic lithium piperidide in boiling ether for example, only a very small amount of aryne formation accompanies the direct (addition–elimination) displacement of chlorine in 4-chloropyridine, whereas the less reactive 3-chloro- and 3-bromo-pyridine appear to react more or less exclusively by the aryne (elimination–addition) mechanism. With the more reactive 3-fluoropyridine however, direct displacement is the main reaction *[37,57]*. Den Hertog and his co-workers *[38,39,58]* have similarly found that the 2-halogenopyridines react with amide ion by direct displacement, whereas substitution with rearrangement indicates that the 3-halogenopyridines and some of their ethoxy derivatives react by the aryne mechanism.

Comparison of the azines, in which ring nitrogen is in place of carbon (methine = CH), with benzenoid systems is readily made, and is most clearly seen in its simplest form in the comparison of pyridine with benzene derivatives.

Ploquin *[59]*, by using the Wheland–Pauling method *[13]*, computed π-electron densities for pyridine, *viz.* N, 1.38; $C_2 = C_6$, 0.863; $C_3 = C_5$, 0.989; C_4, 0.918; suggesting for nucleophilic substitution the reactivity order, 2- > 4- > 3-pyridine derivative > benzene. The data of Orgel *et al.* *[60]* also suggest the same order. Longuet-Higgins and Coulson *[16]* quote values somewhat different from those of Ploquin, *viz.* N, 1.586; $C_2 = C_6$, 0.849; $C_3 = C_5$, 0.947; C_4, 0.822; suggesting higher reactivity relative to benzene and the order 4- > 2-position in pyridine.

Jaffé *[61]* considered both pyridine and pyridine-*N*-oxide, and pointed out that the static charge (or π-electron density) method cannot correctly predict maximum electrophilic and nucleophilic reactivity for the same point, as occurs for the 4-position of the *N*-oxide *[cf. 62]*. By using the localisation energy method, which is essentially the calculation of the energy to change the system from the aromatic configuration to the transition-state configuration without considering the reagent explicitly, he obtained the reactivity orders: 4- > 2- > 3-pyridine derivative > benzene; and 4- > 3-pyridine-*N*-oxide > benzene > 2-pyridine-*N*-oxide; 4-pyridine-*N*-oxide > 4-pyridine derivative > benzene. He commented on the clearly incorrect result for the 2-position in pyridine-*N*-oxide.

Brown and Heffernan *[63a]*, by using what they consider are better parameters for hetero-atoms than those used by earlier workers, calculated both charge densities and localisation energies and included estimates for nitrogen in the onium form (pyridinium salts). Their localisation energy results give the order: 2- > 4- > 3-pyridine > benzene; 2- > 4-pyridinium > benzene > 3-pyridinium; and charge density results are similar. They are correlated qualitatively with experimental results, except for the anomalous position of the 3-pyridinium salts, and the inversion of reactivity of 2- and 4-pyridine derivatives, in the absence of acid catalysis. In relation to experiment, M.O. calculations particularly underestimate reactivity at the 3-position.

The use of nuclear magnetic resonance spectroscopy in consideration of substituent effects is common, and studies have been made of azines *[e.g. 64,65]*. Katritzky and Swinbourne *[64b]* have studied the chemical shifts of the α-proton in a series of β-arylacrylic acids and attempted to use this to estimate Hammett σ-values for hetero-atoms as "substituents", *i.e.* to compare benzenoid and aza-aromatic systems, but with only limited success.

Simple qualitative theory applied to assessment of stabilities of transition states/intermediate complexes (Fig. 72) leads to predicted reactivity orders, 4- > 2- > 3-pyridine > benzene; 2- or 4- > 3-pyridine-*N*-oxide > benzene; 2- > 4- > 3-pyridinium > benzene; pyridinium > pyridine-*N*-oxide > pyridine > benzene.

It is known that relative reactivities at the 2- and 4-position, and at the 3-position depend on the reagent used *[45,66,67]*, and also differ in pyridines, their *N*-oxides and salts. Liveris and Miller *[22]* investigated these three series in the reactions of their monochloro derivatives with methoxide and *p*-nitrophenoxide ions. Their results are summarised in Tables 53 to 55, which include a comparison with chlorobenzene and chloronitrobenzenes *[5,51,68]*. In Table 56 there are additional data, *viz.* the combined results of Chapman and Russell-Hill *[66]* and of Coppens *et al.* *[56]*, by which may be compared reactions of chloropyridines and their *N*-oxides with those of the chloronitrobenzenes. A further comparison with chlorobenzene is made by using the results of Amstutz and his co-workers *[10,65]*, allowing for the minor effect of solvent change, as recorded in Table 57 for the similar 2-chloroquinoline *[10,66,69]*. The reactions of the above two reagents with the chloropyridines and their *N*-oxides are compared in Table 58.

References p. 306

TABLE 53

REACTIVITY OF CHLOROBENZENE, CHLORONITROBENZENES, CHLOROPYRIDINES, THEIR *N*-OXIDES, AND *N*-METHYLPYRIDINIUM SALTS, WITH OMe^- IN MeOH AT 50° ($\rho = 8.47$) [22]

	Kinetic data at 50°							
	k_2 (l·mole^{-1}·sec^{-1})	Rate ratios[d,e]	Substituent constant[f] (σ^-)	$\Delta E^{\ddagger}$ (kcal·mole^{-1})	$\log_{10} B$	$\Delta S^{\ddagger}$ (e.u.)	$T\Delta S^{\ddagger}$ (kcal·mole^{-1})	$\Delta G^{\ddagger}$ (kcal·mole^{-1})
Chlorobenzene[a] (2-, 3- and 4-Cl)	$1.20 \cdot 10^{-16}$	1	0 (0)	39.9_5	11.1	− 9.90	− 3.2	43.1_5
Chloronitrobenzene 2-Cl[b]	$2.52 \cdot 10^{-6}$	$2.10 \cdot 10^{10}$ (2.83)	1.219 (1.301)	23.6_5	10.4	− 13.1	-4.2_5	27.9
Chloronitrobenzene 3-Cl	$6.76 \cdot 10^{-11}$	$5.64 \cdot 10^{5}$	0.679 (0.573)	31.5	12.0	− 5.77	-1.8_5	33.3_5
Chloronitrobenzene 4-Cl	$8.47 \cdot 10^{-6}$	$7.05 \cdot 10^{10}$ (9.48)	1.270 (1.269)	24.0_5	11.2	− 9.42	-3.0_5	27.1
Chloropyridine 2-Cl	$3.31 \cdot 10^{-8}$	$2.76 \cdot 10^{8}$ (0.0372)	0.996 (0.878)	28.9	12.1	− 5.33	− 1.7	30.6
Chloropyridine 3-Cl	$1.09 \cdot 10^{-11}$	$9.12 \cdot 10^{4}$	0.586 (0.596)	32.8_5	11.2_5	− 9.21	− 3.0	35.8_5
Chloropyridine 4-Cl	$8.91 \cdot 10^{-7}$	$7.43 \cdot 10^{9}$ (1)	1.165 (1.165)	25.2	11.0	− 10.3	− 3.3	28.5

Chloropyridine-*N*-oxide 2-Cl	$6.40 \cdot 10^{-4}$	$5.30 \cdot 10^{12}$ (714)	1.502 (1.567)	20.3	10.5_5	− 12.4	− 4.0	24.3
Chloropyridine-*N*-oxide 3-Cl	$1.16 \cdot 10^{-6}$	$9.67 \cdot 10^{9}$	1.178 (1.225)	24.6	10.7	− 11.7	− 3.8	28.4
Chloropyridine-*N*-oxide 4-Cl	$1.00 \cdot 10^{-3}$	$8.33 \cdot 10^{12}$ (1010)	1.526 (1.674)	19.0	9.8_5	− 15.6	$- 5.0_5$	24.0_5
N-Methyl-chloropyridinium salts[c] 2-Cl	$1.53_5 \cdot 10^{5}$	$1.28 \cdot 10^{21}$	2.492 (2.079)	13.9_5	14.6	+ 6.11	+ 2.0	11.9_5
N-Methyl-chloropyridinium salt 3-Cl	$3.14 \cdot 10^{-3}$	$2.62 \cdot 10^{13}$	1.584 (1.147)	25.6	14.8	+ 7.03	$+ 2.2_5$	23.3_5
N-Methyl-chloropyridinium salt 4-Cl	$5.08 \cdot 10^{3}$	$4.23 \cdot 10^{19}$	2.317 (2.164)	13.0	12.4_5	− 3.73	− 1.2	14.2

[a] Data by extrapolation of results for a series of 1-Cl-4-X-benzenes *[68f]*. [b] All values for *m*-chloronitrobenzene were calculated from values for *o*- and *p*-chloronitrobenzene by using the same relationship as between *m*-, and *o*- and *p*-fluoronitrobenzene *[68b,c,e]*. In both series σ and ρ values are very similar *[5b]*. [c] Calculated from data for the reaction with *p*-nitrophenoxide ion, using the OAr^-/OMe^- relationship found with 1-chloro-2,4-dinitrobenzene *[68a]*, which also has a high level of reactivity. [d] Reactivity relative to hydrogen as substituent = 1, also called the substituent rate factor, S.R.F. or *f [51]*. [e] Ratios in parentheses are relative to 4-Cl-pyridine = 1. [f] Hammett substituent constant based on k_2 (directly related to $\Delta G^{\ddagger}$). Values in parentheses, for comparison, are based on changes in $\Delta E^{\ddagger}$ alone (approx. equal to $\Delta H^{\ddagger}$ in these reactions in solution).

TABLE 54

RING POSITION RATE RATIOS FOR REACTIONS OF OMe^- IN MeOH AT 50° WITH CHLORONITROBENZENES, CHLOROPYRIDINES, CHLOROPYRIDINE-*N*-OXIDES, AND *N*-METHYL-CHLOROPYRIDINIUM SALTS *[22,68]*

Position of Cl relative to ring N or $C(NO_2)$ group	Rate ratios							
	Chloronitro-benzenes		Chloro-pyridines		Chloro-pyridine-*N*-oxides		*N*-methyl-chloropyridinium salts	
2-	$3.73 \cdot 10^4$	1	$3.02 \cdot 10^3$	1	$5.52 \cdot 10^2$	1	$4.89 \cdot 10^7$	1
3-	1	—	1	—	1	—	1	—
4-	$1.25 \cdot 10^5$	3.36	$8.14 \cdot 10^4$	26.9	$8.62 \cdot 10^2$	1.56	$1.62 \cdot 10^6$	0.0331

TABLE 55

RING ATOM RATE RATIOS FOR REACTIONS OF OMe^- IN MeOH AT 50° WITH CHLOROPYRIDINES, CHLOROPYRIDINE-*N*-OXIDES, AND *N*-METHYL-CHLOROPYRIDINIUM SALTS (INCLUDING COMPARISON WITH CHLORONITROBENZENES) *[22,68]*

Position of Cl relative to ring N or $C(NO_2)$ group	Ring atom			
	$C(NO_2)$	N	$\overset{+}{N}$–$\overset{-}{O}$	$\overset{+}{N}$–Me
2-	$7.62 \cdot 10^1$	1	$1.93 \cdot 10^4$	$4.64 \cdot 10^{12}$
3-	6.20	1	$1.06 \cdot 10^5$	$2.87 \cdot 10^8$
4-	9.50	1	$1.12 \cdot 10^3$	$5.70 \cdot 10^9$

The data for methoxide ion as reagent are fuller and can be discussed in more detail. The order of activation energy ($\Delta E^{\ddagger}$) in all three pyridine series is 4- < 2- < 3-position. The order 4-, 2- < 3-position is associated with the placing of the negative charge of the azabenzenide system on nitrogen in the former, with resultant lowering of energy. The order of 4- < 2-position is in accord with the view that the transition state has lower energy when the negative charge of the arenide system can be placed at the centre rather than the end of the delocalised system, and corresponds to the greater stability of *para*- than *ortho*-quinonoid compounds *[70,71]*. Differences in activation entropy ($\Delta S^{\ddagger}$) due to electrostatic and associated solvation effects *[22]* lead

TABLE 56

COMPARATIVE REACTIVITY OF CHLOROBENZENE, CHLORONITROBENZENES, CHLOROPYRIDINES AND CHLOROPYRIDINE-N-OXIDES AT 80° WITH PIPERIDINE IN MeOH ($\rho = 5.83$) *[56,66]*

	Kinetic data at 80°							
	k_2 ($l \cdot mole^{-1} \cdot sec^{-1}$)	Rate ratios	Substituent constant[d]	$\Delta E^{\ddagger}$ ($kcal \cdot mole^{-1}$)	$\log_{10} B$	$\Delta S^{\ddagger}$ (e.u.)	$T\Delta S^{\ddagger}$ ($kcal \cdot mole^{-1}$)	$\Delta G^{\ddagger}$ ($kcal \cdot mole^{-1}$)
Chlorobenzene (2-, 3- and 4-Cl)	$2.44 \cdot 10^{-13}$	1[a]	0	28.6	5.1	− 37.7	− 13.3	41.9
Chloronitrobenzene 2-Cl	$1.44 \cdot 10^{-5}$	$5.90 \cdot 10^{7}$[a] 8.79[b]	1.330	18.1	6.4	− 31.6	− 11.2	29.3
Chloronitrobenzene 3-Cl	—	1.61[c]	—	—	—	—	—	—
Chloronitrobenzene 4-Cl	$6.14 \cdot 10^{-6}$	$2.51 \cdot 10^{7}$[a] 3.74[b]	1.270	17.7	5.8	− 34.4	− 12.1	29.8
Chloropyridine 2-Cl[e]	$1.49 \cdot 10^{-7}$	$6.11 \cdot 10^{5}$[a] 0.0909[b]	0.993	23.0	7.4	− 27.0	− 9.5_5	32.5_5
Chloropyridine 3-Cl	—	1[c]	—	—	—	—	—	—
Chloropyridine 4-Cl	$1.64 \cdot 10^{-6}$	$6.73 \cdot 10^{6}$[a] 1[b]	1.170	17.0	4.8	− 38.9	− 13.7	30.7
Chloropyridine-N-oxide 2-Cl	$4.05 \cdot 10^{-4}$	$1.66 \cdot 10^{9}$[a] 247[b]	1.582	15.0	5.9	− 33.9	− 12.0	27.0
Chloropyridine-N-oxide 3-Cl	—	6.94[c]	—	—	—	—	—	—
Chloropyridine-N-oxide 4-Cl	$1.11 \cdot 10^{-4}$	$4.55 \cdot 10^{8}$[a] 67.6[b]	1.487	13.6	4.5	− 40.3	− 14.2	27.8

[a] Relative to chlorobenzene = 1. [b] Relative to 4-chloropyridine = 1. [c] Relative to 3-chloropyridine = 1. [d] Hammett's substituent constant obtained by using the ρ-value 5.83, based on rate data for chlorobenzene and p-chloronitrobenzene with σ-values of zero and 1.270 respectively. [e] Ref. *66* for the reaction of EtOH gives values at 80°, k_2 $1.76 \cdot 10^{-7}$, rate ratio ([a]), $7.21 \cdot 10^{5}$; σ, 1.007; $\Delta E^{\ddagger}$, 19.9; $\log_{10} B$, 5.55; $\Delta S^{\ddagger}$, − 35.9; $T\Delta S^{\ddagger}$, − 12.7; $\Delta G^{\ddagger}$, 32.6 and these seem much more in accord with the other data in the Table than the results of Coppens *et al.* *[56]* recorded in the main Table.

TABLE 57

REACTIVITY OF 2-CHLOROQUINOLINE WITH PIPERIDINE IN A NUMBER OF SOLVENTS AT 50° *[10,45,66]*

	Kinetic parameter					
Solvent	k_2[a] (l·mole^{-1}·sec^{-1}) (ratio)	$\Delta E^{\ddagger}$ (kcal·mole^{-1}) ($\Delta\Delta E^{\ddagger}$)	$\log_{10} B$ ($\Delta \log_{10} B$)	$\Delta S^{\ddagger}$ (e.u.)	$T\Delta S^{\ddagger}$ (kcal·mole^{-1})	$\Delta G^{\ddagger}$ (kcal·mole^{-1}) ($\Delta\Delta G^{\ddagger}$)
Piperidine	$3.52 \cdot 10^{-6}$ (1)	13.8 (0)	3.88 (0)	− 42.9	− 13.8_5	27.6_5 (0)
Ethanol	$1.81 \cdot 10^{-6}$ (0.515)	15.6 (+ 1.8)	4.81 (+ 0.93)	− 38.9	− 12.5_5	28.1_5 (+ 0.5)
Toluene	$2.10 \cdot 10^{-6}$ (0.597)	12.9 (− 0.9)	3.04) (− 0.84)	− 46.8	− 15.1	28.0 (+ 0.3_5)
Petroleum ether	$5.10 \cdot 10^{-7}$ (0.145)	14.9 (+ 1.1)	3.78 (− 0.10)	− 43.4	− 14.0	28.9 (+ 1.2_5)

[a] Values of k_2 (and k_2 ratio) at 80° are $2.19 \cdot 10^{-5}$ (1) in piperidine and $1.43_5 \cdot 10^{-5}$ (0.655) in ethanol.

to the free energy of activation ($\Delta G^{\ddagger}$) or rate order being slightly different, affecting only the relative reactivity at the 2- and 4-positions. The rate (reverse of $\Delta G^{\ddagger}$) order is 4- > 2- ≫ 3-position in the pyridines; 4- ~ 2- ≫ 3-position in the pyridine-*N*-oxides, and 2- > 4- ≫ 3-position in the pyridinium salts.

In the chlorobenzenes the order of $\Delta E^{\ddagger}$ values is 2- ~ 4 ≫ 3-position, but the rate order based on $\Delta G^{\ddagger}$ values is 4- > 2- ≫ 3-position. It is significant that the ratio of reactivity at the 4- to that at the 2-position is greater for the pyridines, for which the expected smaller stability of *ortho*- than of *para*-quinonoid transition states is further reduced by the adverse electrostatic interaction in the 2-position: $\overset{-}{\mathrm{N}}$–C—$\overset{\delta-}{\mathrm{O}}$Me. This additional interaction should be less for an *o*-nitro group, being partly counteracted by the positive charge on the nitro group nitrogen; while retention of a relatively small twist in the transition state, as in the ground state *[68a]*, can further minimise the unfavourable interaction. It is significant that the 4- to 2-position ratio is similar in the dipolar attached $-\overset{+}{\mathrm{N}}\mathrm{O_2}^{-}$ group and

TABLE 58

COMPARISON OF REACTIONS OF PIPERIDINE[a] ($\rho = 5.83$ AT 80°) AND METHOXIDE ION[b] ($\rho = 8.47$ AT 50°) WITH THE CHLOROPYRIDINES AND CHLOROPYRIDINE-*N*-OXIDES IN METHANOL[c]

	Substituent rate factor (S.R.F. or *f*) *[51]*		Hammett's substituent constant σ^-		$\Delta\Delta E^{\ddagger}$ (kcal·mole^{-1})		$\Delta\Delta S^{\ddagger}$ (e.u.)		$\Delta\Delta G^{\ddagger}$ (kcal·mole^{-1})	
	$C_5H_{11}N$	OMe^-	$C_5H_{11}N$	OMe^-	$C_5H_{11}N$	OMe^-	$C_5H_{11}N$	OMe^-	$C_5H_{11}N$	OMe^-
Chloropyridine 2-Cl[d]	$6.11\cdot10^5$ ($7.21\cdot10^5$)	$2.76\cdot10^8$	0.993 (1.007)	0.996	− 5.6 (− 8.7)	-11.0_5	+ 10.7 (+ 1.8)	+ 4.57	-9.3_5 (− 9.3)	-12.5_5
Chloropyridine 3-Cl	—	$9.12\cdot10^4$	—	0.586	—	− 7.1	—	+ 0.69	—	− 7.3
Chloropyridine 4-Cl	$6.73\cdot10^6$	$7.43\cdot10^9$	1.170	1.165	− 11.6	-14.7_5	− 1.2	− 0.40	− 11.2	-14.6_5
Chloropyridine-*N*-oxide 2-Cl	$1.66\cdot10^9$	$5.30\cdot10^{12}$	1.582	1.502	− 13.6	-19.6_5	+ 3.8	− 2.50	− 14.9	-18.8_5
Chloropyridine-*N*-oxide 3-Cl	—	$9.67\cdot10^9$	—	1.178	—	-15.3_5	—	− 1.80	—	-14.7_5
Chloropyridine-*N*-oxide 4-Cl	$4.55\cdot10^8$	$8.33\cdot10^{12}$	1.487	1.526	− 15.0	-20.9_5	− 2.6	− 5.7	− 14.1	− 19.1

[a] Ratios and differences at 80°. [b] Ratios and differences at 50°. [c] All relative to chlorobenzene for which k_2 with piperidine at 80° $= 2.44\cdot10^{-13}$ ($5.57\cdot10^{-14}$ at 50°); k_2* with methoxide at 50° $= 1.20\cdot10^{-16}$; substituent rate factor (S.R.F. or *f*) *[51]* = unity; $\sigma^- = 0$; $\Delta E^{\ddagger}$, piperidine = 28.6, $\Delta E^{\ddagger}$, methoxide = 39.95; $\Delta S^{\ddagger}$ (80°) piperidine = −37.7; $\Delta S^{\ddagger}$ (50°) methoxide = −9.90; $\Delta G^{\ddagger}$ (80°) piperidine = 41.9; $\Delta G^{\ddagger}$ (50°) methoxide = 43.15. [d] Values in parentheses from ref. *66* for EtOH, which however seem more consistent with the rest of the piperidine results, and are preferred. * l·mole^{-1}·sec^{-1}.

dipolar ring $\overset{+}{N}$–O^-. The converse electrostatic effect in the pyridinium salts, plus solvation factors, discussed by Liveris and Miller *[22]*, lead to a low 4- to 2-position rate ratio there.

It is interesting to note that the relative reactivity at the 4- and 2-, and the 3-position decreases in the order, pyridinium salts > pyridines > pyridine-*N*-oxides. The order of the first two is simply related to the difference in electron-withdrawing power of the ring nitrogen, whereas the inferior position of the *N*-oxides is ascribed to a mesomeric electron release conferring a deactivating component on reactivity at the 4- and 2- but not at the 3-position in that system.

The most striking feature of the reactions of these azines however is the very potent effect of substituting the more electronegative nitrogen atom for the ring carbon (with attached hydrogen as a methine unit), and its very marked further increase as the electronegativity of the nitrogen is increased in the *N*-oxides and pyridinium salts. Ingold *[6b]* has discussed the relative electronegativity of N and CH in terms of carrying the proton of CH from a peripheral to the central location in relation to the electron octet. In the 4-position, for example, the $\Delta\Delta E^{\ddagger}$ value in kcal·mole^{-1} for the change from ring methine to ring nitrogen is 14.7_5, as compared with 15.9 for an attached *p*-nitro group; and increases to 20.9_5 for the change from ring methine to $\overset{+}{N}$–O^-, and 26.9_5 to $\overset{+}{N}$–Me. Even in the 3-position, $\Delta\Delta E^{\ddagger}$ values are large, being about half the above. The effect of a ring $\overset{+}{N}$–Me group is even larger than that of the most activating attached group so far investigated kinetically, *viz.* the diazonium group *[68d,72]*, and one can predict even greater activation with ring O^+ (pyrylium salts), and other ring $\overset{+}{N}$–X groups where X is electron-attracting, *e.g.* X = cyano, 2,4-dinitrophenyl, although complicating nucleophilic additions and ring-fission reactions may be expected.

The kinetic data have been speculatively correlated with the following Pauling electronegativity indices *[73]* for ring atoms using sp^2 bonding orbitals: C–H, 2.8; $C^{\delta+}(\overset{+}{N}Me_3)$, 2.9_1; N, 3.3; $\overset{+}{N}(O^-)$, 3.5_2; $\overset{+}{N}$(Me), 3.8_3.

The results of Amstutz, Coppens, Chapman and their co-workers *[9,56,66,69]* taken together in Table 56 are similar but less complete. An expected difference is the enhancement of reactivity at the

α-position to the carbon bearing a nitro group, or to ring nitrogen or *N*-oxide, with a neutral reagent (of the type $\ddot{Y}$–H) as compared with an anionic reagent. This has already been explained as due to intramolecular hydrogen-bonding and electrostatic effects stabilising the transition states. The former must be the more important, since this reactivity change does not occur with tertiary amines *[54]*. Nevertheless the converse electrostatic destabilising effect in reactions of methoxide ion with 2-chlorocarboxylates has been demonstrated *[74]*.

Activation in the azines appears to be substantially less with piperidine than methoxide. Thus in positions 2- and 4- relative to ring nitrogen, rate ratios compared with chlorobenzene are about 10^6 instead of 10^9 in the pyridines; and about 10^9 instead of 10^{13} in the pyridine-*N*-oxides. For both reagents however activation is almost entirely $\Delta E^{\ddagger}$ dependent, as is seen by similarity in the patterns of $\Delta\Delta E^{\ddagger}$ and $\Delta\Delta G^{\ddagger}$ values. The difference in magnitude of activation is reflected by the Hammett reaction constants (ρ) which are 5.83 for piperidine (at 80°) and 8.47 for methoxide (at 50°), and only a minor part of the difference is due to the variation of ρ with temperature. A possible explanation is that the transition states of the methoxide reactions involve a greater degree of bond formation and nearer approach to arenide structures than the piperidine reactions. However it must be recalled that ρ-values for piperidine and methoxide are quite similar in reactions in the reactive benzene series (*e.g.* see Tables 9 and 10 of Chapter 4, pp. 77 ,78) so that more data on this aspect of the azine reactions would be valuable.

The lower reactivity at the 2- and 4-position with piperidine as compared with methoxide should be matched also in substitution at the 3-position. This may be inferred from the qualitative data available, but not confirmed quantitatively. The Hammett substituent constants (σ) are seen to be similar with both reagents. Chapman and Russell-Hill *[66]* have also measured the reactivity of 2- and 4-chloropyridine with ethoxide ion in ethanol, and their results are similar to those of Liveris and Miller *[22]* who used methoxide ion in methanol.

(*ii*) *Monocyclic diazines*

From the preceding discussion it is to be expected that the polyaza-benzenes should be still more susceptible to attack by nucleophilic reagents, and there are numerous supporting references *[e.g. 1,2]*. There is a great deal of qualitative and semi-quantitative information but kinetic data are scarce.

Chan and Miller *[75]* have made a preliminary study of the reactivity of all six monochlorodiazines with *p*-nitrophenoxide ion in methanol. They added an excess of *p*-nitrophenol to avoid any concurrent methanolysis *[68e]* but this necessarily leads to mild acid-catalysis due to hydrogen-bonding by the phenol to azine nitrogen (*cf.* many references quoted in ref. 2). This is not regarded as enough to cause a major increase in reactivity but is likely to affect those positional orders of reactivity which differ little intrinsically. For example, with this acid catalysis and an anionic reagent, the reactivity at position 4- compared with that at position 2- to nitrogen will be decreased. A relevant parallel example is the reaction of 2,4,6-tribromopyridine with phenoxide ion. In water mainly the 4-bromine is displaced, whereas in phenol mainly the 2-bromine is displaced *[76]*. While the effects of the phenol catalysis must be borne in mind, important conclusions are obtainable, and as far as is known this study represents the only full set of data for comparing the reactivity of each position in monocyclic aromatic diazines.

The six compounds are 3- and 4-chloropyridazine; 2-, 4-, and 5-chloropyrimidine; and 2-chloropyrazine. The reagent is well-known to have similar reactivity patterns to those of alkoxides *[68a–d]*, though substantially less reactive (a factor of about 10^4) than methoxide ion. The results are given in Table 59, which for convenient reference also includes results for the reaction of *N*-methylchloropyridinium salts with *p*-nitrophenoxide ion, together with estimates for reaction with chlorobenzene and chloropyridines, made by using the *p*-nitrophenoxide–methoxide relationships found with 1-chloro-2,4-dinitrobenzene.

It is noteworthy that even the most reactive chlorodiazines are substantially less reactive than the very reactive *N*-methylchloropyridinium salts and since a quaternised diazine would be more reactive, this indicates that the catalysis due to hydrogen-bonding by *p*-nitrophenol in no way approaches that due to a large measure of cationisation. This is supported by consideration of the results of Hill and Krause *[77]* for methanolysis (OMe^- in MeOH) of 3-chloro-

TABLE 59

COMPARATIVE REACTIVITY OF CHLOROBENZENE, CHLOROPYRIDINES, *N*-METHYLCHLOROPYRIDINIUM SALTS, AND CHLORODIAZABENZENES WITH *p*-NITROPHENOXIDE ION (WITH 10-FOLD EXCESS ArOH) IN MeOH AT 50° (ρ= 8.47)[a] *[75]*

	Kinetic data								
	k_2 (l·mole^{-1}·sec^{-1})	Rate ratios[d]	σ^{-e} (σ^- (calcd.)[f])	$\Delta E^{\ddagger}$ (kcal·mole^{-1})	$\log_{10} B$	$\Delta S^{\ddagger}$ (e.u.)	$T\Delta S^{\ddagger}$ (kcal·mole^{-1})	$\Delta G^{\ddagger}$ (kcal·mole^{-1})	$\Delta\Delta G^{\ddagger}$ (kcal·mole^{-1})
Chlorobenzene (2-, 3- and 4-Cl[b])	$1.1 \cdot 10^{-20}$	1	0	44.6	10.2	−14.0	−4.5	49.1	0
Chloropyridine 2-Cl[b]	$3.0 \cdot 10^{-12}$	$2.7 \cdot 10^{8}$	1.00	33.5	11.2	−9.4	−3.0	36.5	−12.6
Chloropyridine 3-Cl	$9.9 \cdot 10^{-16}$	$9.0 \cdot 10^{4}$	0.58	37.5	10.4	−13.1	−4.2	41.7	−7.4
Chloropyridine 4-Cl	$8.1 \cdot 10^{-11}$	$7.4 \cdot 10^{9}$	1.16	29.8	10.2	−14.0	−4.5	34.3	−14.8
N-Methylchloropyridinium salts 2-Cl	$1.4 \cdot 10^{1}$	$1.3 \cdot 10^{21}$	2.49	18.6	14.3	4.8	1.5	17.1	−32.0
N-Methylchloropyridinium salts 3-Cl	$2.8 \cdot 10^{-7}$	$2.6 \cdot 10^{13}$	1.58	30.2	13.9	2.9	0.9	29.3	−19.8
N-Methylchloropyridinium salts 4-Cl	$4.6 \cdot 10^{-1}$	$4.2 \cdot 10^{19}$	2.32	17.6	11.6	−7.6	−2.5	20.1	−29.0
Chloropyridazine (1,2-diaza-) 3-Cl	$1.3 \cdot 10^{-6}$	$1.2 \cdot 10^{14}$	1.66 (1.58)	24.4	10.6	−12.1	−3.9	28.3	−20.8
Chloropyridazine (1,2-diaza-) 4-Cl	$2.1 \cdot 10^{-6}$	$1.9 \cdot 10^{14}$	1.69 (1.74)	24.3	10.8	−11.4	−3.7	28.0	−21.1
Chloropyrimidine (1,3-diaza-) 2-Cl	$6.9 \cdot 10^{-4}$	$6.3 \cdot 10^{16}$	1.98 (2.00)	20.5	10.7	−11.7	−3.8	24.3	−24.8
Chloropyrimidine (1,3-diaza-) 4-Cl[c]	$1.2 \cdot 10^{-5}$	$1.1 \cdot 10^{15}$	1.78 (2.16)	25.4	12.2	−4.8	−1.5	26.9	−22.2
Chloropyrimidine (1,3-diaza-) 5-Cl	$1.3 \cdot 10^{-7}$	$1.2 \cdot 10^{13}$	1.54 (1.16)	25.8	10.5	−12.4	−4.0	29.8	−19.3
Chloropyrazine (1,4-diaza-) 2-Cl	$4.5 \cdot 10^{-6}$	$4.1 \cdot 10^{14}$	1.72 (1.58)	23.4	10.5	−12.6	−4.1	27.5	−21.6

[a] We assume the Hammett reaction constant ρ is the same as with OMe^- in MeOH = 8.47 at 50° (Table 53). [b] Calculated from data for OMe^- in MeOH by using the reagent relationship obtained with 1-chloro-2,4-dinitrobenzene *[68a]*, and thus without any acid-catalytic effect of *p*-nitrophenol. [c] The values of $\Delta E^{\ddagger}$, and $\log B$ and other derived data seem somewhat inconsistent: a lower value of $\Delta E^{\ddagger}$ and of $\log B$ would be expected, but values of k_2 and $\Delta G^{\ddagger}$ seem more reliable. [d] Relative to chlorobenzene = 1, These are substituent rate factor, S.R.F. or *f* values *[51]*. [e] Hammett substituent constants (σ^-) based on rate ratios and a value of $\rho = 8.47$ at 50°. [f] σ^--values based on the σ^--values in the pyridine series (Table 53). [g] In these reactions in solution $\Delta E^{\ddagger}$ is taken as approximately equal to $\Delta H^{\ddagger}$.

pyridazine. The rate ratio (compared with the rate for chlorobenzene *[68g]*) is $5.53 \cdot 10^{12}$ at 50°, equivalent to a σ-value of 1.505, whereas the values with *p*-nitrophenoxide in the presence of a ten-fold excess of *p*-nitrophenol are $1.20 \cdot 10^{14}$ and 1.66. The value of σ calculated from chloropyridine reactions with methoxide ion is 1.58. The quite typical values (for reaction of an anion with a neutral substrate) of $\Delta S^{\neq}$ for the *p*-nitrophenoxide ion reactions also support this conclusion. Further, from the data quoted by Albert *[17c]*, it is probable that the two chloropyridazines are more basic than any of the others and thus will be more responsive to catalysis by *p*-nitrophenol. It may thus be concluded that the acid catalysis in these reactions is not likely to be equivalent to an increase in σ of more than 0.2 units.

The main conclusion from Chan and Miller's preliminary results, supported by the data of Hill and Krause *[77]* for 3-chloropyridazine, is that to a reasonable degree of approximation, reactivity is that expected from *independent* contributions of each ring nitrogen atom. Any deficiency is much less than that shown when there are two similar *attached* activating groups in a benzene ring. The latter is illustrated by σ-values for the nitro groups of *o*- and *p*-chloronitrobenzene which for displacement of chlorine by methoxide ion are equal to 1.219 and 1.270, (Table 53) adding to 2.489, whereas that for the 2,4-dinitro group as a unit taken from rates for 1-chloro-2,4-dinitrobenzene *[68a]* is only 1.816.

From the pyridine data *[22]* the following reactivity order would be predicted: *N*-methyl-2- > *N*-methyl-4-chloropyridinium > 4-chloro- > 2-chloro-pyrimidine > 4-chloro- > 3-chloro-pyridazine = 2-chloro-pyrazine = *N*-methyl-3-chloropyridinium > 5-chloropyrimidine. The corresponding experimental or calculated σ-values for these systems are 2.49, 2.32, 2.16, 2.00, 1.74, 1.58, 1.58, 1.58, 1.16. The results with *p*-nitrophenoxide ion (Table 59) are in accord with this except for the inversion of the order of 2- and 4-chloropyrimidine, ascribable, at least in part, to the more favourable effect of the acid catalysis on positions α to ring nitrogen. The same effect is presumed to result in the smaller-than-predicted difference between 3- and 4-chloropyridazine.

The Arrhenius parameters for 4-chloropyrimidine differ somewhat from the normal pattern. This casts some doubt on the degree of precision of the derived rate data for this compound and thus on whether the difference between the experimental and calculated σ-values for this compound is really significant. It seems better to withhold comment until further data are available.

In more general terms it appears that as with pyridines, the very large substituent effects depend essentially on $\Delta E^{\neq}$, with $\Delta\Delta E^{\neq}$ values ranging up to 24.1 kcal·mole^{-1} for 2-chloropyrimidine, and may be compared with values of 26.0 and 27.0 kcal·mole^{-1} for *N*-methyl-2-chloro- and *N*-methyl-4-chloropyridinium compounds. The larger differences in $\Delta\Delta G^{\neq}$ for the last pair, as compared with 2-chloropyrimidine, are due to the high $\Delta S^{\neq}$ values for reactions of an anion and a cation. The typical values of $\Delta S^{\neq}$ in the *p*-nitrophenoxide reactions also indicate that there is relatively minor cationisation under the reaction conditions used by Chan and Miller.

Additional experimental data available support the conclusions given above. Chapman and Russell-Hill *[66]* have measured the rate of reaction of ethoxide ion with 2-chloropyrimidine and it is in a reasonable relationship with the above results *[75,77]*. Chapman and Rees *[44b]* have inferred the reactivity order 4- > 2-chloropyrimidine with secondary amines, from reactions with methyl derivatives. It is significant in relation to the effects of mild acid catalysis suggested above that in 2,4-dichloropyrimidine the 4-chlorine is replaced by methoxide ion in methanol *[78,79]* (no acid catalysis), whereas with methanol alone the 2- and 4-chlorine have similar reactivity *[78]*. Cheeseman *[80]* has reported high reactivity of 2-chloropyrazine, and the still higher reactivity of 2-chloropyrazine-*N*-oxide is known *[81,82]*. In the pyridazine series facile alkoxide exchange has been reported *[83]*. There are numerous reports of reactions of polysubstituted derivatives *[2]*.

Albert *[17a]* quotes calculated charge distributions in the three diazines, and these are shown in Fig. 74, with results for 1,3,5-triazine (electron deficiency shown with a positive sign). These, and the values calculated by Orgel *et al.* *[60]*, are correlated only approximately with kinetic data.

(*iii*) *Monocyclic triazines*

There is a great deal of information, though kinetic data are scarce, about *s*-triazines (1,3,5-triaza-) but little about *as*-triazines (1,2,4-triaza-) and *v*-triazines (1,2,3-triaza-benzene). In the absence of acid

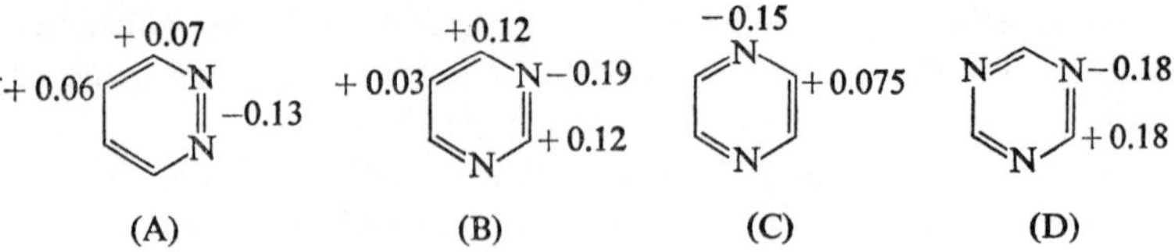

Fig. 74. Calculated charge distribution in pyridazine (A), pyrimidine (B), pyrazine (C) and *s*-triazine (D) *[17a]*.

catalysis one would predict the positional reactivity order 2-, 4-, or 6-*s*-triazine > 5-*as*-triazine = 4-(or 6-)*v*-triazine > 3-*as*-triazine > 5-*v*-triazine > 6-*as*-triazine = 4-pyrimidine, and estimates of σ-values for the ring system can be given. The small differences between 5-*as*-triazines, 4- (or 6)-*v*-triazines, and 3-*as*-triazines are quite likely to be reduced or inverted by acid catalysis, and similarly with 5-*v*-triazine and 6-*as*-triazine.

Shepherd and Fedrick *[2]* have given in detail an essentially similar qualitative scheme based on reactivity *ortho*, *meta* and *para* to nitrogen, assuming also that reactivity *ortho* to a ring nitrogen is about equivalent to that *meta* to two ring nitrogens. This may be compared with estimates based on data of Liveris and Miller *[22]* for reactions of methoxide ion with the pyridines. Their σ-values, *viz. ortho*, 0.996; twice *meta*, 1.172, *para*, 1.165 suggest that the mildly greater reactivity at a position *para* than that at a position *ortho* to a ring nitrogen is equivalent to reactivity *meta* to two ring nitrogens. Clearly this difference is only a minor one and affected in any case by acid catalysis. Regrettably the data for piperidine as reagent (Table 56) do not include a σ-value for a position *meta* to a ring nitrogen.

There seems to be no information available about susceptibility of *v*-triazines to nucleophilic attack, and little about *as*-triazines. Piskala *et al. [84]*, however, reported briefly on the nucleophilic reactivity of 3,5,6-trichloro-1,2,4-triazine and, as predicted above, showed that the 5-position is the most reactive, but gave no comparisons with other ring systems. As Shepherd and Fedrick [2] have pointed out, results from systems with several replaceable groups, even though the same group occurs in different positions, must be interpreted with caution, unless the structure is completely symmetrical, since the reciprocal activation or deactivation of the substituents is unlikely to be equal. An example of this was given in Chapter 4 (p. 100).

In the *s*-triazine series most is known about poly-substituted compounds, particularly cyanuric chloride, from which one class of reactive dyestuffs is formed *[85]*, but this information does include data for compounds in which only one group is highly mobile. Thus 2-chloro-4,6-diphenyl- and 2-chloro-4,6-dimethyl-*s*-triazine react very readily with a number of anionic and neutral nucleophiles *[86,87]*. Similarly 2-phenoxy-*s*-triazine reacts readily with aliphatic amines and aniline *[88]*. Both Illuminati *[1]*, and Shepherd and Fedrick *[2]*, give Tables indicating the high reactivity of cyanuric chloride, and reactivity at high levels, though decreasing, as chlorines are replaced by deactivating groups: kinetic data are available in a few cases *[47,48, 89]*. For example 2,4,6-trichloro-*s*-triazine (cyanuric chloride) reacts with aniline in benzene some $2 \cdot 10^4$ faster than 2,4,6-trichloropyrimidine. If data for piperidine are taken as a guide to reactions of aniline *i.e.* assuming similar ρ-values for the reactions of aniline and of piperidine, these data suggest a rather smaller activation by the third than by the two preceding (pyrimidine) ring nitrogen atoms. Available data *[46,89–91]* suggest that substituent effects are similar to those in the benzene series.

It is possible to obtain a reasonable link through to chlorobenzene, and thus obtain a measure of total activation by all the substituents. Horrobin *[46]* quotes data for the hydrolysis (water as reagent and solvent) of cyanuric chloride, from which a bimolecular rate constant (k_2) at 40° of $4.8 \cdot 10^{-2}$ l·mole–^{1}·sec^{-1} may be estimated. Murto *[92]* has measured k_2 for solvolysis of 1-fluoro-2,4-dinitrobenzene by water as $3.6 \cdot 10^{-9}$, and for reaction with methoxide in methanol as $4.5 \cdot 10^1$ l·mole^{-1}·sec^{-1} at 40°. Miller and his co-workers *[33a,68f]* have measured k_2 for 1-fluoro-2,4-dinitrobenzene, and estimated that for chlorobenzene in reaction with methoxide ion in methanol. At 40° their values are $4.3 \cdot 10^1$ and $1.7 \cdot 10^{-17}$ l·mole^{-1}·sec^{-1} respectively. Assuming the same ratios for methanolic methoxide ion to water (as reagent and solvent), with cyanuric chloride as for 1-fluoro-2,4-dinitrobenzene, the value of k_2 for the reaction of cyanuric chloride with methoxide ion in methanol at 40° is estimated as $5.9 \cdot 10^8$ l·mole^{-1}·sec^{-1}, thus giving a rate ratio, cyanuric chloride/chlorobenzene, of $3.5 \cdot 10^{25}$. For a ρ-value of 9.15 at 40° *[68f]* this is equivalent to a σ-value for the 1,3,5-triaza system, plus two *meta*-chlorine atoms of 2.8. Allowing *ca.* 0.5 as the σ-value for two *meta*-

References p. 306

chlorines *[6]* the value of σ for activation by the 1,3,5-triaza system in 2-chloro-1,3-5-triazine is 2.3. The $\Delta\sigma$ value of about 0.5, as compared with 2- and 4-chloropyrimidine, is equivalent to a rate ratio of 10^4 to 10^5, and is similar to that reported above with aniline as reagent.

It appears that in the very highly activated 1,3,5-triazine system there is a substantial saturation effect in activating power of the same type as that recorded in comparison of halogenobenzenes, *o*- and *p*-halogenonitrobenzenes, and 1-chloro-2,4-dinitrobenzene; whereas in the diazines this effect is present in a minor degree only.

(*iv*) *Monocyclic tetrazines*

The only known aza-benzenoid system with more than three ring nitrogens is *s*-tetrazine (1,2,4,5-tetrazabenzene). The calculated σ-value for reactivity at the equivalent 3- and 6-position is virtually identical with that at the equivalent 2,- 4-, and 6-position of *s*-triazine (1,3,5-triazabenzene), *viz*. 3.16; and this value is presumably in excess of the actual value in *s*-tetrazine as in *s*-triazine, and to about the same extent. Some nucleophilic reactions are known *[93,94]* but there are no quantitative data suitable for comparison with other systems.

While discussion of the reactivity of these azabenzenes with nucleophiles has been in terms of nucleophilic substitutions, it should not be forgotten that the more highly activated systems also undergo nucleophilic addition and nucleophilic ring-fission reactions. Such reactions although of much interest are outside the scope of this book.

(*b*) *Monocyclic π-excessive and derived π-deficient heteroaromatic systems*

The π-deficient *[17a]* heteroaromatics such as have been described above react readily with nucleophilic reagents as expected, especially when the position of substitution is such that the negative charge of the hetarenide system can be placed on the hetero-atom. In contrast, in π-excessive *[17b]* heteroaromatics which contain a 5-membered ring and only one hetero-atom*, electron density is high on ring carbon

*Polyhetero-atom 5-membered ring systems in neutral or positive forms should not be regarded as π-excessive compared with benzene [*cf.* *17c*].

atoms, and for nucleophilic substitution the negative charge of the hetarenide system cannot be placed on the hetero-atom. This implies low reactivity towards nucleophiles, though reactions are facilitated by copper catalysis as with less reactive benzenoid species. Indeed, though the charge can be placed on a carbon atom neighbouring a hetero-atom (Fig. 72), the reactivity in nucleophilic substitution of for example monohalogenopyrroles, even in the more reactive 2-position, should be much less than in the 3-position of pyridine, though the σ-inductive effect of the hetero-atom may be enough to bring reactivity to about the level of that of the halogenobenzenes. There is no very direct evidence, but the results of Amstutz and his co-workers *[21,95]* are in general accord with this view. They showed that the reactivity of 2-halogenofurans (Cl, Br, or I) with piperidine, and in one case with methoxide ion, is a little higher than that of the halogenobenzenes. Similarly Motayama *et al. [96]* have shown that 2-bromo-5-nitrothiophen is a little more reactive than *p*-bromonitrobenzene. The pyrrole system should be less reactive than either the furan or thiophen system, and this reactivity is often further reduced because most powerful nucleophiles are bases, in the presence of which the pyrrole would form the still less reactive pyrrole anion. Stevens *[97]* comments on the low reactivity of halogenopyrroles towards nucleophilic substitution.

The π-electron densities calculated for furan and pyrrole by Orgel *et al. [60]*, and by Brown and Coller *[19]*, support the π-excessive description, and low reactivity, of which the very scarcity of data is itself an indication. On the basis of electronegativity and involvement of the hetero-atom in π-delocalisation the predicted ring-system reactivity order is pyrrole anion < pyrrole < furan < thiophen. Directly confirmatory comparative rate data are not available, but the general level of reactivity is well-known.

Insertion of a second hetero-atom into such ring systems should result in a large increase in susceptibility to nucleophilic attack by virtue of the reduction in π-electron density on ring carbon atoms *[60]*, and especially because, for substitution at certain positions, the negative charge of the hetarenide system can be placed on a hetero-atom. The change thus corresponds in character to the difference between nucleophilic substitution in benzene and pyridine systems. Recent calculations of π-electron densities *[63b,c]* in azasubstituted

References p. 306

pyrroles suggest the order imidazole > pyrazole > pyrrole, with favoured positions 2-imidazole, 3-pyrazole, and 2-pyrrole in the neutral forms; and 4-(5-)imidazole > 4-pyrrole > 3-pyrrole in anionic forms.

Figure 75 illustrates how 2-substitution in imidazole and thiazole, and 3-(5-)substitution in pyrazole are favoured by the placing of the negative charge of the hetarenide transition state on nitrogen.

Amstutz *et al.* *[45b]* have shown that 2-chloro- and 2-bromothiazole are somewhat more reactive than 2-chloro- and 2-bromopyridine, respectively, whereas Sprague and Land *[98a]* have commented on the low reactivity of 4- and 5-halogenothiazoles. Though precise data are not available, Jacobs *[98b]*, and Schipper and Day *[98c]* have commented on the low reactivity of halogenopyrazoles and halogenoimidazoles. Their comments most probably refer to conditions where the reactivity is that of the anionic forms, though even the neutral forms should be less reactive than the thiazoles. It is of interest that Gorbacheva *et al.* *[99]* have shown that hexamethyleneimine reacts only slowly with 5-chloro-3-methyl-4-nitropyrazole at 110°, presumably with the pyrazole rather than its conjugate base, with which it is in equilibrium in these conditions. This is supported by their demonstration that 5-chloro-3-methyl-1-phenyl-4-phenylazopyrazole reacts more readily, despite the azophenyl group being less activating than the nitro group *[62]*, since with a 1-phenyl group the pyrazole does not form a conjugate base. Koehler *[100]* has shown that the 5-bromine atom of 2,5-dibromo-4-nitroimidazole is replaced by the weakly basic but strongly nucleophilic thiophenoxide ion *[7c]*. This presumably illustrates the greater activating power of a nitro group than of a ring nitrogen, *i.e.* as represented in Fig. 76, transition state A has higher energy than transition state B.

Still higher reactivity is clearly predictable in systems with more hetero-atoms; and there are many confirmatory reports, including a

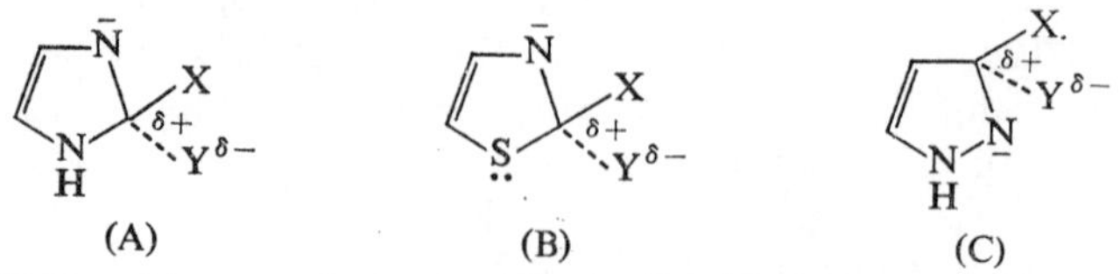

(A) (B) (C)

Fig. 75. Transition states for nucleophilic substitution by Y^- in some hetero-atom-substituted pyrroles: A, 2-X-imidazole; B, 2-X-thiazole(1,3); C, 3-X-pyrazole.

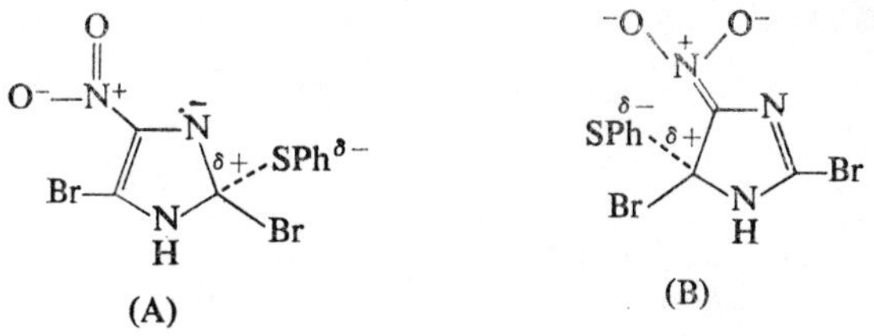

(A) (B)

Fig. 76. Transition states for replacement of 2-bromine (A) and 5-bromine (B) of 2,5-dibromo-4-nitroimidazole by thiophenoxide ion.

few with rate data. Thus Goerdeler and Heller *[101]* have shown that in reaction with piperidine in ethanol, 5-chloro-1,2,4-thiadiazole is somewhat more reactive than 1-chloro-2,4-dinitrobenzene or 2-chloropyrimidine. Rate ratios at 30° are 1:0.73:0.0178. They also demonstrated the much lower reactivity of a 3-chlorine atom, and though the comparison also involved the presence and change in position of a phenyl group, the electronic effect of this group is slight *[6]* and the rate ratio for the two positions ($1:7.45 \cdot 10^{-5}$) may be regarded as a good approximation to a value for the direct comparison. Approximate ratios for the two chlorothiadiazoles and the dinitro and pyrimidine compound are thus $1:1:10^{-2}:10^{-4}$. Calculations of the π-electron densities in the thiadiazole system by Zahradnik and Koutecky *[102a]* are in general accord with the data quoted above. It is possible to infer from the comparisons made between the thiazole and thiadiazole systems, that the thiophen system is more susceptible to nucleophilic substitution than the benzene system, and this is expected from the prediction that the thiophen should be more reactive than the furan system, and the experimental evidence that the latter is a little more reactive than benzene *[21,95]*.

In view of earlier discussion on the increase in reactivity consequent on increasing the electronegativity of a ring hetero-atom, such increases are to be expected in these systems as a result of similar changes. In this general context the recently discovered dithiolium salts *[102b, 103–107]* are of interest. Both 1,2- and 1,3-dithiolium salts are known (Fig. 77A and B) and in these compounds the two sulphur atoms are equivalent.

The most reactive positions are expected to be the 3-(5-)position of 1,2-dithiolium salts and the 2-position of 1,3-dithiolium salts, and the

References p. 306

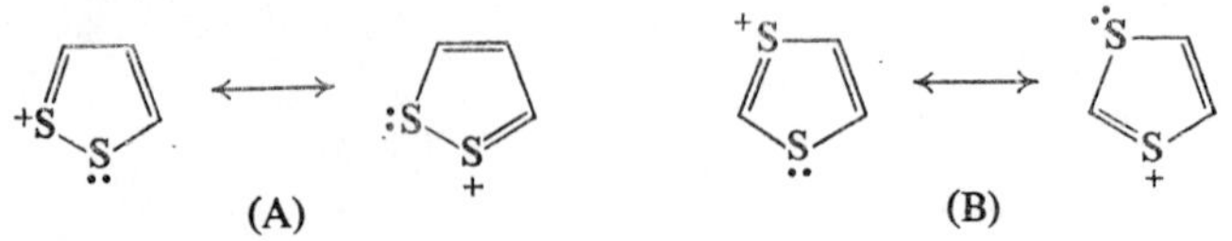

Fig. 77. Dithiolium salts (A, 1,2-; B, 1,3-). The equivalence of the two sulphur atoms is indicated by showing canonical forms.

general evidence is in accord with this *[105]*, though addition and ring-fission reactions cause complications. The replacement of a 3-methylthio group in 1,2-dithiolium salts by amines has been reported *[103]*. An unusual example, indicating high reactivity, is the replacement of a 2-methylthio group in 1,3-dithiolium salts by a dimethylaminophenyl group *[103]*; the nucleophilic atom being the carbon atom in the benzene ring *para* to the dimethylamino group. These salts and their derivatives have recently been reviewed *[107]*.

(*c*) *Fused polycyclic heteroaromatic systems*

More information is available for systems with fused benzene rings than for the simpler monocyclic systems, while the number of possible systems greatly increases when hetero-atoms are present in two or more rings. However steric and inter-annular electronic factors introduce complications and it is preferable to draw conclusions from monocyclic systems and to see how these must be modified in fused systems, including the modification already considered in fused benzenoid systems (Chapter 4, pp. 92, 110).

Amstutz *et al.* *[45]* have measured the mobility with piperidine of bromine or chlorine in positions 2 to 8 in the monohalogenoquinolines. By comparison with their data *[69]* for the corresponding 1- or 2-chloro- or bromo-naphthalenes, the effects of the ring nitrogen in the 1-position on all other positions in the ring system are readily gauged. Details are given in Table 60, which also includes data for 1-chloroisoquinoline, 9-chloroanthracene, and 9-chloroacridine.

In the halogenoquinoline compounds, one would predict the greatest effect on halogen in the same ring as the nitrogen, in the order 4- > 2- ≫ 3-position, with π-electron density and the stabilisation of the arenide system by locating negative charge on nitrogen as controlling factors. From discussions and results for fused benzenoid systems (Chapter 4, pp. 92, 110), the relatively weak conjugation between the rings is expected to result in a markedly reduced effect of the nitrogen on reactivity of halogen in the adjacent benzene ring. It is however greatest when halogen is at the 5- or 7-position, for then in the transition state the negative charge of the hetarenide system can be placed on nitrogen (Fig. 78A and B) whereas it is not possible for substitution of halogen at the 6- or 8-position relative to the nitrogen (Fig. 78C and D). One would however also predict σ-inductive stabilisation in the order 8- > 7- > 6- > 5-, so that the 6-position is the least reactive, and the order 8- to 5- and 7-position depends on the relative magnitudes of the considerable σ-inductive and weakened inter-annular conjugative effects. The results of Table 60 are in accord with this simple discussion.

The experimental order 2- > 4- ≫ 3-position involves a slight inversion of the order of the two most active positions. This could easily be a consequence of a mild steric effect of the second ring on substitution at the 4-position, absent for substitution at the 2-position, and/or a hydrogen-bonding type catalysis (see above). It is significant as regards the steric factor that the value of $\Delta S^{\ddagger}$ for substitution of 1-chloro-2,4-dinitronaphthalene with methoxide ion is low *[72d]*,

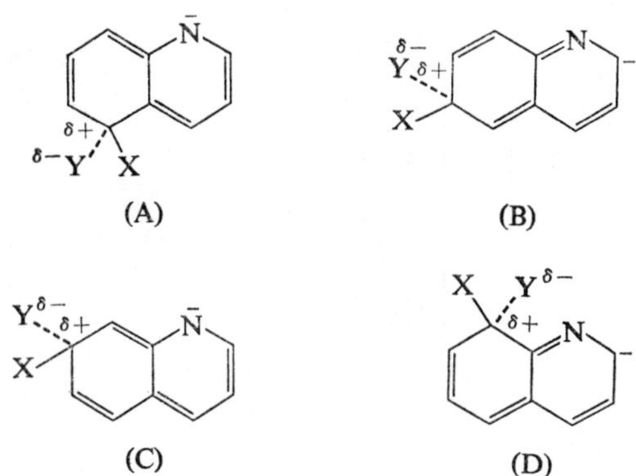

Fig. 78. Transition states for substitution by a nucleophile Y⁻ of 5-, 6-, 7-, and 8-X-quinoline (A, B, C, and D respectively).

TABLE 60

REACTIVITY OF 1- AND 2-BROMO- AND 1- AND 2-CHLORO-NAPHTHALENE, 2- TO 8-BROMO- OR CHLORO-QUINOLINE, 1-CHLOROISOQUINOLINE, 9-CHLOROANTHRACENE, AND 9-CHLOROACRIDINE AT 50° WITH PIPERIDINE AS REAGENT, AND AS SOLVENT EXCEPT WHERE SHOWN *[10,45,66,69]*

	Kinetic data							
	k_2 (l·mole^{-1}·sec^{-1})	Substituent rate factor S.R.F. or f *[51]*	Approx. σ^- of ring N[b]	$\Delta E^{\ddagger}$ (kcal·mole^{-1}) ($\Delta\Delta E^{\ddagger}$)	$\log_{10} B$ ($\Delta\log_{10} B$)	$\Delta S^{\ddagger}$ (e.u.) ($\Delta\Delta S^{\ddagger}$)	$T\Delta S^{\ddagger}$ (kcal·mole^{-1}) [$\Delta(T\Delta S^{\ddagger})$]	$\Delta G^{\ddagger}$ (kcal·mole^{-1}) ($\Delta\Delta G^{\ddagger}$)
Naphthalene 1-Br	$5.62 \cdot 10^{-13}$	1	0	25.0 (0)	4.65 (0)	– 39.4 (0)	– 12.7 (0)	37.7 (0)
Naphthalene 1-Cl	$1.02 \cdot 10^{-13}$	1	0	25.4 (0)	4.18 (0)	– 41.6 (0)	– 13.4_5 (0)	38.8_5 (0)
Naphthalene 2-Br	$1.15 \cdot 10^{-12}$	1	0	25.0 (0)	4.96 (0)	– 38.0 (0)	– 12.3 (0)	37.3 (0)
Naphthalene 2-Cl	$3.11 \cdot 10^{-13}$	1	0	23.1 (0)	3.11 (0)	– 46.5 (0)	– 15.0 (0)	38.1 (0)
Quinoline 2-Cl[a]	$3.52 \cdot 10^{-6}$	$1.13 \cdot 10^{7}$	1.18	13.8 (– 9.3)	3.88 (0.77)	– 42.9 (3.6)	– 13.9 (1.1)	27.7 (– 10.4)
Quinoline 3-Br	$3.20 \cdot 10^{-10}$	$2.78 \cdot 10^{2}$	0.41	21.6 (– 3.4)	5.11 (0.15)	– 37.3 (0.7)	– 12.0_5 (0.25)	33.6_5 (– 3.6_5)
Quinoline 4-Cl	$6.10 \cdot 10^{-7}$	$5.98 \cdot 10^{6}$	1.13	16.1 (– 9.3)	4.67 (0.49)	– 39.3 (2.3)	– 12.7 (0.7_5)	28.8 (– 10.0_5)

Quinoline 5-Br	$7.29 \cdot 10^{-11}$	$1.29 \cdot 10^{2}$	0.35	22.0	4.74	− 39.0	− 12.6	34.6
				(− 3.0)	(0.09)	(0.4)	(0.1)	(− 3.1)
Quinoline 6-Br	$5.67 \cdot 10^{-11}$	$4.93 \cdot 10^{1}$	0.28	23.9	5.91	− 33.6	-10.8_5	34.7_5
				(− 1.1)	(0.95)	(4.4)	(1.4_5)	(-2.5_5)
Quinoline 7-Br	$6.55 \cdot 10^{-10}$	$5.70 \cdot 10^{2}$	0.46	21.6	5.41	− 35.9	− 11.6	33.2
				(− 3.4)	(0.45)	(2.1)	(0.7)	(− 4.1)
Quinoline 8-Br	$5.66 \cdot 10^{-10}$	$1.01 \cdot 10^{3}$	0.50	23.3	6.50	− 30.9	− 10.0	33.3
				(− 1.7)	(1.85)	(8.5)	(2.7)	(− 4.4)
Isoquinoline 1-Cl[a]	$3.83 \cdot 10^{-6}$	$3.76 \cdot 10^{7}$	1.26	13.5	3.18	− 45.9	− 14.8	28.3
	(piperidine)			(− 11.9)	(− 1.0)	(4.3)	(1.3_5)	(-10.5_5)
	$1.97 \cdot 10^{-6}$	—	—	14.5	4.11	− 41.9	− 13.5	28.0
	(ethanol)				—	—	—	—
Anthracene 9-Cl	$2.04 \cdot 10^{-11}$	1	0	20.1	2.90	− 47.4	−15.3	35.4
	(piperidine)			(0)	(0)	(0)	(0)	(0)
Acridine 9-Cl[a]	$8.75 \cdot 10^{-6}$	$4.3 \cdot 10^{5}$	0.94	15.4	5.24	− 36.6	-11.8_5	27.2_5
	(piperidine)			(− 4.7)	(2.30)	(10.8)	(3.4_5)	(-8.1_5)
	$5.22 \cdot 10^{-6}$	—	—	14.5	4.40	− 40.5	− 13.1	27.6
	(toluene)							

[a] Comparative data are relative to the corresponding halogenonaphthalene, except that 9-chloroacridine is compared with 9-chloroanthracene. [b] Values of σ^- assume $\rho = 6.0$ (*cf.* value of 5.83 with halogenopyridines at 80°; Table 58, p. 249).

and that the lower reactivity of 4- than 2-chloroquinoline (each in relation to the appropriate parent chloronaphthalene) is also $\Delta S^{\ddagger}$ dependent. The Arrhenius parameters for the two chloronaphthalenes may however not be very reliable and do not fit such a simple pattern; though the values for the two bromonaphthalenes do.

The rate ratios and σ-values resemble those found in the pyridine series (Table 56), in addition to data for the 3-position relative to the ring nitrogen with piperidine as reagent, giving a $\sigma(m\text{-N})$ value of 0.41. In the monocyclic series (pyridines) this was available with methoxide ion but not with piperidine, and the value, 0.586, is substantially higher. For substitution at the 3-position in pyridine the transition state can be expressed by several canonical forms of somewhat similar stability equivalent to an azabenzenide transition state with the negative charge greatest on the 2- or 6-position (Fig. 79, for a nucleophile Y^-).

For substitution at the 3-position in quinoline the canonical form with the negative charge of the azanaphthalene system at the 4-position has substantially greater π-delocalisation energy than that with the negative charge on the 2-position, for the former (Fig. 80A) retains one fully-benzenoid ring. Since activation by the 3-nitrogen is regarded as a σ-inductive effect of the electronegative element stabilis-

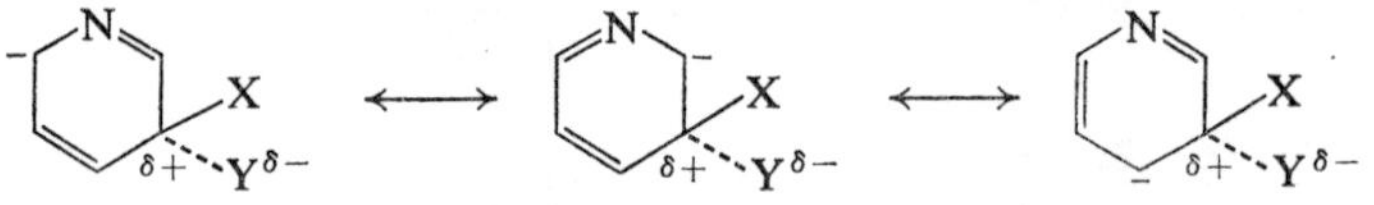

Fig. 79. Transition state representations for substitution of a 3-X-pyridine by a nucleophile Y^-.

(A) (B)

Fig. 80. Transition state representations for substitutions of 3-X-quinoline by a nucleophile Y^-.

ing the negative charge of the hetarenide system placed on neighbouring carbon, this difference between the azabenzene and azanaphthalene systems can explain the smaller σ value for 3-nitrogen in the latter.

The results for displacement of halogen in the benzenoid ring fit very well the weakened conjugative order 5- and 7- > 6- and 8- modified by a σ-inductive order 8- > 7- > 6- > 5-position. The order 7- > 5- > 6-position is a corollary, and the only uncertainty is whether the σ-inductive effect of nitrogen at the 8-position is sufficient only to put it in the range of 5- and 7-position activation, or above it.

Experimentally, reactivity at the 8-position is just greater than that at the 7-position. Particularly significant however are the $\Delta E^{\ddagger}$ values which are in the order 5-, 7- < 6- and 8-position reflecting the conjugative stabilisation available for substitution at the 5- and 7-positions. The lower $\Delta G^{\ddagger}$ value for substitution at the 8-position is thus a consequence of a less negative $\Delta S^{\ddagger}$ value. Similarly the greater part of the difference in $\Delta G^{\ddagger}$ values for substitution at the 7- and 5-positions is due to a less negative $\Delta S^{\ddagger}$ value for the 7-position.

By a simply-based localisation energy procedure, Chapman *[12]* has calculated localisation energies for nucleophilic substitution in azanaphthalenes in terms of a quantity δ which he takes as 22.4 kcal·mole^{-1} from experimental data and equates with values of activation energy. The comparison of $\Delta\Delta E^{\ddagger}$ values extracted from his Table 8 is here given as Table 61 and demonstrates a satisfactory agreement.

Chapman and Russell-Hill *[66]* report results for 1-chloroisoquinoline as well as 2-chloroquinoline with piperidine in ethanol, and comparison with the results of Amstutz *et al.* *[69]* also gives the effect of the change in solvent from piperidine to ethanol for the latter

TABLE 61

COMPARISON OF CALCD. *[12]* AND EXPTL. *[45,69]* $\Delta\Delta E^{\ddagger}$ VALUES (kcal·mole^{-1}) (HALOGENOQUINOLINES–HALOGENONAPHTHALENES)

Halogen position in quinoline	2	3	4	5	6	7	8
$\Delta\Delta E^{\ddagger}$ calcd.	−11.2	−1.8$_5$	−8.1$_5$	−2.0$_5$	−1.8$_5$	−2.8	−1.3$_5$
$\Delta\Delta E^{\ddagger}$ exptl.	− 9.3	−3.4	−9.3	−3.0	−1.1	−3.4	−1.7

compound, which can be applied to the 1-chloroisoquinoline data. It has thus been possible to estimate the kinetic results shown in Table 59 for this compound. These indicate that the activation by a nitrogen atom α- to the chlorine is similar in 1-chloroisoquinoline and 2-chloroquinoline, with σ-values of 1.26 and 1.18 respectively. With ethoxide ion as reagent they demonstrated similar reactivity for 4-chloroquinoline and 1-chloroisoquinoline (*i.e.* 1-chloro-2- and 4-azanaphthalene) again indicating similar σ-values for 2- and 4-nitrogen. Rate data for the reference compound, 1-chloronaphthalene, were not obtained, but approximate data for 2-chloronaphthalene were given and presumably the difference between the two compounds is not large *[cf. 45,69]*.

They also showed that 3-chloroisoquinoline is much less reactive than 1-chloroisoquinoline, although in each case chlorine is α to ring nitrogen. It may be estimated approximately that the rate ratio 1-chloroisoquinoline to 1-chloronaphthalene, is about $10^{9.4}$ and the σ-value for the ring nitrogen thus about 1.1. For the 3-chloro compound the values are about $10^{4.9}$ and 0.58. This difference is the result of the predictably more stable transition state in the former case, in which one ring can be fully benzenoid (Fig. 81A) whereas in the latter it cannot (Fig. 81B) and thus is specially unfavourable relative to the initial state.

Significantly the $\Delta E^{\neq}$ value for 3-chloroisoquinoline is almost 10 kcal·mole^{-1} higher than for 1-chloroisoquinoline. The difference in $\Delta G^{\neq}$ is less because of partly counteracting differences in $\Delta S^{\neq}$.

The data comparing reactions of piperidine with 9-chloroacridine and 9-chloroanthracene demonstrate a smaller activating effect γ to ring nitrogen in this system. It presumably reflects the smaller aromatic character of the central ring in anthracene. Comparative rate data for

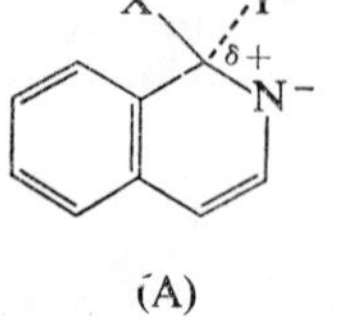

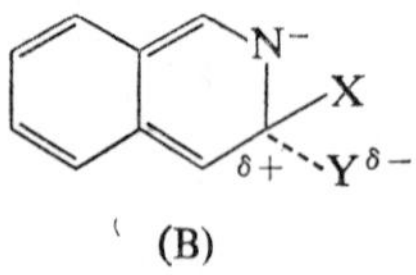

(A) (B)

Fig. 81. Comparison of transition states for substitution of 1- and 3-chloroisoquinoline (A and B respectively) by a nucleophile Y^-.

TABLE 62

REACTIONS OF SOME CHLORODIAZANAPHTHALENES AT 50° WITH PIPERIDINE (SOLVOLYSIS) OR IN ETHANOL AS SOLVENT *[66]*

	Kinetic data									
	k_2		Substituent rate factor	Substituent constant						
	Solvent	(l·mole^{-1}·sec^{-1})	S.R.F. or f *[51]*	σ^{-} [b]	σ^{-} (calcd.) [c]	σ^{-} (calcd.) [d]	$\Delta E^{\ddagger}$ (kcal·mole^{-1})	$\Delta S^{\ddagger}$ (e.u.)	$T\Delta S^{\ddagger}$ (kcal·mole^{-1})	$\Delta G^{\ddagger}$ (kcal·mole^{-1}) ($\Delta\Delta G^{\ddagger}$)
Naphthalene 1-Cl	Piperidine	$1.02 \cdot 10^{-13}$	1	0			25.4	−41.6	−13.45	38.8_5 (0)
Naphthalene 2-Cl	Piperidine	$3.11 \cdot 10^{-13}$	1	0			23.1	−46.5	−15.0	38.1 (0)
Diazanaphthalenes [a]										
1-Chlorophthalazine	EtOH	$1.44 \cdot 10^{-4}$	—				11.8	−42.0	-13.5_5	25.3_5
(2,3-diaza-)	Piperidine	$2.80 \cdot 10^{-4}$	$2.74 \cdot 10^{9}$	1.57	1.66	1.58	—	—	—	24.8_5 (−14.0)
2-Chloroquinazoline	EtOH	$2.82 \cdot 10^{-3}$	—				11.1	−37.8	−12.2	23.3
(1,3-diaza-)	Piperidine	$5.48 \cdot 10^{-3}$	$1.76 \cdot 10^{10}$	1.71	1.98	2.00	—	—	—	22.8 (−15.3)
4-Chloroquinazoline	EtOH	2.02	—				7.0	−37.5	−12.1	19.1
(1,3-diaza-)	Piperidine	3.92	$3.84 \cdot 10^{13}$	2.26	1.78	2.16		—	—	18.6 (-20.2_5)
2-Chloroquinoxaline	EtOH	$3.86 \cdot 10^{-4}$	—				11.3	−40.9	−13.2	24.5
(1,4-diaza-)	Piperidine	$7.50 \cdot 10^{-4}$	$2.41 \cdot 10^{9}$	1.56	1.72	1.58		—	—	24.0 (−14.1)

[a] Values at 50° in EtOH are converted to values in piperidine by using the relationship found with 2-chloroquinoline (see Table 57).
[b] By assuming $\rho = 6.0$. [c] From values for monocyclic diazines and *p*-nitrophenoxide ion (Table 59, p. 253). [d] From values for monocyclic azines with methoxide ion (Table 53, p. 244).

TABLE 63

REACTIONS OF SOME CHLORODIAZANAPHTHALENES AT 50° WITH ETHOXIDE ION IN ETHANOL *[66]*[a]

	Kinetic data								
	k_2 (l·mole^{-1}·sec^{-1})	Substituent rate factor (S.R.F. or f)	Substituent constant			$\Delta E^{\ddagger}$ (kcal·mole^{-1})	$\Delta S^{\ddagger}$ (e.u.)	$T\Delta S^{\ddagger}$ (kcal·mole^{-1})	$\Delta G^{\ddagger}$ (kcal·mole^{-1}) ($\Delta\Delta G^{\ddagger}$)
			σ^-	σ^- (calcd.)[c]	σ^- (calcd.)[d]				
Naphthalene 1-Cl[b]	$\sim 10^{-14}$	1	0	0	0	—	—	—	~ 40
Naphthalene 2-Cl	$\sim 2._7 \cdot 10^{-14}$	1	0	0	0	~ 39	-2.0	-0.6	~ 39.6
Diazanaphthalenes									
1-Chlorophthalazine (2,3-diaza-)	$2.53 \cdot 10^{-2}$	$2._5 \cdot 10^{12}$	$1._4$–$1._5$	(1.66)	(1.58)	16.5	-16.9	-5.4_5	21.9_5 (~ -18)
4-Chlorocinnoline (1,2-diaza-)	$5.88 \cdot 10^{-2}$	$5._9 \cdot 10^{12}$	$1._5$	(1.69)	(1.74)	15.8	-17.3	-5.6	21.4 (~ -18)
2-Chloroquinazoline (1,3-diaza-)	$4.32 \cdot 10^{-2}$	$1._6 \cdot 10^{12}$	$1._4$–$1._5$	(1.98)	(2.00)	16.8	-15.9	-5.1_5	21.9_5 (~ -18)
2-Chloroquinoxaline (1,4-diaza-)	$9.38 \cdot 10^{-2}$	$3._5 \cdot 10^{12}$	$1._5$	(1.72)	(1.58)	15.4	-18.2	-5.9	21.4 (~ -18)

[a] We assume $\rho = 8.5$; *cf.* for azachlorobenzenes, $\rho = 8.47$ at 50° with methoxide ion. [b] Assuming about the same k_2 ratio and $\Delta\Delta G^{\ddagger}$ for 1- compared with 2-chloronaththalene as were found with piperidine (Table 62). [c] From values for monocyclic diazines with *p*-nitrophenoxide ion (Table 59, p. 253). [d] From values for monocyclic azines with methoxide ion (Table 53, p. 244).

6-halogenophenanthridines and the corresponding phenanthrenes would be of interest in this connection.

These workers have also reported results for a number of diazanaphthalenes *[66]*, which can be compared with 1- or 2-chloronaphthalene as parent compounds *[45,69]*, and data for the reactions with piperidine are given in Table 62. Data for reactions with ethoxide ion are given in Table 63, though the information concerning 1- and 2-chloronaphthalene is not very precise. The results are parallel with those for the azabenzenes. As usual the large increase in reactivity due to the ring nitrogen atoms are $\Delta E^{\ddagger}$ dependent. It is interesting to note however that 2-chloroquinazoline is less reactive and 4-chloroquinazoline more reactive than would be expected on the basis of data from monocyclic azines. This difference is the result of a predictably less stable transition state in the former case (Fig. 82A) than in the latter (Fig. 82B).

From Fig. 82A it can be seen that the 3-nitrogen is less effective than the 1-nitrogen since placing the negative charge of the hetarenide system on the former means neither ring is benzenoid, whereas when placed on the latter, one ring can be fully benzenoid. A corresponding explanation has been given for the low reactivity of 3-chloroisoquinoline. The unusually high reactivity of 4-chloroquinazoline, in which the negative charge of the hetarenide system can be placed on

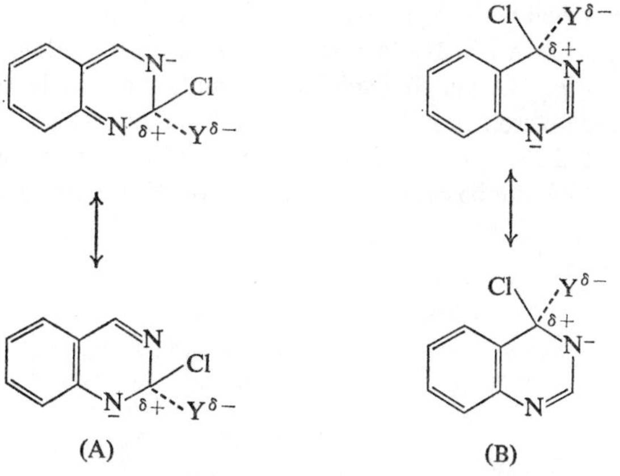

(A)

(B)

Fig. 82. Comparison of transition states for reactions of 2- and 4-chloroquinazoline with a nucleophile Y^- (A and B respectively).

References p. 306

both nitrogens, presumably reflects a smaller difference than in monocyclic compounds in delocalisation energy of arene and arenide system *[cf. 7b,c]*. Table 63 gives similar results to those in Table 62 except for the inclusion of 4-chlorocinnoline which is a little less reactive than expected. With ethoxide ion all four substrates reported have rather negative $\Delta S^{\ddagger}$ values, but since there is no $\Delta S^{\ddagger}$ value for 1-chloronaphthalene, and that for 2-chloronaphthalene may not be very reliable, it is not certain that the low values apply only to the azanaphthalenes.

The 4-position reactivity order, quinazoline > cinnoline > quinoline has been semi-quantitatively reported by Kenneford *et al.* *[108]* in a wide variety of reactions, *e.g.* with methoxide in methanol, with water as reagent and solvent; and with replaced groups such as chlorine and amino groups.

For alkoxide exchange the semi-quantitative reactivity order, 2-, 4-quinazoline > 4-cinnoline ~ 1-phthalazine > 2-quinoxaline > 2-quinoline has been reported *[109]*. This fits the data of Table 61, augmented by Table 62.

Corresponding to the low reactivity of 3-chloroisoquinoline and 2-chloroquinazoline, and for the same reasons, it is reported that 3-halogenocinnolines too have unusually low reactivity *[110]*.

There are numerous semi-quantitative reports, based on temperature of reaction and yield, quoted by Shepherd and Fedrick *[2]* which correspond well with the above general discussion.

Recent work with 2-chlorobenzothiazole and 2-chlorobenzimidazole (see Fig. 83A and B) leads to information about the two parent monocyclic heteroaromatic systems. Harrison and Ralph *[111]* have demonstrated the very low reactivity of 2-chlorobenzimidazole with reagents which are bases, so that the heterocyclic compound is in the

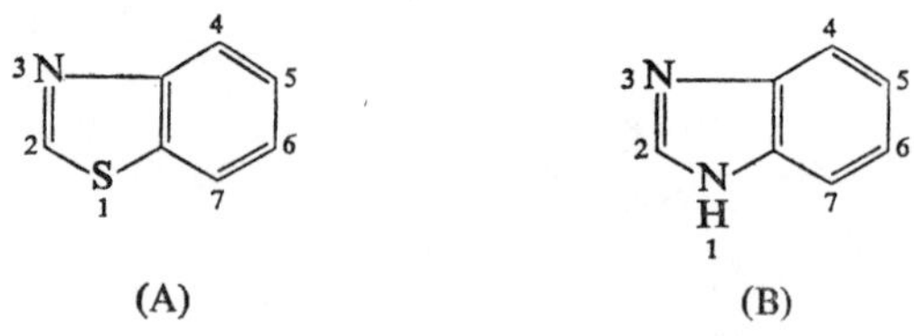

Fig. 83. A, benzothiazole; B, benzimidazole.

form of the conjugate base. Even nitro-substituted compounds did not react in the conditions reported. In contrast, the chlorine of 2-chloro-1-methylbenzimidazole was readily replaced. Steric hindrance with bulky alkoxides was also demonstrated with this compound. While there is no kinetic comparison, the reactivity of 2-chlorobenzothiazole reported by Ricci *et al.* *[112]* is greater than that of 1-methyl-2-chlorobenzimidazole, as would be expected from the relative electron-releasing power of nitrogen and sulphur, a factor involved in the π-electron densities of the 5-membered rings.

Before considering the diazanaphthalenes with one nitrogen in each ring (pyridopyridines or naphthyridines) it is convenient to compare the diazanaphthalenes with both nitrogens in the same ring not only with corresponding naphthalenes as above, but also with corresponding benzenes. This gives a measure of the activating power of the fused benzene ring, and though this is essentially the same as a comparison of chlorobenzene with chloronaphthalenes which was discussed in Chapter 4 (pp. 92, 110), it will give a basis for comparison of the effect of a fused pyridine ring when sufficient rate data become available for such compounds. From the data of Table 56, the second order rate constant (k_2) for reaction of piperidine with chlorobenzene at 50° is $5.83 \cdot 10^{-15}$ l·mole^{-1}·sec^{-1} and the value of $\Delta E^{\ddagger}$ is 28.6 kcal·mole^{-1}. Thus a *meta–para* fused ring, as in 2-chloronaphthalene, activates by a factor of *ca.* 53, with $\Delta\Delta E^{\ddagger}$ –5.5 kcal·mole^{-1}; and an *ortho–meta* fused ring, as in 1-chloronaphthalene, activates by a factor of 17.5, with $\Delta\Delta E^{\ddagger}$ –3.2 kcal·mole^{-1}.

There are also a few results of a more direct nature. Data from Tables 56 and 58, making an approximate allowance for change of solvent, permit direct comparison of 2- and 4-chloroquinoline with 2- and 4-chloropyridine. The rate ratios at 50° are about 12 for the 2-position and 0.2 for the 4-position. While a lower value for the 4-position would be expected, a presumed steric effect large enough to counteract the mild activation by the fused ring is surprising. Indeed both values seem low.

From the data of Chapman and Russell-Hill *[66]* and those in Table 58, a comparison of 2- and 4-chloroquinazoline with 2- and 4-chloropyrimidine is possible. Rate ratios are about 2.6 and 500 respectively. The low reactivity of the former has been explained above, but the high value for the 4-position, even allowing for its

References p. 306

rather approximate nature, is surprising in view of the results for naphthalene/benzene and quinoline/pyridine comparisons. The very small value of the former is paralleled in reaction with ethoxide ion *[66]*, for which the rate ratio at 50° is only 1.8. The data of Miller and Williams *[72d]* for 1-chloro-2,4-dinitrobenzene and 1-chloro-2,4-dinitronaphthalene are also relevant, and lead to a rate ratio at 50° equal to 2.77 for an *ortho–meta* fused ring. By use of the data of Amstutz *et al.* *[45b]* the reactivity with piperidine of 2-chlorothiazole and 2-chlorobenzothiazole may be compared. At 50° the rate ratio, due to the *meta–para* fused benzene ring, is 116 and $\Delta\Delta E^{\ddagger}$ is –1.9 kcal·mole^{-1}.

In summary and as a rough guide, a benzene ring fused *ortho–meta* activates by a factor of about 10^0–10^1, and when fused *meta–para* about 10^1–10^2 (in some cases a little more). The rather wide variation in the effect of the fused ring is unfortunate since the overall effect of a fused pyridine ring ought to be weighted and assessed in relation to the effect of a fused benzene ring. This is done below, though necessarily approximately.

As regards the naphthyridines (pyridopyridines) we may concern ourselves with the 5-, 6-, 7- and 8-nitrogen atom in their effect on one or other of the halogenoquinolines or halogenoisoquinolines, cognisance being taken of the unfavourable arrangement of 2,3-leaving group and activating group in naphthalene-type systems; and we may also consider activation as by a pyridine ring.

From the data of Amstutz *et al.* *[69]* the second order rate constant at 50° for reaction of bromobenzene with piperidine as reagent and solvent may be estimated at $7.5 \cdot 10^{-13}$ l·mole^{-1}·sec^{-1}. Comparison with data for 5-, 6-, 7- and 8-bromoquinoline in Table 59 then gives some measure of the effect of a pyridine ring. The ratio for 5-bromoquinoline, with an *ortho–meta* fused pyridine ring and nitrogen in a conjugatively favourable position, is 97, which is considerably greater than the range 1–10 for an *ortho–meta* fused benzene ring. For 6-bromoquinoline with a *meta–para* fused pyridine ring, but conjugatively unfavourable position of nitrogen, the ratio is 76, which is within the range 10 to 100 for a *meta–para* fused benzene ring. For 7-bromoquinoline with a *meta–para* fused pyridine ring, conjugatively favourable position of nitrogen, and larger σ-inductive effect than with the 5-bromo compound, the ratio is 870, substantially greater than the range 10 to 100 for a *meta–para* fused benzene ring. Finally for 8-bromoquinoline with an *ortho–meta* fused pyridine ring, and conjugatively unfavourable position of nitrogen, but largest σ-inductive effect, the ratio is 750 which is much greater than the range 1 to 10 for an *ortho–meta* fused benzene ring.

In the pyridines the reactivity order with methoxide ion, ethoxide ion, or piperidine as reagent is 4- > 2- ≫ 3-position. In quinolines with piperidine as reagent the rate order is 2- > 4- ≫ 3-position. Making allowance for differences in the reactivity of the parent 1- and 2-chloronaphthalenes removes much of the difference which leads to the changeover at the 2- and 4-position, but activation by the 2-nitrogen is nevertheless slightly greater. To a first rough approximation therefore, for discussion of reactivity of the pyridopyridines, we may assume the equality of the reactivity of 1-X-isoquinolines, 2-X-quinolines, and 4-X-quinolines as parent compounds. For reasons already given however, 3-X-isoquinolines are much less reactive parent compounds, about the same as 3-X-quinolines and 4-X-isoquinolines even though the latter are activated only in the same ring by *meta*-nitrogen.

Supported by the data in Table 59 for 2- to 8-halogenoquinolines, we may assess the activating power of nitrogen in the second ring on the basis of (*a*) whether or not the negative charge of the hetarenide system can be placed on it, (*b*) its σ-inductive effect on the replaced group X in the neighbouring ring, strongest for X in the 1- or 4-position. It is implied in this simple assessment that steric requirements of entering and replaced groups are not so large that the 5- and 8-position have reduced reactivity, and that there is no quaternisation of, or hydrogen-bonding to, nitrogens.

In systems based on 1-X-isoquinolines, the interannular conjugation order 5-, 7- > 6-, 8-position does not simply match the σ-inductive order 8- > 7- > 6- > 5-position. The resulting reactivity order is then 8- ≥ 7- > 5- > 6-position, so that the reactivity of systems based on 1-X-isoquinolines is 1-X-2,8-diaza- ≥ 1-X-2,7-diaza- > 1-X-2,5-diaza- > 1-X-2,6-diaza-naphthalenes.

By a similar analysis of systems based on 2-X- and 4-X-quinolines the reactivity orders, (*i*) 2-X-1,8-diaza- > 2-X-1,6-diaza- ≥ 2-X-1,5-diaza ~ 2-X-1,7-diaza-naphthalene, and (*ii*) 4-X-1,5-diaza- ≥ 4-X-1,6-diaza- > 4-X-1,8-diaza- > 4-X-1,7-diaza-naphthalene are obtained.

References p. 306

The most reactive of these systems are 1-X-2,7-diaza-, 1-X-2,8-diaza-, 2-X-1,8-diaza-, 4-X-1,5-diaza-, and 4-X-1,6-diaza-naphthalene.

Analysis of systems based on the less reactive 3-X- and 4-X-isoquinoline and 3-X-quinoline leads to reactivity orders, (*i*) 3-X-2,5-diaza > 3-X-2,7-diaza- ⩾ 3-X-2,6-diaza- ~ 3-X-2,8-diaza-naphthalene; (*ii*) 4-X-2,5-diaza- ⩾ 4-X-2,6-diaza- > 4-X-2,8-diaza- > 4-X-2,7-diaza-naphthalene; (*iii*) 3-X-1,5-diaza- > 3-X-1,7-diaza- ⩾ 3-X-1,6-diaza- ~ 3-X-1,8-diaza-naphthalene.

Among the systems based on the less reactive parent compounds, the most reactive are 3-X-1,5-diaza-, 3-X-2,5-diaza-, 4-X-2,5-diaza-, and 4-X-2,6-diaza-naphthalenes. There are no reliable quantitative results to test these predicted orders of reactivity, which differ somewhat from those suggested by Shepherd and Fedrick *[2]*.

Corresponding to the above analyses, it is possible to predict mobility within individual systems.

1,5-Diazanaphthalenes. In this system the 2-, 3- and 4-position are equivalent to the 6-, 7- and 8-position. The predicted order is 4- > 2- > 3-position. This is supported by experimental evidence *[113–115]*, though the relative reactivity at the 2- and 4-position is probably affected by such factors as hydrogen-bonding by a reagent facilitating attack at the 2-position.

1,6-Diazanaphthalenes. In this system the predicted order is 4- > 2- > 5- > 7-, 8- > 3-position. The high reactivity of the 4-position has been reported *[116,117]*.

2,5-Diazanaphthalenes. This is the same system as 1,6-diazanaphthalenes, and the same order renumbered applies. High reactivity of the 1-position (equivalent to the 5-position in the 1,6-diazanaphthalene numbering) has been reported *[118a]*.

1,7-Diazanaphthalenes. In this system the predicted order is 8- > 2- ⩾ 4- > 3- > 5- ⩾ 6-position. The reactivity of the 4-position has been recorded *[116,117,118b]*.

2,8-Diazanaphthalenes. This is the same system as 1,7-diazanaphthalenes and the same order renumbered applies.

1,8-Diazanaphthalenes. In this system the 2-, 3- and 4-position are equivalent to the 7-, 6- and 5-position. The predicted order is 2- > 4- > 3-position. The reactivity at the 2- and 4-position, with some indication of the order 2- > 4-position has been reported *[114,119–121]*.

2,6-Diazanaphthalenes. In this system the 1-, 3-, and 4-position are equivalent to the 5-, 7- and 8-position. The predicted order is 1- $>$ 4- $\geqslant$ 3-position. The existence of this system has only recently been reported *[122]* and replacement of chlorine atoms in the 1- and 3-position by hydrazine is mentioned.

2,7-Diazanaphthalenes. In this system the 1-, 3- and 4-position are equivalent to the 8-, 6- and 5-position. The predicted order is 1- $>$ 3- $>$ 4-position. There is supporting evidence for high reactivity at the 1-position *[118c,123–125]*.

The number of possible systems multiplies enormously with increase in number of hetero-atoms and of fused rings, but some confidence can be placed on estimates based on methods applied earlier. As examples of more complex systems the purine and pteridine systems are considered. The first consists of a π-deficient ring derived from a π-excessive type (see footnote to p. 258) fused to a π-deficient ring. The second consists of two fused π-deficient rings. Both are of biological importance.

The purine system consists of fused pyrimidine and imidazole rings (Fig. 84). Sutcliffe and Robins *[126a]* have considered qualitatively the reactivity of 2,6,8-trichloropurine and suggested the reactivity order 6- $>$ 2- $>$ 8-position. The order in 7- or 9-methyl derivatives, though it varies with the reagent, may be summarised as 8- $\sim$ 6- $>$ 2-position. The shift in reactivity of a substituent at the 8-position is presumably in accord with the inability of the methyl derivatives to react with basic nucleophiles to form the less reactive conjugate bases. In support, 8-chloropurine which can form the conjugate base, is very unreactive with such nucleophiles. However the orientation shifts probably also involve acid-type catalysis, which is believed to occur by protonation of nitrogen in the imidazole ring *[127b]*. This is supported by a definite shift to maximum reactivity at the 8-position in acidic

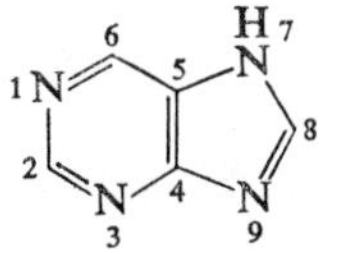

Fig. 84. Purine system (there is a tautomeric form with hydrogen at the 9-position). Numbering as in Chemical Abstracts.

conditions *[126b]*. It should not be forgotten that discussions based on data for unsymmetrical substrates such as 2,6,8-trichloropurine ought to allow for mutual activation by the substituents: this has been pointed out by Pullman in relation to this particular substrate *[127a]*.

Barlin and Chapman *[128]* have recently measured the reactivity, towards piperidine in ethanol, of 2-, 6-, and 8-chloro-9-methylpurine, in which *N*-methylation precludes formation of a heterocyclic conjugate base. The rate ($\Delta G^{\ddagger}$) order is 6- > 8- ≫ 2-chloro compound, but the $\Delta E^{\ddagger}$ order of reactivity is 8- > 6- ≫ 2-chloro compound. Values of $\Delta S^{\ddagger}$ are similar for 2- and 6- and low for the 8-chloro compound. They point out that the (neutral) *N*-methyl group of the imidazole ring may reduce reactivity of the 8-chlorine in that ring towards substitution by nucleophiles compared with that in the parent purine (N–H system). Theoretical calculations of different workers do not agree. Miller *et al.* *[129]* give the order 6- > 2- > 8-position, whereas Mason *[130]* gives the order, 8- ~ 6- > 2-position. Pullman *[127b]* also suggests this order, based on calculation of localisation energies, and further suggests caution in utilising π-electron densities for discussion of reactivity.

Earlier discussions suggest that the imidazole ring is less susceptible to nucleophilic attack than the pyrimidine ring. However in the fused purine system, canonical forms of the reaction intermediates illustrate that for substitution of an 8-X-purine, the negative charge of the hetarenide system can be placed on the 3-, 6-, and 9-nitrogen atoms (Fig. 85A), and in one of these canonical structures with negative charge on the 9-nitrogen atom, the pyrimidine ring remains fully aromatic suggesting particularly powerful enhancement of reactivity by the 9-nitrogen atom. For substitution of a 2-X- or 6-X-purine, however, the negative charge can be placed only on the two pyrimidine nitrogens (Fig. 85B and C). The reactivity order 8-X- > 6-X-, 2-X-purine is thus predicted (the possible effect of a 9-methyl group has been mentioned). The large difference 6-X-≫ 2-X-purine may be predicted however as a further example of the unfavourable 2,3- (in the purine system as 1,2-) leaving group and activating group arrangement.

Fig. 85B demonstrates that in substitution of a 6-X group the negative charge of the hetarenide system can be placed on both 1- and 3-

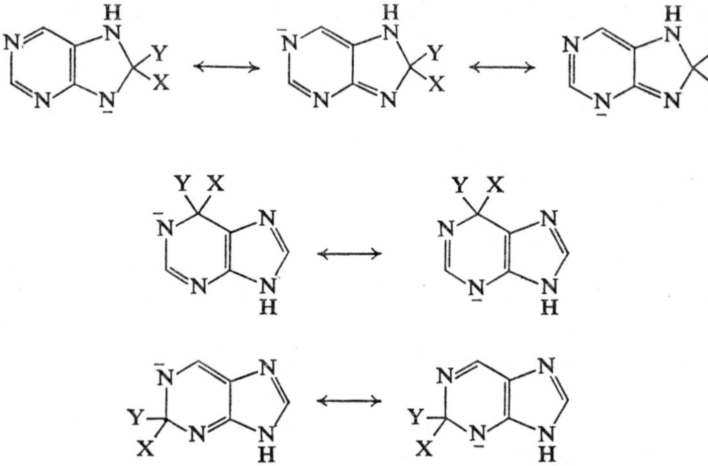

(A)

(B)

(C)

Fig. 85. Canonical forms of intermediate-complexes for substitution of 8-X-, 6-X- and 2-X-purine (A, B, and C respectively) by a nucleophile Y^-.

nitrogen atoms and leave the imidazole ring in its normal quasi-aromatic form, whereas Fig. 85C demonstrates that in substitution of a 2-X-group, the negative charge can be placed on the 3-nitrogen but *not* on the 1-nitrogen and leave the imidazole ring aromatic.

The larger σ-inductive effect of imidazole nitrogen at the 6-position, α to the ring junction, than at the 2-position which is β to it, may also contribute to the 6- $\gg$ 2-position order.

Cresswell and Brown *[131]* have demonstrated the higher reactivity of the purine-1-oxide than the purine system with nucleophiles such as hydroxide ion, water (acid-catalysed) and β-hydroxyethylamine.

The pteridine system (Fig. 86) consists of two fused π-deficient rings. Substitution at any carbon atom can place the negative charge of the hetarenide system on three nitrogen atoms. It should thus be a highly reactive system, but more so for substitution in the pyrimidine

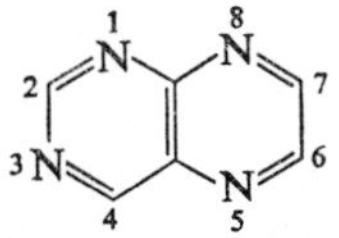

Fig. 86. Pteridine system. Numbering as in Chemical Abstracts.

References p. 306

ring, for intra-annular activation is greater than inter-annular activation. Due note must however also be taken of the diminished activation of the 2-position by the 3-nitrogen (the unfavourable 2,3-activating and replaced group arrangement). The reactivity order based on this and previous discussions is 4- $\gg$ 6-,7- $>$ 2-position, but consideration of S_N reactions in pteridines is complicated by alternative reactions involving ring addition and ring fission, and is particularly obscure if this is followed by recyclisation *[132a]*.

The predicted order needs some further comment and elaboration. Activation at the 4-position utilises fully activation by 1- and 3-nitrogens (α and γ) plus inter-annular activation by the pyrazine ring (5- and 8-)nitrogens and includes a substantial σ-inductive effect from the 5-nitrogen. Activation at the 2-position utilises fully activation by the 1-nitrogen (α) but the activating power of the 3-nitrogen, though also α, is reduced by the unfavourable 2,3-activating and replaced group arrangement to approximately that of a β-nitrogen. The resultant conjugation by the two nitrogens of the pyrimidine ring is thus equivalent only to that at the 6- and 7-positions by the two nitrogens (α- and β-) in the pyrazine ring. However the inter-annular effect of the pyrimidine ring on reactivity at the 6- and 7-position is presumably somewhat larger than that of the pyrazine ring on the 2-position, and not upset by weak σ-inductive effects, so that the overall order is as given above, *viz.* 4- $\gg$ 6-, 7- $>$ 2-position. The difference between the 6-, 7- and 2-position is likely to be rather small, however, and *all* positions are reactive.

The differences in the sum of σ-inductive effects of the two pyrimidine nitrogens acting on the 6- and 7- position may be sufficient to give the order 7- $>$ 6-position.

While precise rate data are lacking, there is clear evidence to support these conclusions. The order 4- $>$ 6- $>$ 2-position in aqueous solvolysis has been demonstrated by Albert and his co-workers *[132a]*. Generally high reactivity at the 6- and 7-positions is also reported but with definite indications of higher reactivity at the 7- than the 6-position *[132b,f]*. The same workers have also demonstrated competitive 3,4-hydration (nucleophilic addition of water across the 3,4-double bond) and substitution in these compounds. The order 4- $>$ 2-position with a wide variety of reagents has been demonstrated *[133,134a]*.

Reactivity orders may be upset in conditions where hydrogen-bonding or more direct acid catalysis occurs. The expected result is protonation of a pyrazine ring nitrogen atom in preference to one in the pyrimidine ring. The pyrazinium ring of the system so formed then has the greater reactivity and substituents at 6- and 7-positions are more easily displaced than at 2- and 4-positions. This effect could be enhanced by substituents for which the resultant increase in the base strength of the pyrazine ring increases pyrazinium-ion formation enough to outweigh any general deactivating effect of the substituent. The report that 6-amino- and 6-dimethylamino-pteridines are more rapidly hydrolysed in $0.01N$ hydrochloric acid than the 2- and 4-isomers *[132]* is readily explicable in these terms.

Shepherd and Fedrick *[2]* have suggested the reactivity order 7- > 4-position, but it is not clear on what basis. The 4-position in pteridines and the (corresponding) 6-position in purines have been compared *[134b,135]* and as expected the former is the more reactive.

The reactivity order pyridinium > pyridine-*N*-oxide > pyridine *[22]* makes it clear that very high reactivity will be found in other hererocyclic systems as a result of corresponding change in coordination of ring nitrogen. An interesting example of this is the formation of a cyanine (polymethine) dye by condensation of 2-chloro-1-ethyl-quinolinium salts with 1-ethylquinaldinium salts *[136]*. The chloro compound is the highly reactive substrate and the activity of the reagent illustrates a parallel consequence of very high activation by quaternary nitrogen, *viz.* nucleophilic reactivity of an α-carbon bearing a hydrogen atom which can be removed as a proton. The reaction is illustrated in Fig. 87.

The isoquinolinium series gives examples of very high susceptibility to reaction with nucleophiles leading to the facile formation of stable adducts *[137]*. This is illustrated for 2-cyanoisoquinolinium salts in Fig. 88, with the reagent adding at the most activated 1-position. In this reaction the electron-withdrawing power of the N^+, having the powerfully electron withdrawing cyano group attached, is even higher than in simple quaternary nitrogen compounds, while the value of $\Delta E^{\ddagger}$ is lowered relative to that of a similar reaction in a monocyclic compound because this addition can proceed leaving one ring fully benzenoid.

Both condensation and addition reactions are examples of widely

References p. 306

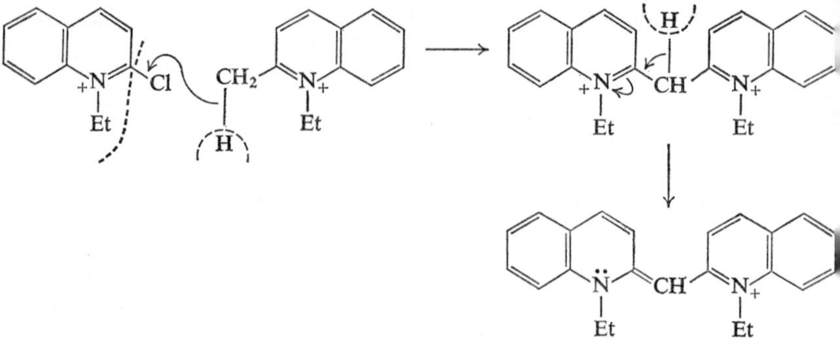

Fig. 87. Condensation of the 2-chloro-1-ethylquinolinium ion with the 1-ethylquinaldinium ion.

known reactions with close theoretical and experimental relationships with nucleophilic substitutions, which are, however, outside the scope of this book.

Quinolizinium salts have the quaternary nitrogen at a ring junction (Fig. 89) and are expected to be less reactive than quinolinium and isoquinolinium systems since the effect of the N^+ is shared between two rings. This is supported by the theoretical calculations of Acheson and Goodall *[138]*, which may be compared with those of Brown and Harcourt *[139]* for quinolinium and isoquinolinium systems. The three ions and pyridinium are compared by Acheson and Goodall. Table 64 gives the results of their calculations of localisation energies for structures suitable for nucleophilic substitution, in terms of a common energy parameter β. Thus, taking the most reactive ring position in each case, the reactivity order is: 1-isoquinolinium > 4-quinolinium > 2-pyridinium > 4-quinolizinium. The calculations of π-electron densities *[138]* lead to a similar prediction. The predicted order is in general accord with qualitative information available, but

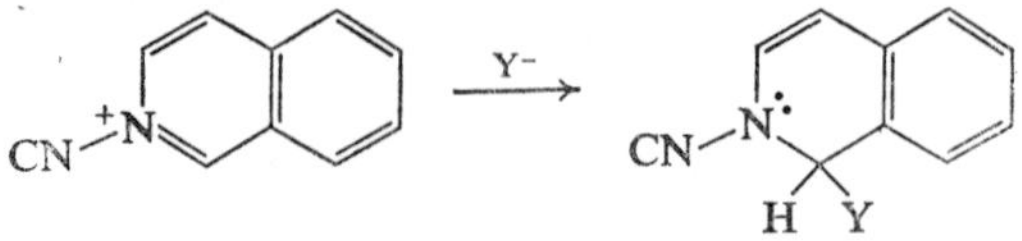

Fig. 88. Addition of a nucleophile Y^- to 2-cyanoisoquinolinium ion.

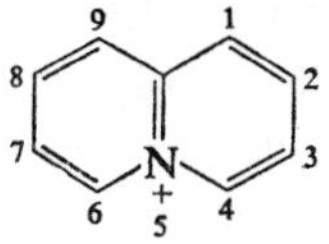

Fig. 89. Quinolizinium ion. Numbering as in Chemical Abstracts.

TABLE 64

LOCALISATION ENERGIES (UNITS OF A COMMON ENERGY PARAMETER β), TO GIVE STRUCTURES SUITABLE FOR NUCLEOPHILIC SUBSTITUTION, OF SOME MONOQUATERNARY AZA-AROMATIC SYSTEMS *[137,138]*

	Ring system										
	Pyridinium			Quino-linium		Isoquino-linium		Quinolizinium			
	2-	3-	4-	2-	4-	1-	3-	1-	2-	3-	4-
Localisation energy (β)	1.86	2.56	2.01	1.96	1.81	1.61	2.02	2.31	2.20	2.50	1.97

there are no precise quantitative results, except that Miller and Sung *[140]* have shown that in reaction with *p*-nitrophenoxide ion in methanol, the 2-chloropyridinium ion is 110 times more reactive at 0° than the 4-chloroquinolizinium ion; the difference is $\Delta E^{\ddagger}$ dependent. Fozard and Jones *[141]* have reported lack of mobility of bromine in 4-bromo-1-hydroxy-2-nitroquinolizinium betaine, but this presumably reflects the very powerful deactivating influence of the O^- group (see Chapter 4, p. 75).

(*d*) *Leaving group mobility*

There is no reason or evidence to suggest any major differences in the patterns of mobility of leaving groups in heterocyclic as compared with benzenoid systems. Thus in the normal S_N2 addition–elimination mechanism of heterocyclic substrates the characteristic order F $\gg$ Cl, Br > I should be found with the first row nucleophiles in protic solvents *[4,5,7,68a]* and in a few other cases (see Chapter 5). Similarly cationic groups should be very easily replaced *[4,5,7]*.

The known greater ease of displacement of fluorine than of the other halogens has already been reported *[26–31]*. The relative reactivity of Cl, Br and I is much less diagnostic and varies both with the degree of activation and reagent, but patterns Cl > Br > I with alkoxides and Cl < Br > I with amines, as in benzenoid compounds, have been recorded *[1,2]*. Recently Ricci *et al.* *[112]* have measured precisely the rates of reactions of 2-halogenobenzothiazoles with methoxide ion in methanol, and from their data for 25° the halogen mobility ratios, F, 1000; Cl, 1; Br, 0.75; I, 0.11_5 and $\Delta\Delta E^{\ddagger}$ values F, –3.5; Cl, 0; Br, –0.7; I, +2.9 may be calculated. These results may be compared with values for 1-halogeno-2,4-dinitrobenzenes *[33,68a]*, *viz.*, mobility ratios, F, 400; Cl, 1; Br, 0.65; I, 0.19; and $\Delta\Delta E^{\ddagger}$, F, -3.9_5; Cl, 0; Br, –0.4; I, +1.5. Even the much lower F/Cl ratio with thiophenoxide ion occurs in the heterocyclic as in the benzenoid series *[7]*.

High mobility of cationic groups is well-documented too. Thus Kloetzer and his co-workers *[142]* have reported on the facile replacement of the trimethylammonio group in the pyrimidine series. The nitro group is another group with high mobility in both benzenoid and heterocyclic series. In 2-chloro-4-nitropyridine and its *N*-oxide, the nitro group is more readily displaced *[143]* than chlorine, even allowing for complicating features, *viz.* the greater activating power of the nitrogen at the 4- than at the 2-position, and the greater activation of the chlorine by the *m*-nitro group than of the nitro group by the *m*-chlorine (see Chapter 4, pp. 118, 122). The reactivity of such groups as azido, alkoxyl, thioalkoxyl and sulphonyl *[e.g. 144–150]* also follow the same pattern (see Chapter 5, p. 165) of mobility as in benzenoid compounds.

Brief references have already been made in Chapter 5 (p. 173) to the displacement of hydride ion, and to the Chichibabin reaction *[8]*, *viz.* the direct amination of pyridine and its derivatives by alkali-metal amides. This reaction has been discussed in detail recently by Abramovitch and his co-workers *[151]* and their results indicate that the reaction proceeds by the addition–elimination and not the elimination–addition (benzyne) S_N2 mechanism. Although lack of data precludes precise calculations by Miller's procedure, it is possible to predict that whereas addition of amide ion to pyridine is a facile reaction proceeding most readily at the 2-, 4- or 6-position, hydride ion displacement is much less facile and is the rate-limiting step for the substitution process. The preferred formation of 2-aminopyridine in the reaction of the parent heterocycle may be ascribed to greater facilitation of hydride displacement at the 2-position than elsewhere by the nitrogen, which is negatively charged in the reaction intermediates.

(e) *Substituent effects*

There is no reason or evidence to suggest major differences in substituent effects, *i.e.* the same Hammett substituent constants can be used in heterocyclic as in benzenoid systems. Evidence given above suggests that values of Hammett reaction constants (ρ), which are markedly less in benzenoid systems with two than in those with one powerful exocyclic activating group, differ less when activation is by one or more electronegative ring atoms in heterocycles. In general terms this may reflect the lower proportion of π-electrons delocalised on ring atoms in the transition states of reactions of exocyclic activated benzenoid than of ring-atom activated heterocyclic substrates.

Within one ring σ_m- and σ_p-values should be used. The situation in polycyclics is less clear but appropriate correlations should be obtainable, depending on whether the negative charge of the hetarenide system can be placed on the substituent group (σ_p) or not (σ_m). Values are modified by use of a transmission coefficient (Chapter 4, p. 90) to allow for reduction in reactivity in inter- as compared with intra-annular transmission of substituent effects: a phenomenon well known in heterocyclic as well as in benzenoid systems. Numerous examples are quoted by Illuminati *[1]*, and Shepherd and Fedrick *[2]*.

Hill and Krause *[77]* have measured the mobility of chlorine in a series of 3-Cl-6-R-pyridazines including that with R = H. Their ρ-value is 6.82 at 40.2°, and substituent effects are very similar to those in the benzene series (see relevant sections of Chapter 4). The only difference is that the methylthio group is slightly deactivating whereas in the benzene series it is mildly activating. This may be because of cross-conjugation with the α-nitrogen in the pyridazines, whereas there is no such cross-conjugation in 4-chloro-3-nitrothioanisole, in which its effect as a *para*-substituent in the benzene ring has been measured *[152]*. A good indication of a parallel pattern of activation is the very large number of nitro-activated hetero-aromatic substitutions *[2]*, and the fact that these all indicate similar activating power in both series. There are other references of a similar type *[e.g. 153–157]*. Jaffé and Jones *[158]* have recently discussed generally the applications of the Hammett equation to heterocyclic compounds, and included the effects of exocyclic substituents in S_N reactions.

The effects of *ortho* substituents are also similar in both benzenoid and heterocyclic systems, and may be separated into polar and specific *ortho* effects (see Chapter 4).

Marino and his co-workers *[156d]* have made the interesting comparison of the effects of 2- and 6-methyl and -tert-butyl groups on the rate of substitution of alkyl-4-chloropyrimidines by piperidine in toluene or ethanol at 30°. They obtained the following rate-ratios (k_{Me}/k_{t-Bu}): 6-R, 1.62 in toluene and 2.57 in ethanol; 2-R, 2.36 in toluene and 17.3 in ethanol. The marked increase for the 2-position in ethanol is ascribed to steric hindrance to solvation of the aza-group by the tert-butyl group. When solvated, therefore, the steric requirements of the aza-group are so enhanced that one cannot indiscriminately assume lack of steric interaction between an aza-group and a substituent *ortho* to it.

6. Reactivity of Heterocyclic Aromatic Systems in the Elimination–Addition Mechanism

It is convenient first to consider the usual requirements for the elimination–addition mechanism to operate in benzenoid compounds (Chapter 3). These are (*i*) the presence on neighbouring ring atoms of a hydrogen atom, and a group X which is sufficiently electronegative to be displaced readily with its bonding electrons; this being facilitated also by a weak bond between the ring atom and X, together with (*ii*) the use of strongly basic reagents; or (*iii*) the presence on neighbouring ring atoms of two such groups X (or X and X′) as described in (*i*) above, together with (*iv*) a reactive metal or similar reagent; and in either case (*v*) low reactivity of the substrate with the reagent via the normal addition–elimination mechanism.

Conditions (*i*) and (*ii*) correspond to the formation of alkynes by elimination of HX from vinyl–X compounds; and conditions (*iii*) and (*iv*) correspond to the formation of alkynes from 1,2-di-X-alkenes. Condition (*v*) is an obvious prerequisite, so that for example, halogeno-polynitro compounds are not known to react via the aryne mechanism; and fluoro compounds, at least with first row reagents in protic solvents, are less susceptible to reaction via this mechanism than corresponding compounds of the heavier halogens—the strong C–F bond being additionally disadvantageous.

Corresponding reactions may be expected in the less reactive heterocyclic systems, and the subject of elimination–addition reactions in heterocyclic systems has recently been reviewed *[32,159]*. The term hetarynes is used in these for heterocyclic analogues of benzynes (*o*-dehydrobenzenes). It is of interest that what appears to be the first reference *[160]* to an aryne was in fact to a hetaryne, *viz.* benzofuran-2,3-yne, with the triple bond in a π-excessive system, which has low reactivity with nucleophiles by the addition–elimination mechanism.

In accord with the brief discussion above, the formation of hetarynes occurs most readily in such species as 3-halogenopyridines (Cl, Br, I compounds) and 3-halogeno-furans and -thiophenes. With lithium piperidide in ether (at 35°) it was found for example, that 3-chloro-pyridine reacts via the elimination–addition (aryne) mechanism; whereas 3-fluoro- and 4-chloro-pyridine react almost exclusively via the addition–elimination mechanism *[57]*. In reaction with amide ion in liquid ammonia (at –33°) both 3- and 4-chloropyridine react via the aryne mechanism. The change may be in part a differential solvent and temperature effect on the two mechanisms, but in the main is a consequence of the greater base strength and smaller nucleophilic strength of amide than piperidide ions, corresponding to a somewhat similar discrepancy in basic and nucleophilic strengths of hydroxide and alkoxide ions (Chapter 6, p. 201).

3-Chloropyridine could react via formation of 2,3- or 3,4-pyridyne, but the absence of 2-substituted pyridines among the products suggests that only 3,4-pyridyne is formed. Kauffmann and Boettcher *[37,159a]* suggest that the negative charge formed on C–2, before any elimination of 3-halogen, is strongly attracted by the nitrogen and is

References p. 306

thus unable to facilitate expulsion of halogen to form 2,3-pyridynes. The 2-halogenopyridines react with amide ion giving only 2-aminopyridine and have been assumed to react by the addition–elimination mechanism. However Jones and Beveridge *[161]* on the basis of M.O. calculations have suggested that 2,3-pyridyne would give only 2-aminopyridine with amide ion, so that the situation is somewhat obscure.

Zoltewicz and Smith *[162a]* have shown that the non-formation of 2,3-pyridyne is a consequence of the very much lower mobility of the 2- than the 4-hydrogen in a 3-halogenopyridine. Abramovitch and his co-workers *[162b]* have compared this situation with that in the corresponding pyridine-*N*-oxide and pyridinium systems in which the 2-hydrogen has the highest mobility. Their explanation of the difference is essentially that of Kauffmann and Boettcher *[37,159a]*.

There is some evidence for the existence of 2,3-pyridyne, however, for in the reaction of 3-bromo-2-chloropyridine with lithium amalgam in the presence of furan *[39]* the pyridyne is trapped to form some 5,8-epoxy-5,8-dihydroquinoline which is then reduced to quinoline (Fig. 90A). Similarly 3,4-pyridyne has been trapped with furan to give finally some isoquinoline (Fig. 90B).

There is evidence from the formation of some rearranged product that even a halogenopyridine-*N*-oxide may react in part by the aryne mechanism *[39]*. The possible existence of a pyridynium ion (Fig. 91) has been suggested by Kauffmann *[159a]*.

Recently evidence has been put forward for the formation of 4,5-pyrimidynes and substituted forms by reaction of 5-halogenopyrimidines (the least reactive of the halogenopyrimidines) with amide ion in liquid ammonia *[159b]*.

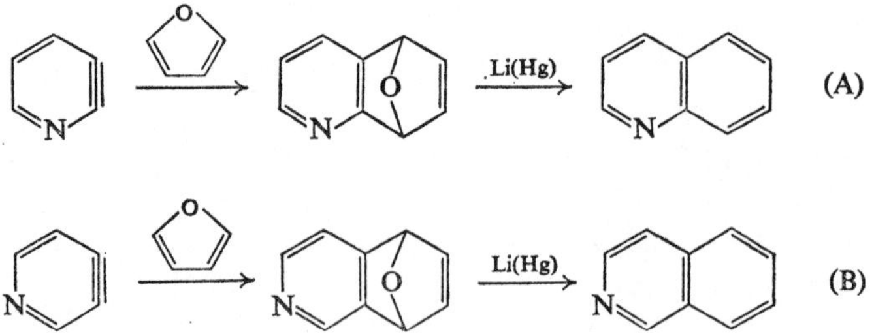

Fig. 90. Formation of quinoline and isoquinoline from 2,3-pyridyne (A) and 3,4-pyridyne (B).

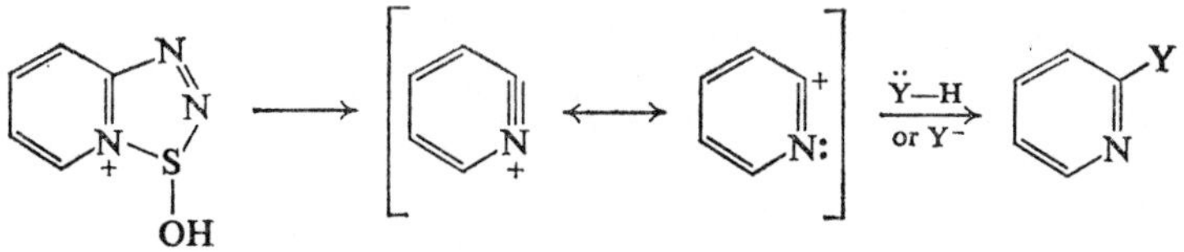

Fig. 91. Possible intermediacy of a pyridynium ion.

It has been suggested by Den Hertog and Van der Plas *[32]* that the benzothiophen compound corresponding to benzofuran-2,3-yne *[158]* (see above) is an intermediate in reactions described by Komppa and Weckman *[163]*. The simpler 2,3-dehydrothiophen system has been reported by Wittig and Wahl *[164]* who trapped it by reaction with tetraphenylcyclopentadienone.

Other bicyclic hetarynes include 2,3- and 3,4-quinolyne, formed from 3-bromo-2- or -4-chloroquinoline and lithium amalgam *[40,165]*. 3,4-Quinolyne was also reported to be formed by reaction of 3-halogenoquinolines except the fluoro compound, with lithium piperidide in ether. There is evidence also for the formation of 1,5-diazanaphthalene-3,4-yne (1,5-naphthyrid-3,4-yne) *[159c]* by reaction of the 2- or 4-bromo compound with amide ion in liquid ammonia, and probably accompanied by direct substitution via the addition–elimination mechanism.

As regards substituent effects and group mobility the limited available data indicate parallel behaviour in heteroaromatic and benzenoid compounds. There appears to be no body of evidence as yet for the elimination–addition reaction in other non-benzenoid aromatic systems, but see Section 7(e).

7. Reactivity of Inorganic and Other Non-Benzenoid Aromatic Systems

(*a*) *Introduction*

In recent years interest has been focused on several inorganic aromatic systems in which nucleophilic substitution is known to take place. These include halogenoborazines, phosphonitrilic halides and thiazyl halides, three examples of which are given in Fig. 92.

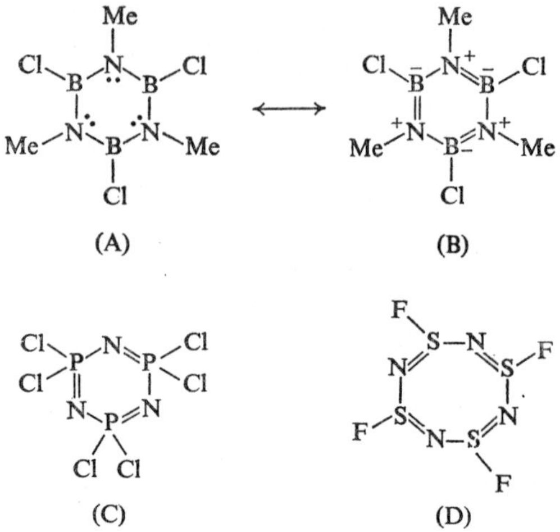

Fig. 92. A and B, structures for 1,3,5-trimethyl-2,4,6-trichloroborazine; C, trimeric phosphonitrilic chloride (2,2,4,4,6,6-hexachlorotriazaphosphorine); D, tetrameric thiazyl fluoride (1,3,5,7-tetrafluorotetrathiatetrazine).

There are also organic non-benzenoid (carbocyclic) aromatic systems in which nucleophilic substitution takes place, the most studied being tropylium (and related compounds) and quinones.

In the borazines aromaticity requires the delocalisation of the unshared pairs of the nitrogen and presumably involves some contribution from a canonical structure with pπ bonds of the type shown in Fig. 92B.

Aromaticity in the other two examples does not require the use of the unshared pair of electrons (not shown) on nitrogen atoms but does involve contribution of p and d electrons to the delocalised π-electron system. In a more general discussion of these and other similar ring systems a first row element provides p electrons and a second row element d electrons to a π-electron system. Another or π'-electron system involves interaction of a pair of unshared electrons on the first row element with vacant d orbitals of the second row element.

Several lengthy reviews have been published on the borazines and phosphonitrilic compounds, especially the latter *[166–174]*. The thiazyl compounds have yet to be reviewed at any length, though scattered references to them may be found *[e.g. 174,175]*. Only a few of the more important features are discussed here.

(*b*) *Borazines*

There is evidence that there is partial delocalisation of the three pairs of electrons, shown in Fig. 92A as unshared pairs on nitrogen, to give aromatic character, but not to the same extent as in benzenoid compounds, and possibly it could involve "island" type delocalisation. Ring atoms are coplanar in *B*-trichloroborazine *[176]*, and the molecule has D_{3h} symmetry *[177]*. Dipolar character of the molecule involving N–B (Fig. 92B) and Cl–B coordination is supported by physical evidence *[178,179]*. Estimates have been made of the contributions of double bond structures, and of π-bond orders, in the borazine system *[179–181]* with similar conclusions. Thus Watanabe *et al.* *[179]* estimate that the contribution of the structure represented by Fig. 92B (R and X = H) is 24%, and the π-bond order 0.45.

Even with a fractional negative charge, boron is more electropositive than carbon, and also forms strong bonds with common nucleophiles, so that more facile S_N reactions than in simple benzenoid compounds are to be expected. The fractional positive charges on the nitrogen atoms should also result in activation relative to benzenoid compounds.

The conversion of the B–Cl group in *B*-trichloroborazines into the B–OH group by hydrolysis *[182]*, and its reversal with acid chlorides *[183]*, resembles that in highly activated halogenobenzenes. Various other reagents *e.g.* alkoxides, thiocyanate, and cyanide ions are also known to replace the chlorine of *B*-trichloroborazines *[184–186]*. Recently the *B*-trifluoro compound has been prepared by reaction of the *B*-trichloro compound with titanium tetrafluoride without solvent *[187]*. This reaction is reminiscent of the formation of fluoro- from chloro-aromatic compounds by halide exchange with potassium fluoride without solvent or in dipolar aprotic solvents *[188]*. The *B*-trifluoro compound is reported as sensitive to hydrolysis but there is no indication of the relative facility of hydrolysis of the fluoro and chloro compounds.

References p. 306

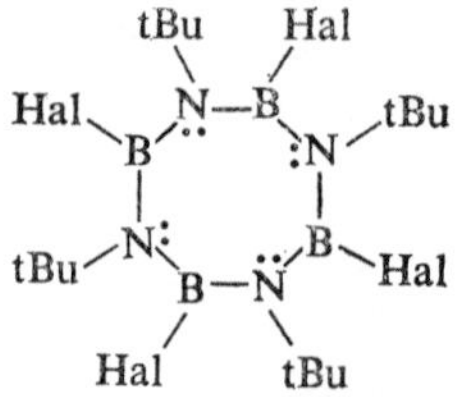

Fig. 93. 1,3,5,7-Tetra-tert-butyl-2,4,6,8-tetrahalogenoborazines.

Turner and Warne *[189]* have prepared cyclic tetramers (Fig. 93). Models suggest that these have a boat form. It is further suggested that extended π-delocalisation is unlikely. In the boat and other possible suggested forms, the boron and nitrogen atoms are highly inaccessible, and the much lower reactivity of halogen in these than in corresponding trimers is ascribed to this steric factor.

It should be recalled that in nucleophilic substitutions in organic compounds unsaturated systems are found in many respects to be intermediate in character between saturated and aromatic systems. Characteristically bond making assumes greater importance than in saturated systems but not to the extent found in aromatic systems with suitable reagents, as indicated by the relative mobility of fluorine having correspondingly intermediate values *[190]*. The available evidence on reactivity of halogenoborazines is insufficient to give definite indications as to whether delocalisation is equivalent to that in unsaturated organic or in aromatic compounds, nor is the physical evidence conclusive. Rate measurements on halogen and other group mobility, and on transmission of substituent effects, ought to be obtainable readily with compounds already known and could lead to well founded conclusions on this aspect of borazine chemistry. The facile formation of *B*-tris(alkylamino) but not of intermediate derivatives from the *B*-trichloro compound *[191–193]*, though these may be made from mono- and di-chloro compounds, is perhaps such an indication of weak transmission of substituent effects and thus is against substantial delocalisation, for the alkylamino groups are expected to be deactivating (*cf.* phosphonitrilic halides).

It is also noteworthy in connection with borazine chemistry that hydrogen is more electronegative than boron and so is much more readily replaced by nucleophiles than when attached to carbon, or other more electronegative elements.

(c) *Phosphonitrilic derivatives*

These compounds have been well reviewed in recent years *[168–174]* and many ring systems, some very large [*e.g.* $(PNF_2)_{17}$], have been reported. The best-known however are the 6-membered planar ring $(PNCl_2)_3$ and 8-membered puckered ring $(PNCl_2)_4$, and derivatives of these, in which ring bonds are of equal length.

Delocalisation in these involves not only a π-system formed by overlap of electrons in nitrogen p and phosphorus d orbitals, but an additional (or π') system, utilising the formally unshared electrons on nitrogen interacting with vacant d orbitals on phosphorus. This has been discussed by Craig and by Dewar and their co-workers *[194,195]*, and in a wider context by Craig *[194a]* and Cruickshank *[196]*. While there is general agreement on the existence of π-type delocalisation in these compounds, there is as yet no agreement as to whether it extends over the whole ring or alternatively to a lesser extent as in P–N–P "islands". In a paper *[194e]* based on recent calculations, Craig suggests that fully cyclic delocalisation plays an important part; but not to the extent to which it occurs in benzenoid systems, so that there is some "island" character to the delocalisation. Craig has also discussed the inapplicability to these $(AB)_n$ systems (A = first row element, B = second row element) of Hückel's rule, which refers to the stabilisation, due to π-delocalisation, of monocyclic systems of alternate single and double bonds which have $(4n + 2)$ electrons in the π-system, and the lack of such stabilisation of those with $4n$ electrons. Measurements of diamagnetic anisotropy, though suggestive of a fully cyclic delocalisation in a $p\pi$–$d\pi$ system, are not conclusive.

Before considering the S_N reactions of phosphonitrilic compounds, it is of interest to consider briefly the S_N reactions of phosphoryl compounds. The evidence suggests that these reactions are bimolecular and involve optical inversion *[197–199]*. It is also known that electron withdrawal facilitates and electron accession hinders reaction. Nevertheless, the halogen mobility order, $F < Cl < I$, with a variety of first-row nucleophiles suggests that bond-breaking is also involved

References p. 306

in the rate-limiting step *[7b,c,190]*. This argument, paralleling that used in carbon compounds, is more persuasive for oxygen than nitrogen nucleophiles, for in comparing bond strengths to phosphorus and carbon it is relevant to note that P–F and P–OR bonds are a little stronger than C–F and C–OR bonds *[7b,200]*, differing by about the same amount in each case, whereas P–N bonds are weaker than C–N bonds. The evidence suggests formation of a transition state involving both bond-making and bond-breaking as rate-limiting step, rather than formation of the second transition state, in an addition–elimination mechanism proceeding via an intermediate complex of substantial stability. There are no precise data for comparison, but it seems that the F/Cl ratio is as small as in acyl halides and may be less.

In the phosphonitrilic halides, S_N reactions are also bimolecular, with F/Cl mobility ratios less than unity *[201,202]*. Other interesting features are that substitution of halogen is stepwise, and that subsequent steps may take place on a different or the same phosphorus atom depending on the group first introduced. The presence of amino groups which are strongly electron-releasing, particularly results in subsequent steps occurring on other phosphorus atoms *[203]*. In contrast, groups such as fluoro *[204]* and methylthio *[205]*, which do not possess such marked electron-releasing power, and, being more electronegative than phosphorus, exert a $-I$ effect, lead to geminal substitution, *i.e.* the second step takes place on the same phosphorus atom. The phenyl group also causes geminal substitution *[205]* and this presumably reflects a greater inductive withdrawal of electrons than mesomeric electron release, when it is attached to phosphorus. This is not surprising since carbon is substantially more electronegative than phosphorus *[73]*.

Shaw and his co-workers *[205]* have shown that substitution of a piperidino group into the trimeric chloride leads to the next piperidino group replacing chlorine at a second phosphorus atom at a rate 50 times slower, and a third piperidino group (replacing chlorine at the third phosphorus atom) about 500 times slower than the first substitution. These facts suggest that delocalisation must exist at least at the "island" P–N–P level. Measurement of the rates of the first and second steps in replacement of chlorine by piperidine in tetrameric phosphonitrilic chloride, which is known *[206]* to give the 1,5-dipiperidino compound, would give definite evidence as to whether the

1,5-disubstitution is due simply to deactivation at 7- and 3-positions by the 1-piperidino group which would occur in "island" delocalisation, or whether in addition the *rate* of a second substitution, at the 5-position, is also less, which would indicate transmission of a deactivating effect to the 5-position, and thus substantial cyclic delocalisation. Steric contributions cannot be disregarded *[205]* but substituent effects in the cases discussed above would appear to be largely polar in origin.

The relatively poor discrimination towards the reagents ethanol and aniline, shown by the ratio of bimolecular rate constants (aniline: ethanol, in ethanol as solvent) being only 148 at 34.5° *[207]*, whereas it is *ca.* 10^9 with picryl chloride *[208]* (*cf.* Table 50, p. 209), has led Bailey and Parker *[207]* to suggest that bond breaking is relatively advanced compared to bond making, *i.e.* reaction is S_N1-like. Paddock *[174]*, considering the reactivity order ethanol < aniline (both in ethanol–benzene) < piperidine (in toluene) < chloride ion (in acetonitrile), has also suggested that bond-breaking is the more important step, indicating at the same time however the importance of a penta-coordinated complex in the transition state. The significance of placing chloride above the other three is much minimised by its being based on reactivity in a dipolar aprotic solvent, in which nucleophilic reactivity of an anion is greatly enhanced compared with that in a protic solvent in which it is strongly solvated *[7b,c,209,210]*. The lack of discrimination between ethanol and the nitrogen bases reflects the great importance of the bond-making factor. Since P–NR_2 bonds are weaker than C–NR_2 bonds, whereas P–OR bonds are stronger than C–OR bonds *[7b,c,200]*, the reactivity of oxygen nucleophiles compared with nitrogen nucleophiles is greatly enhanced in substitution at phosphorus compared with carbon. Corresponding to the relevant figures for the relative reactivity of aniline and ethanol given above, it is of interest to note that, in reaction with amines, trimeric phosphonitrilic chloride has a reactivity comparable with that of 1-chloro-2,4-dinitrobenzene, whereas with alcohols it is more reactive than picryl chloride *[174,206,208,211,212]*.

While the F/Cl mobility ratios with first row nucleophiles in protic solvents, which are less than unity *[201,202]*, suggest that an aliphatic type S_N2 mechanism is operative, the small margin of levels of transition-state energies by which, in aromatic compounds, high F/Cl mobility ratios are expected and found, should not be forgotten (Chapter 5, Section 3). Since in phosphorus compounds the difference in bond-dissociation energies of P–F and P–Cl bonds *[200]* is substantially larger than that for C–F and C–Cl bonds (Chapter 5, Section 3), there is at least a possibility that the low F/Cl ratio is a consequence of S_N reactions of phosphonitrilic halides proceeding by the addition–elimination mechanism via an intermediate complex with the formation of the second transition-state rate-limiting, even with some first-row nucleophiles. Aliphatic and aromatic type S_N2 mechanisms for reactions of trimeric phosphonitrilic chloride with an anionic reagent Y^- are illustrated in Fig. 94A and 94B respectively.

Ignoring the delocalisation in the π'-system (see above) the aliphatic type S_N2 mechanism involves a change at the reaction centre from $sp^3(\sigma)$ $d(\pi)$ bonding to $sp^2(\sigma)$ $d(\pi)$ bonding plus a p_z orbital used for part-bonds to the nucleophile (Y) and leaving group (Cl). In the aromatic type S_N2 mechanism the change is to $sp^2(\sigma)$ $pd(\sigma)$ bonding. In *both* cases the geometrical arrangement about phosphorus at the reaction centre is that of a trigonal bipyramid.

The reported deactivation by a piperidino group indicated both by occurrence of non-geminal substitution, and more specifically by the 50-fold smaller rate constant for the reaction of 2-piperidino-2,4,4,6,6-pentachlorotriazaphosphorine than of the hexachloro compound with piperidine, indicates delocalisation in the ring, but not necessarily greater than that of P–N–P "island" type.

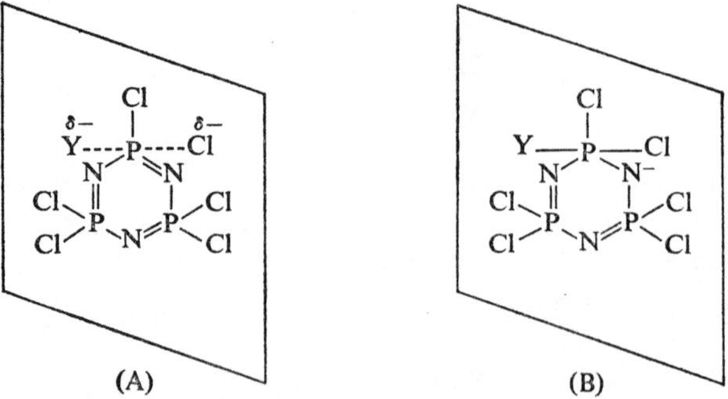

Fig. 94. A, suggested transition state for aliphatic type S_N2 reaction of trimeric phosphonitrilic chloride; B, suggested intermediate complex for aromatic type S_N2 reaction of trimeric phosphonitrilic chloride.

Evidence for or against fully cyclic delocalisation in trimeric phosphonitrilic chloride would however be given by measuring the kinetic effect on replacement of chlorine atoms in the 2-position, of quaternisation of nitrogen in the 5-position, or its conversion to the *N*-oxide.

Shaw and his co-workers *[205]* have shown that with diethylamine in toluene, tetrameric phosphonitrilic chloride is about 10^2–10^3 more reactive than the trimer *[cf. 174,202]*. This might reflect the occurrence of cyclic delocalisation in both, with an extra nitrogen in the latter resulting in additional activation.

(*d*) *Thiazyl and sulphanuric halides*

It is known that the bonds in trithiazyl and α-sulphanuric chlorides (Fig. 95A and B), which exist in chair conformations, are equal in length *[175,213a,214]* indicating a substantial measure of delocalisation. In tetrathiazyl fluoride (Fig. 92D), which has a puckered ring, bonds alternate in length *[214b]*, suggesting that delocalisation is not important in this system *[174,175,194]*. The reason for these differences between trimeric and tetrameric sulphur–nitrogen compounds is not clear. Craig and Paddock *[194b]* suggest that it may be due to the lone pair of electrons on sulphur preventing delocalisation of the lone pair on nitrogen, *i.e.* it is an effect on the π' system.

Little is known of their chemistry in mechanistic terms except that reactions with nucleophiles result in ring fission *[215,216]*.

An interesting compound related both to trimeric phosphonitrilic

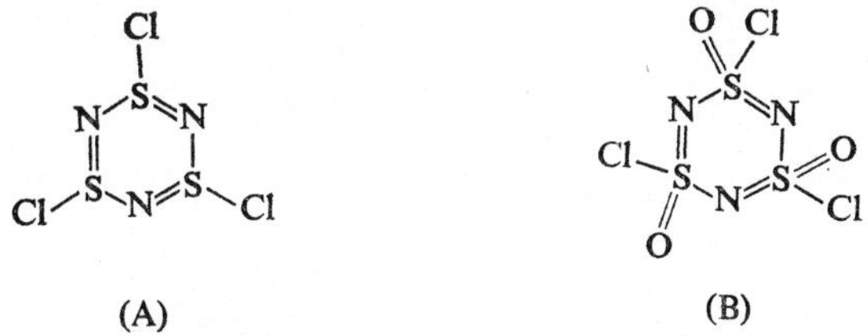

Fig. 95. A, trithiazyl chloride; B, α-sulphanuric chloride.

References p. 306

chloride and α-sulphanuric chloride, with one PCl_2 group instead of an SOCl group, has been reported *[217]*.

(*e*) *Tropylium ions and related compounds*

The marked aromatic stability of 7-membered ring systems such as the tropylium ion and tropone (Fig. 96), which possess a sextet of π-electrons, has been realised for a considerable time *[218]* and has recently been discussed at length by Doering *[219]*. A number of reviews have appeared *[220–224]*.

Ring carbon atoms are markedly electron deficient in the tropylium ion, less so in tropone, and still less in tropolone (2-hydroxytropone) and its anion. The expected order of reactivity with nucleophiles is thus tropylium ion > tropone > tropolone > tropolonate anion. Reactivity towards electrophiles is in the converse order. The pattern resembles that in pyridinium ions, pyridine-*N*-oxide, and 2-hydroxypyridine-*N*-oxide.

Physical measurements on the tropylium ion *[e.g. 224–228]* are in accord with a symmetrical aromatic structure, having C–C bonds of equal length. This has been confirmed by converting labelled (^{14}C) tropilidene (cycloheptatriene) into tropylium bromide and thence to phenyltropilidene–^{14}C, oxidising it to benzoic acid, and showing that this contained just one-seventh of the specific activity of the tropylium ion. This shows the uniform distribution of positive charge over all seven carbon atoms *[229a]*.

The tropylium ion reacts readily with nucleophilic but not with electrophilic reagents. For example, it does not show replacement of more than 1% of its hydrogen by deuterium in $DBr/AlBr_3$ at room temperature. It reacts readily with nucleophiles *[219,230]*, so much

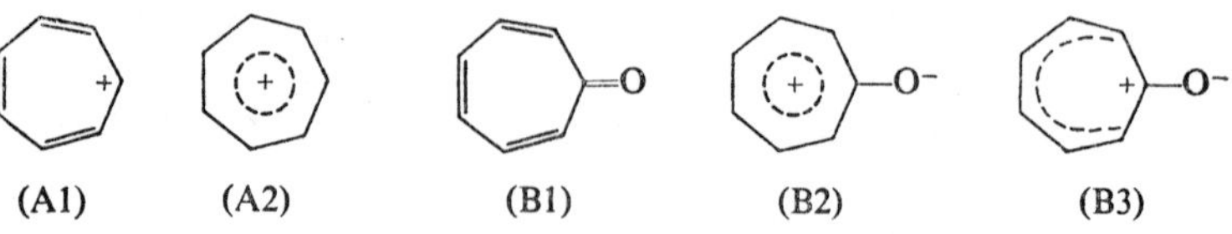

Fig. 96. Formal valence structure and π-delocalised representations of tropylium ion (A1 and A2), and tropone (B1 to B3).

so that the equilibrium, tropylium$^+X^-$ (ionic) $\rightleftharpoons$ X-tropilidene (covalent) lies to the right-hand side unless the nucleophilic power of X^- is low *[229b]*. Reaction by the activated aromatic addition–elimination mechanism seems most likely, though reliable kinetic evidence for this is not yet available.

Whereas all positions in the unsubstituted tropylium ion are equally reactive, this should not be the case with tropone. If there is full delocalisation as in Fig. 96,B2, one would expect reactivity towards nucleophiles to be greatest at the 4- and 5-position, whereas if the positive charge is largely localised on C-1, to which an O^- group is attached, with less delocalisation of the π-electron system as in Fig. 96,B3, then the nearer 2- and 7-positions might be the most susceptible to reaction with nucleophiles. The recent structure determination of 2-chlorotropone *[231]* by X-ray diffraction indicates C1–C2 and C1–C7 bond lengths close to values for C_{sp^2}–C_{sp^2} single bonds, whereas the other C–C bond lengths as a group more nearly approximate to benzene C–C bond lengths, and this support less delocalisation of the π-electron system, with structure B3 as that most nearly approximating to the true structure of tropone.

The kinetic evidence is very limited, but Nozoe *[224]* reports that in 2,4-dimethoxytropone the 2-methoxyl group is more reactive, and the 4-methoxyl group resistant to attack. In 2,3-dibromotropone however the 3-bromine is reported as preferentially displaced *[224]*, but the weight of available evidence favours highest reactivity at the 2- (and 7-) position.

Jutz and Voithenleitner *[232]* have calculated electron densities in (1-)phenyltropylium ion. Their data suggest the reactivity order with nucleophiles, 2- > 4- > 3-position in the tropylium ring (but all very similar), and 4- > 2- > 3-position in the benzene ring. They showed experimentally that this substance reacts readily with phenylmagnesium bromide to form 1,4-diphenyltropylium ion. The inversion of the 2-, 4-position reactivity order was ascribed to more extended conjugation for substitution at the 4-position, and steric hindrance at the 2-position. No experimental evidence was offered for enhanced reactivity in the benzene ring.

Brown *[233]* has calculated both electron densities and localisation energies for nucleophilic substitution in tropone. The former indicate the lowest electron density on C-1, equivalent to substantial localisa-

References p. 306

tion of positive charge at this point, to which the oxygen is attached; and suggests the reactivity order 3- > 4- > 2- (equivalent to 6- > 5- > 7-) position. The latter calculations suggest the reactivity order 2- > 4- > 3-position, and there are experimental data for substituted tropones in accord with both orders. It is of interest to recall that the reactivity order in chloropyridine-*N*-oxides is 2-, 4- > 3-position *[22b]*.

Experimental evidence is often complicated by concurrent substitution with transannular rearrangement, or ring contraction. Pauson *[223]* and Nozoe *[224]* suggest that these, as well as simple substitutions without rearrangement, are addition–elimination reactions. Likely mechanisms, now suggested, are illustrated in Fig. 97.

Pratt and Webster *[234]* report that 2-fluorotropone is hydrolysed in good yield to tropolone, *i.e.* by normal substitution without rearrangement (attack at the 2-position), whereas 2-chlorotropone reacts with ring contraction (see above) to give benzoic acid (initial attack at the 1-position) and *o*-hydroxybenzaldehyde (initial attack at the 3-position). This indicates that the electronegativity of fluorine is facilitating attack at the 2-position, with a F/Cl mobility ratio for direct substitution by a first row nucleophile in a protic solvent thus substantially greater than unity *[cf. 7b,c]*. This is characteristic of activated aromatic addition–elimination S_N2 reactions. In acidic conditions the oxygen of tropones is protonated, thus forming hydroxytropylium salts, with consequent enhancement of reactivity.

In tropolone (2-hydroxytropone) the hydroxyl group is phenolic in type, and with the geometry suitable, thus forms a mobile tautomeric system (Fig. 98). If the tautomeric shift is not blocked by, for example, etherification, the two oxygen functions are effectively equivalent at ordinary temperatures and orientation effects are obscured. The system is amphoteric (Fig. 98). In strongly acid conditions protonation of oxygen occurs with formation of 1,2-dihydroxytropylium (tropolonium) salts which are much more susceptible to nucleophilic attack, *e.g.* halogen exchange reactions *[235]*; whereas in basic conditions the tropolonate anion is formed with consequent marked reduction in reactivity, and is commonly accompanied by ring contraction and vicinal substitution with rearrangement. X-ray structure determinations *[236,237]* suggest more complete ring delocalisation of π-electrons in the tropolonate anion and complete delocalisation in the tropolonium cation (a tropylium derivative).

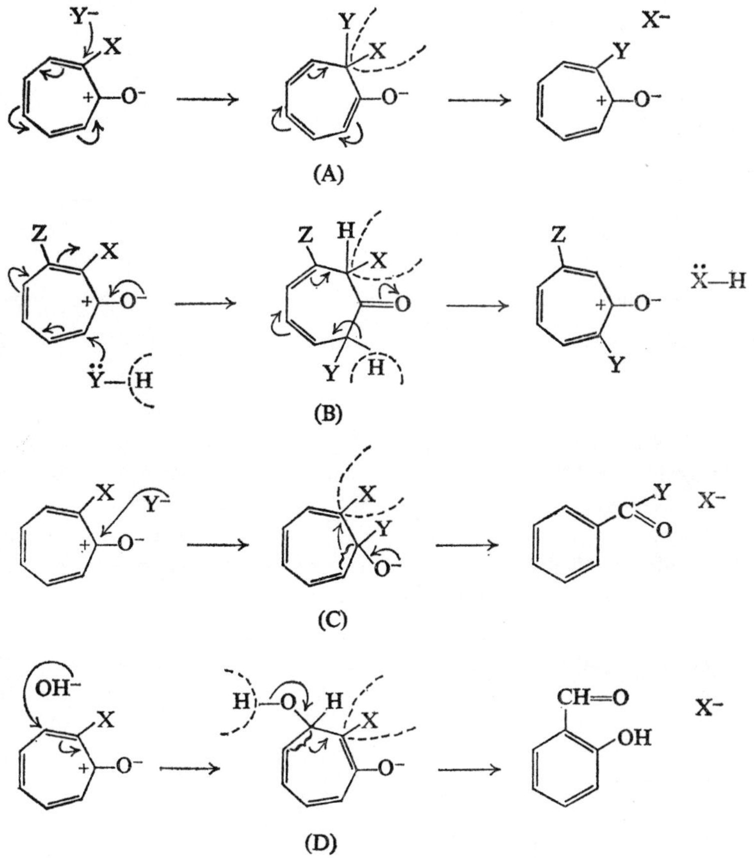

(A)

(B)

(C)

(D)

Fig. 97. Suggested mechanisms for A, substitution of 2-X-tropones without rearrangement to form 2-Y-tropones (attack at 2-position); B, substitution of 2-X-3-Z tropones with transannular rearrangement to form 2-Y-6-Z-tropones, equivalent to 7-Y-3-Z-tropones, (attack at 7-position); C, substitution of 2-X-tropones with ring contraction to form a benzoyl-Y compound (attack at 1-position); D, alkaline hydrolysis of 2-X-tropones with ring contraction to form *o*-hydroxybenzaldehyde (attack at 3-position).

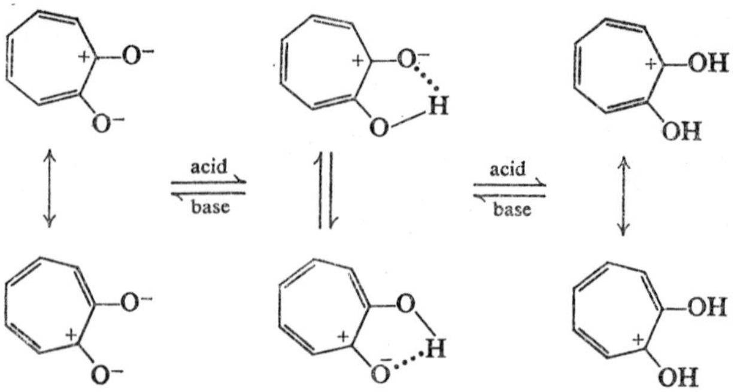

Fig. 98. Tautomeric equilibrium and amphoteric behaviour of tropolone (formal valence structures).

Dewar *[238]*, and Kurita and Kubo *[239]*, have carried out M.O. calculations of reactivity of the tropolone system. Dewar suggests that substitution occurs more readily at the 4-position, *i.e.* in 4-X-2-hydroxytropones. Kurita and Kubo correspondingly calculate that the electron density is least at the 4-position. This is also equivalent to the 6-position since they regard both oxygen functions as equivalent. Confirmatory kinetic data are not available, but a noteworthy parallel with less reactive benzenoid compounds occurs in 3-bromotropolone, in which the bromine is presumably in a less

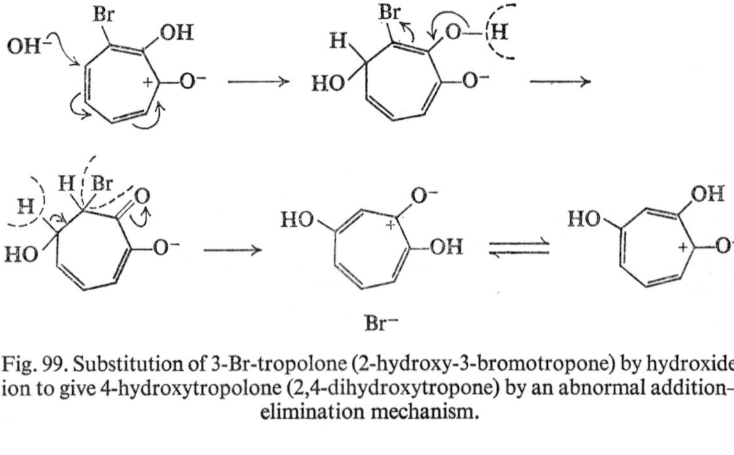

Fig. 99. Substitution of 3-Br-tropolone (2-hydroxy-3-bromotropone) by hydroxide ion to give 4-hydroxytropolone (2,4-dihydroxytropone) by an abnormal addition–elimination mechanism.

reactive position. This substance reacts only slowly with methoxide ion in methanol, even at 150°, but does so by direct substitution *[240]* to form 3-methoxytropolone (2-hydroxy-3-methoxytropone); whereas at about the same temperature the less nucleophilic but more basic hydroxide ion reacts with vicinal rearrangement to form 2,4-dihydroxytropone. Nozoe *[224]* however suggests as an alternative to a benzyne mechanism for the latter type of reaction an "abnormal" addition–elimination mechanism of the type discussed by Kauffmann *[159a,241]*, and based on work by Bordwell *et al.* *[242]*. This is illustrated in Fig. 99. In more vigorous conditions ring contraction (see above) occurs.

(*f*) *Quinones*

The quinones are a well-known and an important class related to benzenoid compounds. They resemble the intermediates in the normal addition–elimination S_N2 mechanism of aromatic nucleophilic substitution, which are often referred to as quinonoid. Such cyclohexadienide (benzenide) intermediates can be of considerable stability, and where there are several suitably oriented conjugative electron-withdrawing groups present, it is likely that their delocalisation energy exceeds that of the corresponding benzenoid systems *[6a,7b,c]*.

The aromatic stability of a simple quinone is regarded as substantially inferior to that of the corresponding dihydroxybenzene, and that of benzene itself; and the *p*-benzoquinones are regarded as a little more stable than the *o*-benzoquinones *[70,71,243,244]*. Quinones therefore exhibit properties such as would be expected for conjugated unsaturated ketones. It is readily seen that all carbon atoms in the ring are electron deficient, so that substituents X, of the type which can readily form X^{-}, should be readily replaced by nucleophiles, and this also is well-known *[e.g. 245–251]*. Intermediate complexes for such reactions can be envisaged with entering and replaced group fully bonded, as is the case with reactions via the activated addition–elimination aromatic S_N2 mechanism, and these are likely to be of considerable stability relative to the initial states *[248–251]*. This is illustrated in Fig. 100. Reactions would then proceed via two transition states and an intermediate complex, and also, as in typical

References p. 306

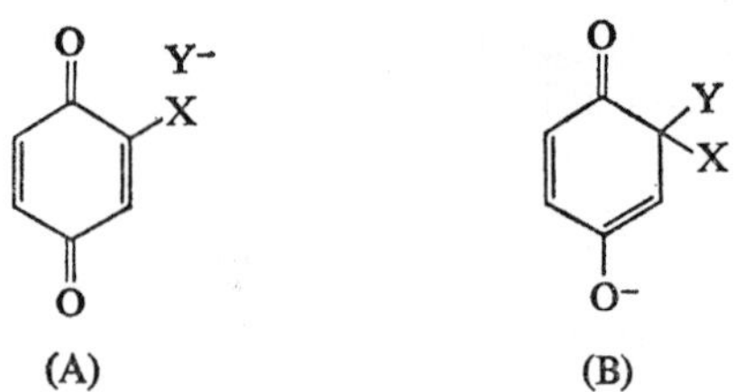

Fig. 100. A, initial state and B, intermediate complex for nucleophilic substitution of a 2-X-*p*-benzoquinone by Y^- in an aromatic type addition–elimination S_N2 mechanism.

aromatic S_N2 reactions, the formation of the first transition state would often be rate-limiting.

There has been little kinetic work on S_N reactions in these systems but available data support these conclusions. In particular measurements by Wallenfels and Friedrich *[250b]* of the reactivity of tetrachloro- and tetrafluoro-benzoquinone (chloro- and fluoro-anil) towards hydroxide ion demonstrate that the F/Cl mobility ratio is large for monosubstitution and, more explicitly, is about 150/1 for disubstitution. This cannot arise from the differential substituent effects of halogen atoms on other halogen atoms in the ring since the total activation by chlorine exceeds that of fluorine *[252]*, and so it must be due to the higher mobility of fluorine in this system. The authors have in fact used the mobility relationship as an argument in favour of the addition–elimination mechanism.

Recently Bishop *et al.* *[253]* have measured halogen mobility in 2-halogeno-3,5,6-trimethyl-*p*-benzoquinones towards sulphite ion, and reported reactivity ratios for Cl, Br, I = 1:3:30 at pH 7. The reactions are acid-catalysed and also complicated by addition of bisulphite ion to the carbonyl group—this is a general problem in reactions of quinones with nucleophiles. An analogous complication in S_N reactions of halogenoaromatic aldehydes and nitriles has been discussed by Miller and his co-workers *[68e,254]*. While the halogen mobility pattern *[253]* corresponds to that normally found with heavy nucleophiles *[7b,c,33]*, and does not therefore invalidate an addition–elimination mechanism (as in activated aromatic S_N2 reactions), Bevan and his co-workers *[255]* have shown the mobility order $F \gg Cl > Br > I$ in reactions of 1-halogeno-2,4-dinitrobenzenes with sulphite ion in aqueous ethanol. In view of the results of Wallenfels and Friedrich described above giving the order F $\gg$ Cl, the normal addition–elimination S_N2 mechanism is favoured, and the inverse Cl, Br, I mobility in the trimethylhalogenoquinones with sulphite ion, is presumed to involve some of the complicating features of these particular substitutions.

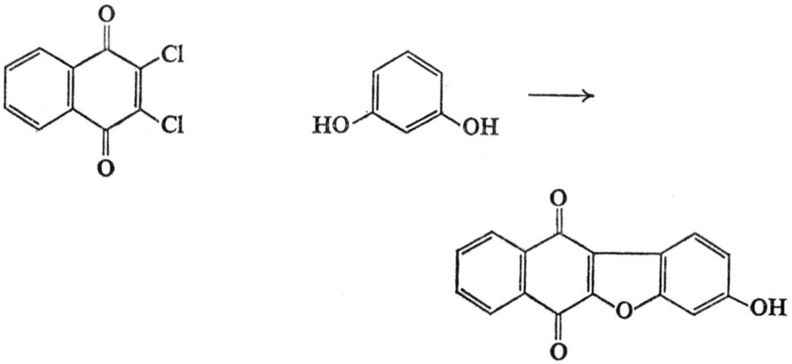

Fig. 101. Formation of 3-hydroxybenzo[*b*]naphthol[2,3-*d*]furan-6,11-dione by reaction of 2,3-dichloro-1,4-naphthoquinone with resorcinol.

Sartori *[256]* has recently reviewed the formation of heterocyclic quinones by methods involving S_N reactions of 2,3-dichloronaphthaquinone. An example shown in Fig. 101, and first reported in 1899 *[257]*, is the formation of 3-hydroxybenzo[*b*]naphtho[2,3-*d*]furan-6,11-dione by its reaction with resorcinol. In this reaction the nucleophilic centres are a carbon atom *ortho* to one hydroxyl and *para* to the other, and the hydroxyl oxygen atom *ortho* to this carbon atom. The reaction with ethyl acetoacetate (in its enolic form) *[258]* resembles this.

There is clearly scope for a great deal of work in the field of S_N reactions of quinones but it seems safe to predict that such reactions will show a close resemblance to activated benzenoid S_N reactions.

REFERENCES

1 G. Illuminati, in A. R. Katritzky (Ed.), *Advances in Heterocyclic Chemistry*, Vol. 3, Academic Press, New York, 1965, p. 285–359.

2 R. G. Shepherd and J. L. Fedrick, in A. R. Katritzky (Ed.), *Advances in Heterocyclic Chemistry*, Vol. 4, Academic Press, New York, 1965, p. 145–423.

3 S. K. K. Jatkar and C. M. Deshpande, *J. Indian Chem. Soc.*, 37 (1960) 11.

4 J. Miller, *Rev. Pure Appl. Chem.* (*Austral.*), 1 (1951) 171.

5 J. Miller, *Australian J. Chem.*, 9 (1956) 61.

6 C. K. Ingold, *Structure and Mechanism in Organic Chemistry*, Cornell Univ. Press, Ithaca, N.Y., 1953, (*a*) p. 61–90; (*b*) p. 9.

7*a* B. A. Bolto and J. Miller, *Australian J. Chem.*, 9 (1956) 74, 304; *b* J. Miller, *J. Am. Chem. Soc.*, 85 (1963) 1628; *c* D. L. Hill, J. Miller and K. C. Ho, *J. Chem. Soc.*, *B*, (1966) 299.

8 A. E. Chichibabin, *J. Phys. Chem.* (*USSR*)., 46 (1914) 1216.

9 N. B. Chapman and R. E. Parker, *J. Chem. Soc.*, (1951) 3301 and subsequent papers.

10 T. E. Young and E. D. Amstutz, *J. Am. Chem. Soc.*, 73 (1951) 4773 and subsequent papers.

11 J. F. Bunnett and R. E. Zahler, *Chem. Rev.*, 49 (1951) 273.

12 N. B. Chapman in *Chemical Society Special Publication, No. 3*, Chem. Soc., London, 1955, p. 155–167.

13 G. W. Wheland and L. Pauling, *J. Am. Chem. Soc.*, 57 (1935) 2086.

14 G. W. Wheland, *J. Am. Chem. Soc.*, 64 (1942) 900.

15 H. C. Longuet-Higgins, *J. Chem. Phys.*, 18 (1950) 283.

16*a* H. C. Longuet-Higgins and C. A. Coulson, *Trans. Faraday Soc.*, 43 (1947) 87; *b* H. C. Longuet-Higgins and C. A. Coulson, *Proc. Roy. Soc.*, A191 (1947) 39; A192 (1947) 16; *c* H. C. Longuet-Higgins and C. A. Coulson, *J. Chem. Soc.*, (1949) 971.

17 A. Albert, *Heterocyclic Chemistry*, Athlone Press, London, 1959 (*a*) p. 39; (*b*) p. 133; (*c*) p. 136–165; (*d*) p. 343–346.

18 B. Bak, *Acta Chem. Scand.*, 9 (1955) 1355.

19 R. D. Brown and B. A. W. Coller, *Australian J. Chem.*, 12 (1959) 152.

20*a* J. R. Morris and F. L. Pilar, *Chem. and Ind.*, (1960) 469; *b* J. R. Morris and F. L. Pilar, *J. Chem. Phys.*, 34 (1961) 389.

21 D. G. Manly and E. D. Amstutz, *J. Org. Chem.*, 22 (1957) 133.

22*a* M. Liveris and J. Miller, *Australian J. Chem.*, 11 (1958) 297; *b* M. Liveris and J. Miller, *J. Chem. Soc.*, (1963) 3486.

23 R. C. Elderfield and M. Siegel, *J. Am. Chem. Soc.*, 73 (1951) 5622.

24*a* H. H. Jaffé, *J. Chem. Phys.*, 20 (1952) 1554; *b* H. H. Jaffé and G. O. Doak, *J. Am. Chem. Soc.*, 77 (1955) 4441.

25 A. Bryson, *J. Am. Chem. Soc.*, 82 (1960) 4871.

26 H. L. Bradlow and C. A. Vanderwerf, *J. Org. Chem.*, 14 (1949) 50.

27 W. K. Miller, S. B. Knight and A. Roe, *J. Am. Chem. Soc.*, 72 (1950) 4765.

28 H. Schroeder, *J. Am. Chem. Soc.*, 82 (1960) 4115.

29 M. Bellas and H. Suschitzky, *J. Chem. Soc.*, (1963) 4007.

30 R. D. Chambers, J. Hutchinson and W. K. R. Musgrave, *J. Chem. Soc.*, (1964) 3736.

31 R. E. Banks, J. E. Burgess, W. M. Cheng and R. N. Haszeldine, *J. Chem. Soc.*, (1965) 575.
32 H. J. den Hertog and H. C. van der Plas in A. R. Katritzky, *Advances in Heterocyclic Chemistry*, Vol. 4, Academic Press, New York, 1965, p. 121–144.
33*a* J. Miller and K. W. Wong, *Australian J. Chem.*, 18 (1965) 117; *b* J. Miller and K. W. Wong, *J. Chem. Soc.*, (1965) 5454.
34 S. C. Chan, K. Y. Hui, J. Miller and W. S. Tsang, *J. Chem. Soc.*, (1965) 3207.
35 D. Spinelli, C. Dell'erba and A. Salvemini, *Ann. Chim.* (*Rome*), 52 (1962) 1156.
36 R. Levine and N. W. Leake, *Science*, 121 (1955) 780.
37*a* T. H. Kauffmann and F. P. Boettcher, *Angew. Chem.*, 73 (1961) 65; *b* T. H. Kauffmann and F. P. Boettcher, *Chem. Ber.*, 95 (1962) 949, 1528.
38 M. J. Pieterse and H. J. den Hertog, *Rec. Trav. Chim.*, 80 (1961) 1377.
39*a* R. J. Martens and H. J. den Hertog, *Tetrahedron Letters*, (1962) 463; *b* R. J. Martens and H. J. den Hertog, *Rec. Trav. Chim.*, 83 (1964) 621.
40 T. H. Kauffmann, F. P. Boettcher and J. Hansen, *Ann. Chem.*, 659 (1962) 102.
41 H. C. van der Plas and G. Geurtsen, *Tetrahedron Letters*, (1964) 2093.
42 E. A. C. Lucken in A. R. Katritzky (Ed.), *Physical Methods in Heterocyclic Chemistry*, Vol. 2, Academic Press, New York, 1963, p. 89.
43 C. K. Banks, *J. Am. Chem. Soc.*, 66 (1944) 1127.
44*a* R. R. Bishop, E. S. Cavell and N. B. Chapman, *J. Chem. Soc.*, (1952) 437; *b* N. B. Chapman and C. W. Rees, *J. Chem. Soc.*, (1954) 1190.
45*a* K. R. Brower, W. P. Samuels, J. W. Way and E. D. Amstutz, *J. Org. Chem.*, 18 (1953) 1648; *b* K. R. Brower, W. P. Samuels, J. W. Way and E. D. Amstutz, *J. Org. Chem.*, 19 (1954) 1830.
46 S. Horrobin, *J. Chem. Soc.*, (1963) 4130.
47*a* B. Bitter and H. Zollinger, *Angew. Chem.*, 70 (1958) 246; *b* B. Bitter and H. Zollinger, *Helv. Chim. Acta*, 44 (1961) 812.
48 H. Zollinger, *Angew. Chem.*, 73 (1961) 1257.
49 N. E. Sbarbarti, T. H. Suarez and J. A. Brieux, *Chem. and Ind.*, (1964) 1754.
50 G. E. Ficken and J. D. Kendall, *J. Chem. Soc.*, (1959) 3988.
51 J. Miller, *J. Chem. Soc.*, (1952) 3550.
52 J. F. Bunnett and R. J. Morath, *J. Am. Chem. Soc.*, 77 (1955) 5051.
53 P. B. D. de la Mare in W. Klyne and P. B. D. de la Mare (Eds.), *Progress in Stereochemistry*, Vol. 2, Butterworths, London, 1958, p. 85.
54*a* S. D. Ross and M. Finkelstein, *J. Am. Chem. Soc.*, 85 (1963) 2603; *b* S. D. Ross, in S. G. Cohen, A. Streitwieser, Jr. and R. W. Taft (Eds.), *Progress in Physical Organic Chemistry*, Vol. 1, Interscience, New York, 1963, p. 31.
55 G. Illuminati and G. Marino, *Chem. and Ind.*, (1963) 1287; *Tetrahedron Letters*, (1963) 1055.
56 G. Coppens, F. Declerck, C. Gillet and J. Nasielski, *Bull. Soc. Chim. Belges*, 70 (1961) 480; 72 (1963) 572.
57 R. Huisgen and J. Sauer, *Angew. Chem.*, 72 (1960) 91.
58 H. J. den Hertog, M. J. Pieterse and D. J. Buurman, *Rec. Trav. Chim.*, 82 (1963) 1173.
59 J. Ploquin, *Compt. Rend.*, 226 (1948) 339.
60 L. E. Orgel, T. L. Cottrell, W. Dick and L. E. Sutton, *Trans. Faraday Soc.*, 47 (1951) 113.

61 H. JAFFÉ, *J. Am. Chem. Soc.*, 76 (1954) 3527.
62 J. MILLER AND A. J. PARKER, *Australian J. Chem.*, 11 (1958) 302.
63 R. D. BROWN AND M. L. HEFFERNAN, *Australian J. Chem.*, (*a*) 9 (1956) 83; (*b*) 12 (1959) 543; (*c*) 13 (1960) 49.
64*a* A. R. KATRITZKY AND J. M. LAGOWSKI, *J. Chem. Soc.*, (1961) 43; *b* A. R. KATRITZKY AND F. J. SWINBOURNE, *J. Chem. Soc.*, (1965) 6707.
65*a* I. C. SMITH AND W. G. SCHNEIDER, *Can. J. Chem.*, 39 (1961) 1158; *b* T. SCHAFFER AND W. G. SCHNEIDER, *Can. J. Chem.*, 41 (1963) 972.
66 N. B. CHAPMAN AND D. Q. RUSSELL-HILL, *J. Chem. Soc.*, (1956) 1963.
67 J. MILLER AND H. W. YEUNG, unpublished work.
68*a* A. L. BECKWITH, G. D. LEAHY AND J. MILLER, *J. Chem. Soc.*, (1952) 3552; *b* G. P. BRINER, P. G. LUTZ, M. LIVERIS AND J. MILLER, *J. Chem. Soc.*, (1954) 1265; *c* B. A. BOLTO, J. MILLER AND V. A. WILLIAMS, *J. Chem. Soc.*, (1955) 2926; *d* G. D. LEAHY, M. LIVERIS, J. MILLER AND A. J. PARKER, *Australian J. Chem.*, 9 (1956) 382; *e* M. LIVERIS, P. G. LUTZ AND J. MILLER, *J. Am. Chem. Soc.*, 78 (1956) 3375; *f* J. MILLER AND K. Y. WAN, *J. Chem. Soc.*, (1963) 3492.
69 A. RICHARDSON, K. R. BROWER AND E. D. AMSTUTZ, *J. Org. Chem.*, 21 (1956) 890.
70 H. E. ZIMMERMANN, *Tetrahedron*, 16 (1961) 169.
71 R. R. BATES, R. H. CARNIGHAN AND C. E. STAPLES, *J. Am. Chem. Soc.*, 85 (1963) 3032.
72*a* B. A. BOLTO AND J. MILLER, *Chem. and Ind.*, (1953) 640; *b* B. A. BOLTO, M. LIVERIS AND J. MILLER, *J. Chem. Soc.*, (1956) 750; *c* B. A. BOLTO AND J. MILLER, unpublished work; *d* J. MILLER AND V. A. WILLIAMS, unpublished work.
73 L. PAULING, *The Nature of the Chemical Bond*, 3rd Edn., Cornell Univ. Press, Ithaca, N.Y., 1960, p. 93.
74 J. MILLER AND V. A. WILLIAMS, (*a*) *J. Chem. Soc.*, (1953) 1475; (*b*) *J. Am. Chem. Soc.*, 76 (1954) 5482.
75 T. L. CHAN AND J. MILLER, *Australian J. Chem.*, 20 (1967) 1595.
76 H. J. DEN HERTOG AND A. P. DE JONGE, *Rec. Trav. Chim.*, 67 (1948) 385.
77 J. H. M. HILL AND J. G. KRAUSE, *J. Org. Chem.*, 29 (1964) 1642.
78 H. YAMANAKA, *Chem. Pharm. Bull.* (*Tokyo*), 7 (1959) 297; *Chem. Abstr.*, 54 (1960) 24782f.
79 G. W. KENNER, C. B. REESE AND A. R. TODD, *J. Chem. Soc.*, (1955) 855.
80 G. W. H. CHEESEMAN, *J. Chem. Soc.*, (1960) 242.
81 A. E. ERICKSON AND P. E. S. POERRI, *J. Am. Chem. Soc.*, 68 (1946) 400.
82 G. PALAMIDESSI AND L. BERNARDI, *Gazz. Chim. Ital.*, 93 (1963) 343.
83 P. COAD, R. A. COAD AND J. HYEPOCK, *J. Org. Chem.*, 29 (1964) 1751.
84 A. PISKALA, J. GUT AND F. ŠORM, *Chem. and Ind.*, (1964) 1752.
85 O. A. STAMM, *J. Soc. Dyers. Col.*, 80 (1964) 416.
86 J. EPHRAIM, *Ber.*, 26 (1893) 2226, 2227.
87*a* C. GRUNDMANN, H. ULRICH AND A. KREUTZBERGER, *Chem. Ber.*, 86 (1953) 181; *b* H. SCHROEDER AND C. GRUNDMANN, *J. Am. Chem. Soc.*, 78 (1956) 2447; *c* H. SCHROEDER, *J. Am. Chem. Soc.*, 81 (1959) 5658.
88 R. HIRT, H. NIDECKER AND R. BERCHTOLD, *Helv. Chim. Acta*, 33 (1950) 1365.
89 M. GOI, *Yûki Gôsei Kagaku Kyôkaishi*, 18 (1960) 327, 332, 337; *Chem. Abstr.*, 54 (1960) 19702g.
90 K. MATSUI, K. HAGIWARA, A. HAYASHI, I. SAKAMOTO AND Y. SOEDA, *Yûki Gôsei Kagaku Kyôkaishi*, 18 (1960) 53, 97, 175, 184; *Chem. Abstr.*, 54 (1960) 5687e, 8843c, 11042e, 11043d.

91 H. Koopman, *Rec. Trav. Chim.*, 81 (1962) 465.
92 J. Murto, *Acta. Chem. Scand.*, 18 (1964) 1043.
93 V. A. Grakauskas, A. J. Tomasewski and J. P. Horwitz, *J. Am. Chem. Soc.*, 80 (1958) 3155.
94 H. J. Marcus and A. Remanick, *J. Org. Chem.*, 28 (1963) 2372.
95 R. J. Petfield and E. D. Amstutz, *J. Org. Chem.*, 19 (1954) 1944.
96 R. Motoyama, S. Nishimura, Y. Morakami, K. Hari and E. Imoto, *Nippon Kagashu Zasshi*, 78 (1957) 950; *Chem. Abstr.*, 54 (1960) 4224c.
97 T. S. Stevens in E. H. Rodd (Ed.), *Chemistry of Carbon Compounds* Vol. IV, Part A, Elsevier, Amsterdam, 1957, p. 35.
98*a* J. M. Sprague and A. H. Land, in R. C. Elderfield (Ed.), *Heterocyclic Compounds*, Vol. 5, Wiley, New York, p. 545; *b* T. L. Jacobs, in R. C. Elderfield (Ed.), *Heterocyclic Compounds*, Vol. 5, Wiley, New York, p. 102; *c* E. S. Schipper and A. O. Day, in R. C. Elderfield (Ed.), *Heterocyclic Compounds*, Vol. 5, Wiley, New York, p. 208.
99 L. I. Gorbacheva, I. I. Grandberg and A. N. Kost, *Zh. Obshch. Khim.*, 34 (1964) 650.
100 F. Koehler, *J. Prakt. Chem.*, 21 (1963) 50.
101 J. Goerdeler and K. H. Heller, *Chem. Ber.*, 97 (1964) 225.
102*a* R. Zahradnik and J. Koutecky, *Collection Czech. Chem. Commun.*, 26 (1961) 156; *b* R. Zahradnik and J. Koutecky, *Tetrahedron Letters*, (1961) 631.
103*a* B. Böttcher and A. Lüttringhaus, *Ann.*, 557 (1947) 89; *b* U. Schmidt, A. Lüttringhaus and F. Hübinger, *Ann.*, 631 (1960) 138.
104*a* E. Klingsberg, *Chem. and Ind.*, (1960) 1568; *b* E. Klingsberg, *J. Am. Chem. Soc.*, 83 (1961) 2930; 84 (1962) 2491, 3410.
105 D. Leaver, W. A. H. Robertson and D. M. McKinnon, *J. Chem. Soc.*, (*a*) (1962) 5104; (*b*) (1965) 32.
106*a* E. Campaigne and R. D. Hamilton, *J. Org. Chem.*, 29 (1964) 171; *b* C. Paulmier, *Bull. Soc. Chim. France*, (1965) 2643.
107 H. Prinzbach and E. Futterer in A. R. Katritzky (Ed.), *Advances in Heterocyclic Chemistry*, Vol. 7, Academic Press, New York, 1966, p. 39–151.
108 J. R. Keneford, J. S. Morley, J. C. E. Simpson and P. H. Wright, *J. Chem. Soc.*, (1950) 1104.
109 K. Adachi, *Yakugaku Zasshi*, 75 (1955) 1426.
110*a* K. Schofield and T. Swain, *J. Chem. Soc.*, (1950) 392, 394; *b* E. J. Alford and K. Schofield, *J. Chem. Soc.*, (1953) 1811.
111 D. Harrison and J. T. Ralph, *J. Chem. Soc.*, (1965) 236.
112 A. Ricci, M. Fòa, P. E. Todesco and P. Vivarelli, *Tetrahedron Letters*, (1965) 1935.
113 V. Petrow and B. Sturgeon, *J. Chem. Soc.*, (1949) 1157.
114 J. T. Adams, C. K. Bradsher, D. S. Breslow, J. T. Amore and C. R. Hauser, *J. Am. Chem. Soc.*, 68 (1946) 1317.
115 W. Czuba, *Rec. Trav. Chim.*, 82 (1963) 1988.
116 A. Albert, *J. Chem. Soc.*, (1960) 1790.
117 H. Rapoport and A. D. Batcho, *J. Org. Chem.*, 28 (1963) 1753.
118 N. Ikekawa, *Chem. Pharm. Bull.* (*Tokyo*), 6 (1958) (*a*) 263, (*b*) 401, (*c*) 269.
119 A. Dornow and J. von Loh, *Arch. Pharm.*, 290 (1957) 136.
120 A. Mangini and M. Colonna, *Gazz. Chim. Ital.*, 73 (1943) 323.
121 S. Carboni, A. da Settimo and G. Pirisino, *Ann. Chim.* (*Rome*), 54 (1964) 677.

122 G. Giacomello, F. Gualtieri, F. M. Riccieri and M. L. Stein, *Tetrahedron Letters*, (1965) 1117.
123 L. Birkofer and C. Kaiser, *Chem. Ber.*, 90 (1957) 2933.
124 B. M. Ferrier and N. Campbell, *J. Chem. Soc.*, (1960) 3513.
125 J. M. Bobbitt and R. E. Doolittle, *J. Org. Chem.*, 29 (1964) 2298.
126*a* E. Y. Sutcliffe and R. K. Robins, *J. Org. Chem.*, 28 (1963) 1622; *b* R. K. Robins, *J. Am. Chem. Soc.*, 80 (1958) 6671.
127*a* B. Pullman, *J. Org. Chem.*, 29 (1964) 508; *b* B. Pullman, *J. Chem. Soc.*, (1959) 1621.
128 G. B. Barlin and N. B. Chapman, *J. Chem. Soc.*, (1965) 3017.
129*a* R. L. Miller and P. G. Lykes, *Tetrahedron Letters*, (1962) 493; *b* R. L. Miller, P. G. Lykes and H. N. Schmeising, *J. Am. Chem. Soc.*, 84 (1962) 4623.
130 S. F. Mason in *The Chemistry and Biology of Purines, Ciba Foundation Symposium*, Little, Brown and Co., Boston, Mass., 1957, p. 72.
131 R. M. Cresswell and G. B. Brown, *J. Org. Chem.*, 28 (1963) 2560.
132*a* A. Albert and W. L. F. Armarego in A. R. Katritzky (Ed.), *Advances in Heterocyclic Chemistry*, Vol. 4, Academic Press, New York, 1965, p. 1–42; *b* A. Albert, D. J. Brown and G. Cheeseman, *J. Chem. Soc.*, (1951) 474; (*c*) (1952) 1620; (*d*) (1954) 3832; (*e*) (1964) 1666; (*f*) (1965) 27.
133 J. W. Daly and B. E. Christensen, *J. Am. Chem. Soc.*, 78 (1956) 225.
134*a* J. J. McCormack and H. G. Mautner, *Abstr. of Papers, 117th Meeting of the Am. Chem. Soc.*, New York, 1963, p. 16–O; *b* J. J. McCormack and H. G. Mautner, *J. Org. Chem.*, 29 (1964) 3337.
135 E. C. Taylor, R. J. Knopf, J. A. Cogliano, J. W. Barton and W. Pflederer, *J. Am. Chem. Soc.*, 82 (1960) 6058.
136 H. Larive and R. J. Dennilauler, *U.S. Patent*, 3,149,105; *Chem. Abstr.*, 61 (1964) 14830h.
137 M. D. Johnson, *J. Chem. Soc.*, (1960) 200.
138 R. M. Acheson and D. M. Goodall, *J. Chem. Soc.*, (1964) 3225.
139*a* R. D. Brown and R. D. Harcourt, *J. Chem. Soc.*, (1959) 3451; *b* R. D. Brown and R. D. Harcourt, *Tetrahedron*, 8 (1960) 23.
140 J. Miller and M. L. Sung, Unpublished work.
141 A. Fozard and G. Jones, *J. Chem. Soc.*, (1964) 3030.
142*a* W. Kloetzer, *Monatsh. Chem.*, 87 (1956) 131; 526; 536; *b* W. Kloetzer and J. Schantl, *Monatsh. Chem.*, 94 (1963) 1190; *c* W. Kloetzer and H. Bretschneider, *Monatsh. Chem.*, 87 (1956) 134; *d* H. Bretschneider and W. Kloetzer, *Monatsh. Chem.*, 87 (1956) 120.
143*a* Z. Talik, *Bull. Acad. Sci. Polon.* (*Chim.*), 9 (1961) 567; *b* Z. Talik, *Roczniki Chem.*, 36 (1962) 1313.
144 E. Ott and E. Ohse, *Ber.*, 54 (1921) 179.
145 C. V. Hart, *J. Am. Chem. Soc.*, 50 (1928) 1929.
146 F. R. Benson, L. W. Hartzel and E. A. Otten, *J. Am. Chem. Soc.*, 76 (1954) 1861.
147 T. Itai and S. Kamiya, *Chem. Pharm. Bull.* (*Tokyo*), 11 (1963) 1059.
148*a* W. E. Taft and R. G. Shepherd, *J. Med. Pharm. Chem.*, 5 (1962) 1335; *b* R. G. Shepherd, W. E. Taft and H. M. Krazinski, *J. Org. Chem.*, 26 (1961) 2764.
149 C. W. Noell and R. K. Robins, *J. Am. Chem. Soc.*, 81 (1959) 5997.
150 R. O. Clinton and C. M. Suter, *J. Am. Chem. Soc.*, 70 (1948) 491.
151 R. A. Abramovitch, F. Helmer and J. G. Saha, *Canad. J. Chem.*, 43 (1965) 725.

152 N. J. Daly, G. Kruger and J. Miller, *Australian J. Chem.*, 11 (1958) 290.
153 R. C. Elderfield and M. Siegel, *J. Am. Chem. Soc.*, 73 (1951) 5622.
154*a* K. H. Schaarf and P. E. Spoerri, *J. Am. Chem. Soc.*, 71 (1949) 2043; *b* G. Karmas and P. E. Spoerri, *J. Am. Chem. Soc.*, 79 (1957) 680.
155 A. Bryson, *J. Am. Chem. Soc.*, 82 (1960) 4871.
156*a* G. Illuminati and G. Marino, *J. Am. Chem. Soc.*, 80 (1958) 1421; *b* E. Baciocchi, G. Illuminati and G. Marino, *J. Am. Chem. Soc.*, 80 (1958) 2270; *c* M. L. Belli, G. Illuminati and G. Marino, *Tetrahedron*, 19 (1963) 345; *d* M. Calligaris, P. Linda and G. Marino, *Tetrahedron*, 23 (1967) 813.
157 N. E. Sbarbarti, *J. Org. Chem.*, 30 (1965) 3365.
158 H. H. Jaffé and H. L. Jones, in A. R. Katritzky (Ed.), *Advances in Heterocyclic Chemistry*, Vol. 3, Academic Press, New York, 1964, p. 209–261.
159*a* T. Kauffmann, *Angew. Chem.*, 77 (1965) 557; *Intern. Ed.*, 4 (1965) 543; *b* T. Kauffmann, J. Hansen, K. Udluft and R. Wirthwein, *Angew. Chem.*, 76 (1964) 590; *c* T. Kauffmann and K. Udluft, *Angew. Chem.*, 75 (1963) 89.
160 R. Stoermer and B. Kahlert, *Ber.*, 35 (1902) 1633.
161 H. L. Jones and D. L. Beveridge, *Tetrahedron Letters*, (1964) 1577.
162*a* J. A. Zoltewicz and C. L. Smith, *J. Am. Chem. Soc.*, 88 (1966) 4766; *b* R. A. Abramovitch, G. M. Singer and A. R. Vinutha, *Chem. Commun.*, (1967) 55.
163 G. Komppa and S. Weckman, *J. Prakt. Chem.*, 138 (1933) 109.
164 G. Wittig and V. Wahl, *Angew. Chem.*, 73 (1961) 492.
165 W. Czuba, *Rec. Trav. Chim.*, 82 (1963) 997.
166 E. K. Mellon and J. J. Lagowski, *Advan. Inorg. Chem. Radiochem.*, 5 (1963) 259.
167 J. C. Sheldon and B. C. Smith, *Quart. Revs.*, 14 (1960) 200.
168 L. F. Audrieth, R. Steinman and A. D. F. Toy, *Chem. Rev.*, 32 (1943) 109.
169 J. R. van Wazer, *Phosphorus and its Compounds*, Vol. 1, Interscience, New York, 1958, p. 309–344.
170 N. L. Paddock and H. T. Searle, *Advan. Inorg. Chem. Radiochem.*, 1 (1959) 347.
171 I. A. Gribova and U. Ban-Yuan, *Russian Chem. Rev.* (*Engl. transl.*) 30 (1961) 1.
172 R. A. Shaw, B. W. Fitzsimmons and B. C. Smith, *Chem. Rev.*, 62 (1962) 247.
173 C. D. Schmulbach in F. A. Cotton (Ed.), *Progress in Inorganic Chemistry*, Vol. 4, Interscience, New York, 1962, p. 275–379.
174 N. L. Paddock, *Quart. Rev.*, 18 (1964) 168.
175 C. W. Allen, *J. Chem. Educ.*, 44 (1967) 38.
176 D. L. Coursen and J. L. Hoard, *J. Am. Chem. Soc.*, 74 (1952) 1742.
177 A. Stock and R. Wierl, *Z. Anorg. Chem.*, 203 (1931) 228.
178 W. C. Price, R. D. Fraser, T. S. Robinson and H. C. Longuet-Higgins, *Disc. Faraday Soc.*, 9 (1950) 131.
179 H. Watanabe, K. Ito and M. Kubo, *J. Am. Chem. Soc.*, 82 (1960) 3294.
180 J. Goubeau and H. Keller, *Z. Anorg. Chem.*, 272 (1953) 303.
181 R. A. Spurr and S. Chang, *J. Chem. Phys.*, 19 (1951) 518.
182*a* R. G. Jones and C. R. Kinney, *J. Am. Chem. Soc.*, 61 (1939) 1378; *b* C. R. Kinney and C. L. Mahoney, *J. Org. Chem.*, 8 (1943) 526.
183 R. K. Bartlett, H. S. Turner, R. J. Warne, M. A. Young and W. S. McDonald, *Proc. Chem. Soc.*, (1962) 153.
184 M. J. Bradley, G. E. Ryschkewitsch and H. H. Sisler, *J. Am. Chem. Soc.*, 81 (1959) 2635.

185 G. L. BRENNAN, G. H. DAHL AND R. SCHAEFFER, *J. Am. Chem. Soc.*, 82 (1960) 6248.
186 D. T. HAWORTH AND L. F. HOHNSTEDT, *J. Am. Chem. Soc.*, 81 (1959) 842.
187 K. NIEDENZU, H. BEXER AND H. JENNE, *Chem. Ber.*, 96 (1963) 2649.
188 G. C. FINGER AND C. W. KRUSE, *J. Am. Chem. Soc.*, 78 (1956) 6034, and subsequent papers.
189 H. S. TURNER AND R. J. WARNE, *J. Chem. Soc.*, (1965) 6421.
190 R. E. PARKER in M. STACEY, J. C. TATLOW AND A. G. SHARPE (Eds.) *Advances in Fluorine Chemistry*, Vol. 3, Butterworths, London, 1963, pp. 63–91.
191 J. H. SMALLEY AND S. F. STAFIEJ, *J. Am. Chem. Soc.*, 81 (1959) 582.
192 K. NIEDENZU AND J. W. DAWSON, *J. Am. Chem. Soc.*, 81 (1959) 3561.
193 W. GERRARD, H. L. HUDSON AND E. F. MOONEY, *J. Chem. Soc.*, (1962) 113.
194*a* D. P. CRAIG, in *Theoretical Organic Chemistry, Proc. of the Kekulé Symposium*, Chemical Society, London, 1959, p. 20; *b* D. P. CRAIG AND N. L. PADDOCK, *Nature*, 181 (1958) 1052; *J. Chem. Soc.*, (1962) 4118; *c* D. P. CRAIG, *J. Chem. Soc.*, (1959) 997; *d* D. P. CRAIG, M. L. HEFFERNAN, R. MASON AND N. L. PADDOCK, *J. Chem. Soc.*, (1961) 1376; *e* D. P. CRAIG AND K. A. R. MITCHELL, *J. Chem. Soc.*, (1965) 4682.
195 M. J. S. DEWAR, E. A. L. LUCKEN AND M. A. WHITEHEAD, *J. Chem. Soc.*, (1960) 2423.
196 D. W. J. CRUICKSHANK, *J. Chem. Soc.*, (1961) 5486.
197 H. MCCOMBIE, B. C. SAUNDERS AND G. J. STACEY, *J. Chem. Soc.*, (1945) 380, 921; (1948) 695.
198 I. DOSTROVSKY AND M. HALMANN, *J. Chem. Soc.*, (1953) 502, 508, 511, 516.
199 M. GREEN AND R. F. HUDSON, *Proc. Chem. Soc.*, (1959) 227; (1962) 307.
200 S. B. HARTLEY, W. S. HOLMES, J. K. JACQUES, M. F. MOLE AND J. C. MCCOUBREY, *Quart. Rev.*, 17 (1963) 204.
201 F. SEEL AND J. LANGER, *Z. Anorg. Chem.*, 295 (1958) 316.
202 T. MOELLER AND S. G. KOKALIS, *J. Inorg. Nucl. Chem.*, 25 (1963) 1397.
203*a* M. BECKE-GOEHRING, K. JOHN AND E. FLUCK, *Z. Anorg. Allgem. Chem.*, 302 (1959) 103; *b* M. BECKE-GOEHRING AND K. JOHN, *Z. Anorg. Allgem. Chem.*, 304 (1960) 126.
204*a* A. C. CHAPMAN, D. H. PAINE, H. T. SEARLE, D. R. SMITH AND R. F. M. WHITE, *J. Chem. Soc.*, (1961) 1768; *b* M. L. HEFFERNAN AND R. F. M. WHITE, *J. Chem. Soc.*, (1961) 1382; *c* G. ALLEN, M. BARNARD, J. EMSLEY, J. L. PADDOCK AND R. F. M. WHITE, *Chem. and Ind.*, (1963) 952.
205*a* A. P. CARROLL AND R. A. SHAW, *Chem. and Ind.*, (1962) 1908; *b* B. CAPON, K. HILLS AND R. A. SHAW, *Proc. Chem. Soc.*, (1962) 390; *J. Chem. Soc.*, (1965) 4059; *c* S. K. RAY AND R. A. SHAW, *J. Chem. Soc.*, (1961) 872; *d* R. A. SHAW AND F. B. G. WELLS, *Chem. and Ind.*, (1959) 152.
206 K. JOHN, T. MOELLER AND L. F. AUDRIETH, *J. Am. Chem. Soc.*, 82 (1960) 5616.
207 J. V. BAILEY AND R. E. PARKER, *Chem. and Ind.*, (1962) 1823.
208 C. W. L. BEVAN AND J. HIRST, *J. Chem. Soc.*, (1956) 254.
209 J. MILLER AND A. J. PARKER, *J. Am. Chem. Soc.*, 83 (1961) 117.
210 A. J. PARKER, *Quart. Rev.*, 16 (1962) 163.
211 I. D. LANTZKE AND J. MILLER, unpublished work.
212 W. GREIZERSTEIN, R. A. BONELLI AND J. A. BRIEUX, *J. Am. Chem. Soc.*, 84 (1962) 1026.
213*a* G. A. WIEGERS AND A. VOS, *Proc. Chem. Soc.*, (1962) 387; *b* G. A. WIEGERS AND A. VOS, *Acta Cryst.*, 14 (1961) 562.

214 A. J. BANNISTER AND A. C. HAZELL, *Proc. Chem. Soc.*, (1962) 282.
215 M. BECKE-GOEHRING, *Advan. Inorg. Radiochem.* 2 (1962) 159.
216 O. GLEMSER, S. AUSTIN AND F. GERHART, *Chem. Ber.*, 97 (1964) 1262.
217 J. C. VAN DE GRAMPEL AND A. VOS, *Rec. Trav. Chim.*, 82 (1963) 286.
218 M. J. S. DEWAR, *Nature*, 155 (1945) 50.
219*a* W. VON E. DOERING AND L. H. KNOX, *J. Am. Chem. Soc.*, 76 (1954) 3203; *b* W. VON E. DOERING AND H. KRAUCH, *Angew. Chem.*, 68 (1956) 661; *c* W. VON E. DOERING, in *Theoretical Organic Chemistry, Proc. of the Kekulé Symposium*, Chemical Society, London, 1959, p. 35–48.
220 J. W. COOK AND J. D. LOUDON, *Quart. Rev.*, 5 (1951) 99.
221*a* A. W. JOHNSON, *J. Chem. Soc.*, (1954) 1331; *b* A. W. JOHNSON, *Sci. Progr. (London)*, 45 (1957) 86.
222 W. BAKER AND J. F. W. MCOMIE in J. COOK (Ed.), *Progress in Organic Chemistry*, Vol. 3, Butterworths, London, 1955, p. 44–80.
223 P. L. PAUSON, *Chem. Rev.*, 55 (1955) 9.
224*a* T. NOZOE, in D. GINSBURG (Ed.), *Non-benzenoid Aromatic Compounds*, Interscience, New York, 1959, p. 339–464; *b* T. NOZOE, in J. COOK AND W. CARUTHERS (Eds.), *Progress in Organic Chemistry*, Vol. 5, Butterworths, London, 1961, p. 132–165; *c* T. NOZOE, *Chemistry of the Tropylium and Related Compounds*, Elsevier, Amsterdam, in preparation.
225 H. J. DAUBEN, F. A. GADECKI, M. HARMON AND D. L. PEARSON, *J. Am. Chem. Soc.*, 79 (1957) 4557.
226*a* N. G. FATELEY AND E. R. LIPPINCOTT, *J. Am. Chem. Soc.*, 77 (1955) 244; *b* R. D. NELSON, N. G. FATELEY AND E. R. LIPPINCOTT, *J. Am. Chem. Soc.*, 78 (1956) 4870.
227*a* M. E. VOL'PIN, S. I. ZHDANOV AND D. N. KURSANOV, *Compt. Rend. Acad. Sci. (URSS)*, 112 (1957) 264; *b* V. I. BELOVA, M. E. VOL'PIN AND YA. K. SYRKIN, *Zh. Obshch. Khim.*, 29 (1959) 693.
228 P. ZUMAN, J. CHODKOWSKI, H. POTEŠIOLVA AND F. ŠANTAVY, *Nature*, 182 (1958) 1535.
229*a* M. E. VOL'PIN, D. N. KURSANOV, M. M. SHEMYAKIN, N. J. MAINUND AND L. H. NEYAMAN, *Chem. Ind.*, (1958) 1262; *b* M. E. VOL'PIN, I. S. AKHREM AND D. N. KURSANOV, *Khim. Nauk i. Prom.*, 2 (1957) 656; (c) *Zh. Obshch. Khim.*, 28 (1958) 330.
230*a* S. I. ZHDANOV, D. N. KURSANOV, V. N. STETKINA AND A. I. SHATENSTEIN, *Bull. Acad. Sci. (URSS) (Chem.)*, (1959) 754; *b* D. N. KURSANOV, *Chem. Abstr.*, 58 (1963) 4398h, 4399g.
231 E. J. FORBES, M. J. GREGORY, T. A. HAMOR AND D. J. WATKIN, *Chem. Commun.*, (1966) 114.
232 C. JUTZ AND F. VOITHENLEITNER, *Chem. Ber.*, 97 (1964) 29.
233 R. D. BROWN, *J. Chem. Soc.*, (1951) 2670.
234 B. C. PRATT AND O. W. WEBSTER, *U.S. Patent* 2,894,989; *Chem. Abstr.*, 55 (1961) 420b.
235 R. MEIER, B. SCHAB AND L. KNIPP, *Experientia*, 10 (1954) 74.
236*a* Y. SASADA AND I. NITTA, *Acta Cryst.*, 9 (1956) 205; *b* Y. SASADA AND I. NITTA, *Bull. Chem. Soc. Japan*, 30 (1957) 62.
237 R. SHIONO, *Acta Cryst.*, 14 (1961) 42.
238 M. J. S. DEWAR, *Nature*, 166 (1950) 790.
239 Y. KURITA AND Y. KUBO, *Bull. Chem. Soc. Japan*, 24 (1951) 13.
240 T. NOZOE AND Y. KITAHARA, *Proc. Japan Acad.*, 30 (1954) 204.
241 T. KAUFFMANN, A. RISBERG, J. SCHULTZ AND R. WEBER, *Tetrahedron Letters*, (1964) 3563.

242 F. G. Bordwell, B. B. Lampert and W. H. McKellin, *J. Am. Chem. Soc.*, 71 (1949) 1702.
243 E. Berliner, *J. Am. Chem. Soc.*, 68 (1946) 49.
244 M. G. Evans, J. Gergely and J. de Heer, *Trans. Faraday Soc.*, 45 (1949) 312.
245 L. F. Fieser, *J. Am. Chem. Soc.*, 48 (1926) 2922.
246 J. W. Dodgson, *J. Chem. Soc.*, (1930) 2498.
247*a* R. H. Thomson, *J. Org. Chem.*, 13 (1948) 377, 870; *b* J. W. Macleod and R. H. Thomson, *J. Org. Chem.*, 25 (1960) 36.
248 M. J. S. Dewar, *Electronic Theory of Organic Chemistry*, Oxford Univ. Press, London, 1949, p. 195.
249 A. H. Crosby and R. E. Lutz, *J. Am. Chem. Soc.*, 78 (1956) 1233.
250*a* K. Wallenfels and W. Draber, *Chem. Ber.*, 90 (1957) 2819; *b* K. Wallenfels and K. Friedrich, *Chem. Ber.*, 93 (1960) 3070.
251 Y. Hopff and Y. Y. Schweizer, *Helv. Chim. Acta*, 45 (1962) 313.
252 K. C. Ho and J. Miller, *Australian J. Chem.*, 19 (1966) 423.
253 C. A. Bishop, R. F. Porter and L. K. J. Tong, *J. Am. Chem. Soc.*, 85 (1963) 3991.
254*a* J. Miller, *J. Am. Chem. Soc.*, 76 (1954) 448; *b* R. L. Heppolette, J. Miller and V. A. Williams, *J. Am. Chem. Soc.*, 78 (1956) 1975; *c* N. S. Bayliss, R. L. Heppolette, L. H. Little and J. Miller, *J. Am. Chem. Soc.*, 78 (1956) 1978.
255 M. A. Adeniran, C. W. L. Bevan and J. Hirst, *J. Chem. Soc.*, (1963) 5868.
256 M. F. Sartori, *Chem. Rev.*, 63 (1963) 279.
257 C. Liebermann, *Ber.*, 32 (1899) 923.
258 G. A. Reynolds, J. A. VanAllen and R. E. Adel, *J. Org. Chem.*, 30 (1965) 3819.

Chapter 8

MEDIUM, CATALYTIC AND STERIC EFFECTS

1. Solvent Effects

(*a*) *Reactions of anions*

Nucleophilic substitution reactions with anionic reagents have commonly been carried out in protic solvents (having an acidic or hydrogen-bonding proton) or in mixed protic and aprotic solvents.* In recent years however the marked effect on reaction rates of carrying out such reactions in pure aprotic solvents and in the absence of solvent, *i.e.* in the melt, has become widely known *[1–12]*.

The importance of this in synthetic aromatic S_N reactions was indicated by the general procedures of Finger and his co-workers *[2]* for replacing chlorine or other heavy halogen atoms or the nitro group by fluorine, through reaction with fluorides in dipolar aprotic solvents or in the melt. This clearly indicates high reactivity (kinetic nucleophilicity) of fluoride ion in these conditions, in contrast to its low reactivity in protic solvents, and this is supported by the fact that potassium, rubidium and cesium fluoride are reactive, whereas lithium and sodium fluoride in which anion–cation ion-pair formation is more facile, are unreactive.

Miller and Parker *[7a]* have compared in considerable detail the effects of protic and dipolar aprotic solvents in activated aromatic nucleophilic substitutions. An essential feature of their discussion is that for the common dipolar aprotic solvents the dipole moment is larger than for protic solvents, and the negative end of the dipole is more fully exposed. Thus cations are very well solvated, probably better than in protic solvents *[cf. 13]*. With the positive end of the dipole well-shielded and lacking the ability to form hydrogen-bonds with anions, aprotic solvents solvate small anions poorly, but solvate large anions to an extent comparable with protic solvents. Protic

* The terms protonic and aprotonic solvents are better, but are not yet in common use.

References p. 361

solvents have the positive end of their dipole well exposed and, with the additionally favourable ability to form hydrogen-bonds, solvate anions well. Often, as with water and the lower alcohols, the negative end of the dipole is also unshielded and cations are then solvated well. A solvent such as liquid sulphur dioxide with both ends exposed is somewhat different in character from other aprotic solvents, but lacks the hydrogen-bonding power of protic solvents. Solvents such as aromatic hydrocarbons have little or no dipolar character but are highly polarisable and their consequent solvating power has largely been underestimated in the past. Nevertheless they are less effective than the dipolar aprotic solvents.

The dielectric constant is not directly relevant to the above discussion but is important in reactions of ions, since in solvents of low dielectric constant ion-pair and ion-aggregate formation occurs with consequent reduction of reactivity. The level at which this becomes serious will clearly depend on the species, and is most marked when small ions are involved, but as a rough guide it is not very important for solvents with dielectric constants above 30, and is expected to operate to a considerable extent as a threshold effect.

Since it is the free anions which are reactive as nucleophiles or bases, *i.e.* they require to be desolvated* or separated out from intimate ion-pairs for reaction, greatly increased reactivity is expected, and is found, in aprotic solvents of sufficiently high dielectric constant to inhibit ion-pairing. The magnitude of the increase will be greatest for those small or multiply charged anions which are most strongly solvated in protic solvents. A further consequence is that characteristic orders of reactivity in protic solvents such as $I^- > Br^- > Cl^- > F^-$, and $SR^- > OR^- > OH^-$, are markedly affected, and this may be sufficient to cause reversal *[7]*.

Miller and Parker *[7a]* showed that the reactivity of sodium or tetraethylammonium azide in reactions with *p*-fluoro- or *p*-iodonitrobenzene is about 10^3–10^5 times greater in dipolar aprotic than in protic solvents, whereas with potassium or lithium thiocyanate or iodide the ratio is about 10^1–10^3 (allowing for incomplete dissociation of the lithium salts).

* Complete desolvation may not be necessary for the initiation of bond-formation, however.

Miller and his co-workers *[7c,d,e]*, in a quantitative approach to aromatic nucleophilic substitution, have included solvation enthalpy and entropy as major factors in calculating energies of activation and reaction, and in discussions of kinetic and thermodynamic nucleophilicity and basicity. Through lack of fundamental data the change from protic to aprotic solvents was considered only qualitatively, but the simple concept that the anions solvated most in protic solvents are most affected by the change to aprotic solvents, and thus gain markedly in reactivity, satisfactorily accounts for observed changes. It is not necessary to postulate that the relative magnitude of solvation of large and small anions is reversed by such a change, though there is some evidence for this *[7]*.

Parker *[7b]* has given at length evidence that these reactivity increases are of quite general occurrence, and has discussed evidence of parallel changes in the properties of anions. Among univalent anions, the largest increases in reactivity are expected for the small anions highly solvated in protic solvents, such as hydroxide, fluoride, and amide ions. Murto and Hiiro *[14a]* have measured the rate of reaction of hydroxide ion in dimethyl sulphoxide (DMSO)–water mixtures, from 0 up to 96.8% DMSO, and demonstrated a rate increase of about 10^5 with both *p*-fluoronitrobenzene and methyl iodide; and in work with methyl iodide *[14b]* they demonstrated an overall increase of 10^6–10^7 in the change to pure DMSO. Slightly less acceleration was found for methoxide and ethoxide ions in the change from alcohols to DMSO. Miller and Parker *[7a]* quoted a report by Langford that hydroxide ion is 10^7 times more basic in sulpholane (tetrahydrothiophen dioxide) than in water. The greatest changes occur at high concentrations of aprotic solvent and are $\Delta E^{\ddagger}$ ($\sim \Delta H^{\ddagger}$) dependent. Changes in $\Delta S^{\ddagger}$ also occur but these are characteristically adverse, as would be expected.

The increased reactivity of fluoride ion in aprotic solvents, and the order $F^- > Cl^-$, is inferred from the procedures of Finger and his co-workers *[2]* though of course halogen mobilities are also involved. It is also inferred with singly charged cationic substrates from the results of Bolton and Miller *[15]*. They showed that the reaction of fluoride ion with 2,4-dinitrophenylpyridinium ion in dipolar aprotic solvents readily produced 1-fluoro-2,4-dinitrobenzene, and that a 2,4-dinitrophenylpyridinium salt was *not* formed by treating pyridine

References p. 361

with 1-fluoro-2,4-dinitrobenzene in conditions where a good yield of the salt was obtained with 1-chloro-2,4-dinitrobenzene.

In view of the evidence that the reactivity increase is a general phenomenon, the work of Weaver and Hutchison *[16]* is relevant to this discussion. In reactions with alkyl tosylates they demonstrated the reactivity order $I^- > Br^- > Cl^-$ in aqueous dioxane and $Cl^- > Br^- > I^-$ in dimethylformamide (DMF). They showed further that a proportion of only 9.1% of water in a mixture with DMF depresses the reactivity of chloride ion some 24-fold, whereas there is only a 2-fold depression in reactivity of the larger iodide ion. For acetone, when using *lithium* salts, they demonstrated the order LiI > LiBr > LiCl. This shows the combined effects of using a small cation and a solvent of lower dielectric constant in facilitating ion-pair formation, which is adverse to reactivity, and which is in the order LiCl > LiBr > LiI. However, Winstein and his co-workers *[17]*, taking account of the effect of ion-pair formation, used tetrabutylammonium halides to demonstrate the reactivity order $Cl^- > Br^- > I^-$ also in acetone.

Kingsbury *[11]*, and Murto and Hiiro *[14a]*, have measured the reactivity of anions in a range of DMSO–water mixtures, the former author including low concentrations of DMSO. While the main increase in reactivity is when the proportion of protic solvent is low and approaches zero, there is a quite noticeable enhancement in rate even with small concentrations of DMSO. Kingsbury also showed that this was essentially independent of the substrate and that Hammett reaction constants were not affected. In the range 0 to 80% DMSO the rate of reaction of methoxide ion with *p*-fluoronitrobenzene was increased about 2000-fold. In the range 0 to 60% DMSO the enhancement of the reactivity of benzyl mercaptide ion was closely similar, in fact slightly larger, than that of methoxide ion and this is perhaps surprising at first sight. One explanation offered by Kingsbury is that the enhancement at the lower DMSO concentrations is an effect on the structure of the protic solvent. Such changes are known and have been referred to, for example, by Heppolette and Miller *[18]* for benzene–methanol mixtures. However with a number of activated substrates, up to 70% of benzene in the mixture has little effect on rates. Addition of the dipolar aprotic solvents, methyl acetate or dioxane, has more effect *[19,20]*, but not as large as that due to DMSO. Kingsbury has alternatively suggested that the change may be due to interaction between the DMSO and the substrate forming a more reactive species. A further alternative, now suggested, is that the admixture of DMSO causes a relative enhancement of solvation of the transition state. While the possibility that *large* anions may be better solvated in dipolar aprotic solvents than in protic solvents *[7]* is relevant, this need not be invoked. It is sufficient to give a rate enhancement, as described by Kingsbury *[11]*, if the decrease of solvation in forming the transition state from the initial state is less in DMSO–alcohol mixtures than in the pure alcohol. This new suggestion would also explain the comparative results with methoxide and benzyl mercaptide ions. What difference there is would be in the observed direction.

Recently Delpuech *[21]* has shown that azide ion reacts with n-butyl bromide in several protic solvents at similar rates with values of $\Delta E^{\ddagger} \geqslant 20$ kcal·mole^{-1}. In aprotic solvents rates are increased by a factor of about 10^3 and values of $\Delta E^{\ddagger}$ are $\leqslant 17$ kcal·mole^{-1}. One solvent, hexamethylphosphoramide $(NMe_2)_3PO$, causes a still greater enhancement and the value of $\Delta E^{\ddagger}$ is about 13.5 kcal·mole^{-1}. The difference however is probably in degree rather than in kind. Delpuech's results not only support the earlier reports that rate enhancement is $\Delta E^{\ddagger}$-dependent, but also show the unimportance of dielectric constant in the system chosen. His slow (protic) solvents have values of dielectric constant from 33 to 183, and the fast (aprotic) solvents from 30 to 90.

Cram and his co-workers *[22]* have demonstrated the marked increases in reactivity of tert-butoxide ion as a nucleophile and as a base, in dimethyl sulphoxide as compared with tert-butanol. They studied its reaction with simple bromobenzenes by the benzyne mechanism, and with simple fluorobenzenes in direct substitution by the addition–elimination mechanism.

In relation to the major changes in reactivity discussed above, effects of admixture of similar types of solvent are much less important, but by no means negligible. Murto *[23]* has discussed in detail hydroxide and alkoxide reactions with 1-fluoro-2,4-dinitrobenzene in alcohol–water mixtures. In these one must first make allowances for the equilibrium $OH^- + ROH \rightleftharpoons OR^- + H_2O$ *[cf. 24]* (see also Section 3), and this is particularly important since alkoxides are considerably more nucleophilic than hydroxide ion in protic solvents (see above and Chapter 6, p. 201). He showed that the "true rates" of reaction of hydroxide and alkoxide ions increase with the alcohol content of the solvent mixture. The increases in the latter rate are slight and not very different for the alcohols, methanol, ethanol and propan-2-ol, whereas the increase in hydroxide ion reactivity is also small in methanol, but larger in ethanol and still larger in propan-2-ol. This would match expected changes in solvating power of the alcohols for a small anion, *viz.* water $\geqslant$ methanol $>$ ethanol $>$ propan-2-ol. The differences are however much smaller than in a change from a protic to an aprotic solvent.

In these reactions between anions and neutral substrates the effect of the solvent change on the neutral substrate has been neglected by comparison with the much larger effect on the anion. However, Parker *[25]* has provided evidence that a neutral aromatic substrate, *p*-iodonitrobenzene, is more solvated, thus less reactive, in dimethyl formamide (DMF) than in methanol and that the 10^4-fold rate increase is due both to less solvation of the reagent, azide ion, and increased solvation of the transition state in DMF as compared with methanol.

The differences between protic and aprotic solvents in solvation of anionic reagents should be paralleled in their effects on displaced anionic groups. This would not however be observed as an effect on the overall rate of the reaction unless the formation of the second transition state is rate-limiting or it becomes so as a result of the solvent change. Such an effect is separate from an increase in rate of formation of the first transition state and intermediate complex when the reagent is an anion. For reactions of thiocyanate and iodide ions with 1-fluoro-2,4-dinitrobenzene it is predicted that formation of the second transition state is rate-limiting in protic solvents. In accord with the above discussion, in these reactions in dipolar aprotic solvents, reagents are consumed faster, but the intermediate complex formed undergoes only slow decomposition, not involving simple displacement of fluoride ion *[7]* (see also Section 3).

The reaction of azide ion with *p*-fluoronitrobenzene might have been expected to show a solvent effect on rate by a shift towards the formation of the second transition state as rate-limiting, but the observations with sodium azide in DMF are now known not to support this *[26]*. There are however some peculiar features about this reaction, since even when the concentration of sodium fluoride formed is far higher than its known solubility in the solvent, it is not precipitated. Fluorescence and viscosity changes on pouring the reaction mixture into water have also been reported *[7]*. Perhaps an intimate association of sodium and fluoride ions and the dipolar nitrophenylazide surrounded by a solvent sheath, is the normal final state in DMF. Some electrochemical investigations of this system and an investigation of this and the methoxide reaction in other aprotic solvents would be of interest.

(*b*) *Reactions of neutral reagents*

A useful survey by Suhr *[27]* of the reactions of piperidine with *inter alia*, the *p*-halogenonitrobenzenes, gives rates in a range of aprotic solvents, and in ethanol, a typical protic solvent. His data, together with a few results from other workers *[28,29]* are given in Table 65,

TABLE 65

RATE CONSTANTS ($10^6 k_2$, l·mole^{-1}·sec^{-1}) AND SOLVENT RATE RATIOS RELATIVE TO ETHANOL AS UNITY, SHOWN IN PARENTHESES, FOR THE REACTION OF PIPERIDINE WITH p-$NO_2C_6H_4$Hal IN VARIOUS SOLVENTS AT 50° *[27]*

	Halogen replaced			
Solvent	F	Cl	Br	I
Dimethyl sulphoxide	27400 (116)	66 (120)	80.2 (115)	18.4 (87.7)
Dimethyl-formamide	7400 (31.2)	22.8 (41.5)	27.1 (38.8)	6.9 (32.8)
Acetonitrile	1150 (4.85)	5.5 (10.0)	6.4 (9.15)	2.1 (10.0)
Benzonitrile	375 (1.58)	3.15 (5.72)	3.85 (5.50)	1.10 (5.25)
Methyl ethyl ketone	223 (0.940)	2.4 (4.4)	3.47 (5.30)	1.01 (4.80)
Ethyl acetate	43 (0.18)	0.32 (0.58)	0.70 (1.0)	0.40 (1.9)
Benzene	3.8 (0.016)	0.11[a] (0.20) 0.16[b] (—)	0.25[a] (0.36)	0.041[a] (0.20)
Ethanol	140 (—) 237[a] (1)	0.55[a] (1)	0.70[a] (1)	0.21[a] (1)

[a] Ref. *28*. [b] Ref. *29*.

and are graphically illustrated in Fig. 102. A clear pattern is visible, with a distinction between solvent effects in replacement of fluorine and of the other halogens. Reactions in the more polar of the aprotic solvents (strongly dipolar and with fairly high dielectric constants) are faster than in ethanol; reactions in benzene (non-polar but polarisable) are slower; and reactions in aprotic solvents of intermediate character are intermediate also in rate. Superimposed on this is a definite indication that the aprotic solvents other than DMF (dimethylformamide) and DMSO (dimethyl sulphoxide) show up less well in displacement of fluorine than of other halogens.

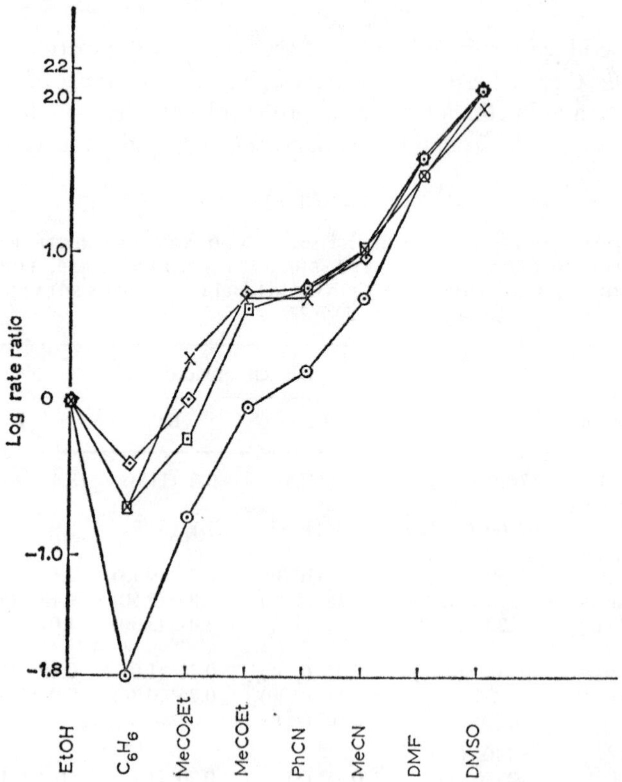

Fig. 102. Solvent rate ratios (EtOH = 1) in reactions of piperidine with *p*-halogenonitrobenzenes at 50° *[27]*. ⊙, F; ⊡, Cl; ◇, Br; ×, I.

The nature of the transition states and intermediate complexes in reactions between neutral reagents and neutral substrates was discussed in Chapter 6 (p. 204), including consideration of their energy–reaction coordinate profiles. Although formally dipolar, these intermediates are equivalent for solvation purposes to a large anion and large cation of the type ArY^- and ArX^+ linked by a polarisable π-electron system, and possessing a solvation energy not less than the sum of those of individual ions ArY^- and ArX^+.

The more polar of the aprotic solvents are expected to solvate cations particularly well, probably better than do protic solvents, since they have larger dipoles with the negative ends well exposed; and also to solvate *large* anions moderately well. Intermediates such as those described above should be formed as well or better in such solvents, in accord with the experimental results. While the lowly position of benzene is significant and expected, there is clear evidence of considerable solvating power, which may be ascribed to the high polarisability of benzene. Its ability and that of other hydrocarbons to act as hydrogen-bond acceptors *[30–32]*, for example, is in accord with this. There is little doubt that reactions in saturated aliphatic hydrocarbons would be much slower, and probably be autocatalytic as polar products are formed, but results are lacking except in reactions where internal solvation can occur and substitute for the external solvent, the distinction being thereby obscured.

The pattern, which demonstrates an inferior place for the less polar solvents in replacement of fluorine, suggests that in them the formation of the second transition state (T.St.2) may be rate-limiting. Correspondingly, in benzene, the F/I mobility ratio at 50° is only 93, compared with 1100 in ethanol and 1500 in DMSO, despite the probability *[28]* that hydrogen–fluorine interactions lower the energy of T.St.2 for the fluoro compound.

Similarly Hammond and Parks *[33]* have shown in the reactions of 1-halogeno-2,4-dinitrobenzenes with *N*-methylaniline that the mobility order F ~ Cl in ethanol becomes F < Cl in nitrobenzene and suggested that this change indicates the influence of the bond-breaking step.

The reactions of piperidine with *p*-chloronitrobenzene and 1-chloro-2,4-dinitrobenzene in benzene suggest that the formation of T.St.2 is kinetically significant in this solvent, for there are some

References p. 361

indications of base catalysis by piperidine leading to a third order term, *[34a,35]* (however, see Section *3a* below).

It is known *[36–40]* that in reactions of many neutral reagents with neutral substrates suitable *ortho* substituents facilitate reaction by intramolecular hydrogen-bonding in the rate-limiting transition states and intermediate complex (see Fig. 103), and this markedly levels out the differential effects of solvents. Favourable electrostatic interactions ("built-in" solvation) may also play a part, but hydrogen-bonding is the key factor, as was demonstrated by Ross *[38]* in comparing the reactivity of di-n-butylamine and 1,4-diaza(2-2-2)bicyclooctane with *o*- and *p*-chloronitrobenzene in benzyl alcohol at 150°. With the former, in which intramolecular hydrogen-bonding as in Fig. 103 is possible, the ratio k_o/k_p is $\geqslant$16, whereas with the latter the ratio is $\leqslant$1/250.

It is relevant to this that the results of Brieux and his co-workers *[34a]* showed that there is little difference in the reactivity of piperidine with *o*-chloronitrobenzene in benzene and ethanol, whereas there is a large difference (reaction in ethanol faster) with *p*-chloronitrobenzene. They obtained corresponding results with 2,4-dichloronitrobenzene in ethanol and benzene *[34b]* by use of radioactive chlorine. A similar levelling effect is seen in the results for reactions of piperidine with 2-chloroquinoline in a range of solvents quoted in Table 57 (p. 248), the *ortho* group being ring nitrogen. At 50°, solvent rate ratios are: piperidine, 1.67; ethanol, 0.862; toluene, 1; light petroleum, 0.243. Even the inclusion of a saturated aliphatic hydrocarbon in the series leads to a reactivity range of only about 7 to 1. Again, Illuminati and Marino *[39]* have compared the

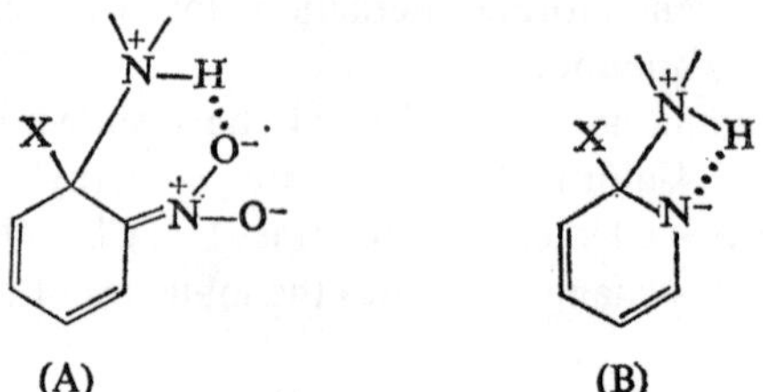

Fig. 103. Favourable intramolecular hydrogen-bonding in reactions of a primary or secondary amine with 1-X-2-nitrobenzenes and 2-X-pyridines.

reactivity of piperidine with 2- and with 4-chloroquinoline in piperidine, methanol and toluene. At 86.5° with 2-chloroquinoline, solvent rate ratios are: piperidine, 7.6; methanol, 6.3; toluene, 1. The values for the 4-chloro-compound are much more discriminating, *viz.* piperidine, 35; methanol, 1100; toluene, 1. Further, Illuminati and Marino showed that the reactivity of 2-chloroquinoline is at a level corresponding to that shown by 4-chloroquinoline in the *best* solvent, methanol. A further consequence of internal solvation in *ortho* compounds is shown by the relative activating power of *ortho*- and *para*-nitro groups referred to above. With anionic reagents and tertiary amines the order is *para*- > *ortho*-nitro; whereas with primary and secondary amines it is *ortho*- > *para*-nitro *[36,38]*. Also relevant is the fact that the *ortho*-carboxylate group is mildly activating in reactions with primary and secondary amines *[34]*, but deactivating in reactions with methoxide ion *[40]*.

The above results and especially the results in Tables 33 and 34 (p. 141) show that the reactivity order, protic solvent > aprotic solvent of low polarity, is due to high values of $\Delta S^{\ddagger}$ more than compensating for higher values of $\Delta E^{\ddagger}$ in protic solvents. These must lose considerable intermolecular hydrogen-bonding energy in order to solvate polar reaction-intermediates but in doing so gain in entropy. Neither of these related factors applies to an aprotic solvent which if sufficiently polar or polarisable to solvate the intermediates does so at the expense of "freezing" solvent molecules with a consequent reduction in entropy.

Data are lacking for consideration of solvent effects in reactions of neutral substrates with cationic reagents, and the scarce data for reactions with potentially anionic substrates are also unsuitable for considering reactivity with charged substrates. Berliner and Monack *[41]*, and Brieux and his co-workers *[34c]* have measured, *inter alia*, the rates of reaction of piperidine with *o*-chloronitrobenzene and 4-chloro-3-nitrobenzoic acid in piperidine and benzene respectively. Whereas with piperidine acting as solvent the acid may be assumed to be converted into the benzoate ion, this may not be the case in benzene in the conditions used by Brieux and his co-workers, and since the free benzoic acid is much more reactive, no reliable solvent effect can be inferred.

2. Salt and Counter Ion Effects

It has been pointed out by Frost and Pearson *[42]* that reactions between ions of unlike charge are characterised by high values of the entropy of activation ($\Delta S^{\ddagger}$), and reactions between ions of like charge by low values. They quoted examples in inorganic and organic systems and explained the phenomenon in general terms as a consequence of increase or decrease in solvation entropy in forming the transition state.

In reactions between ions of unlike charge the transition state has less total charge than the reacting species, and the consequent decreased solvation releases solvent molecules leading to an increase in entropy. In reactions between ions of like charge, the charge in the transition state, though not increased in total, is more closely associated than in the reactants. The consequent increased solvation binds solvent molecules leading to a decrease in entropy, but the magnitude of the entropy change is clearly greater for the reaction between unlike ions than between like ions. From the viewpoint of changing magnitude of charge, the reactions opposite to those between ions of unlike charge, are those between two neutral reactants forming charged transition states.

Effects on energy of activation ($\Delta E^{\ddagger}$), which is approximately equal to enthalpy of activation ($\Delta H^{\ddagger}$) in reactions in solution, are much more specific to the structure of the species involved, depending essentially on transmission of polar effects between different parts of a molecule, and thus do not demonstrate any parallel simple pattern.

As pointed out in earlier chapters, aromatic S_N reactions between anions and neutral substrates in the absence of special steric factors are characterised by having only a small range of $\Delta S^{\ddagger}$ values. With many common anionic nucleophiles in methanol, except when *both* entering and leaving groups are heavy atoms or attached to the reaction centre by heavy atoms, this range is approximately between –5 and –15 e.u., and very often between about –7 and –13 e.u. In contrast, reactions between neutral reagents and neutral substrates, which form charged transition states, are characterised by very negative values of $\Delta S^{\ddagger}$, usually in the range –32 to –42 e.u. This is the opposite pattern to that shown in reactions between ions of unlike charge, for which the values of $\Delta S^{\ddagger}$ should be very much more positive and exceed the values for reactions between an ion and a neutral substrate.

Miller and his co-workers *[43,44]* have demonstrated such high values of $\Delta S^{\ddagger}$ for reactions between univalent anionic nucleophiles and univalent cationic aromatic substrates in cases where the positive charge is distant from, as well as neighbouring the point at which reaction occurs. Some examples are shown in Fig. 104.

The actual values of $\Delta S^{\ddagger}$ in e.u. at 50° are: (A) +16.4; (B) –0.7; (C) +6.8; (D) +2.0; (E) –0.5. These are corrected to zero ionic strength (μ) except for (B)—the value for $\mu = 0$ would be positive. While there is some suggestion of structural and positional differences, all are clearly higher (less negative or more positive) than the anion–neutral-substrate reactions, differences being of the order of 10–20 e.u. and very much higher than the neutral-reagent–neutral-substrate reactions, differences being of the order of 40–50 e.u.

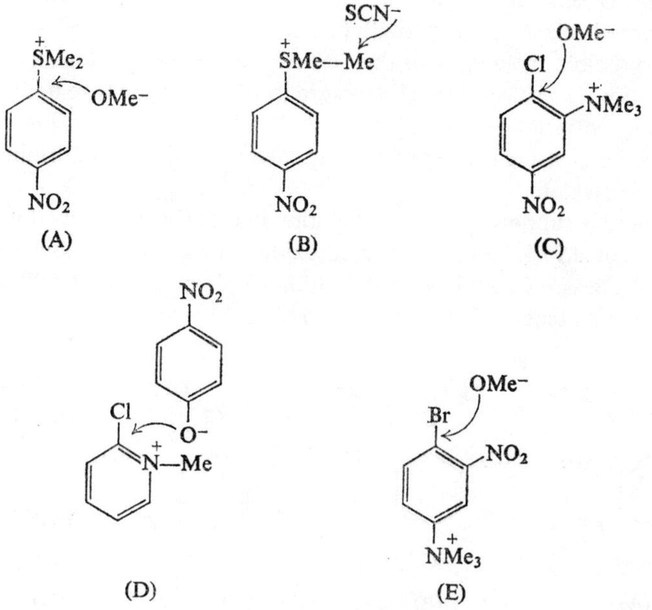

Fig. 104. Some anion–cation reactions possessing high values of entropy of activation ($\Delta S^{\ddagger}$).

References p. 361

The values of $\Delta S^{\ddagger}$ in reactions between anionic nucleophiles and anionic substrates, though expected to be low, should be between the extremes but nearer to values for anion–neutral-substrate reactions than to those for reactions of neutral reagent with neutral substrate, and thus have values $\leqslant -15$ e.u. Apart from an interesting group of exceptions discussed further below, this has also been confirmed by Miller and his co-workers *[43c,45]* in aromatic S_N reactions, examples being illustrated in Fig. 105.

The actual values of $\Delta S^{\ddagger}$ at 50° in e.u. are: (A) –18.6; (B) –23.5; (C) –21.0; (D) –21.1. Values for (B) and (D) are corrected to zero ionic strength (μ) and those for (A) and (C) are estimated values for zero μ, by assuming similar corrections.

The known group of exceptions to the reduction in $\Delta S^{\ddagger}$ for reactions of anions with anionic substrates is when the negative charge is *ortho* to the point of reaction *[46a]*. In these, the values of $\Delta S^{\ddagger}$ are *high*. Examples are given in Fig. 106 (A) and (B). Actual values of $\Delta S^{\ddagger}$ at 50° in e.u. are: (A) +3.1; (B) –4.0. Neither are corrected to zero ionic strength (μ).

A possible explanation is that in such cases the transition state may incorporate a cation-bridge structure in which the ions are not separated by solvent molecules. It is illustrated in Fig. 106 (C). This would result in releasing instead of binding solvent molecules in forming the transition state.

There is support for this explanation in that the markedly different values of $\Delta S^{\ddagger}$ in reactions of methoxide ion with sodium 4-chloro-3-nitrobenzoate ($\Delta S^{\ddagger}$ low) and 2-chloro-5-nitrobenzoate ($\Delta S^{\ddagger}$ high) are not matched in reactions of sodium 4-chloro-3-nitrocinnamate

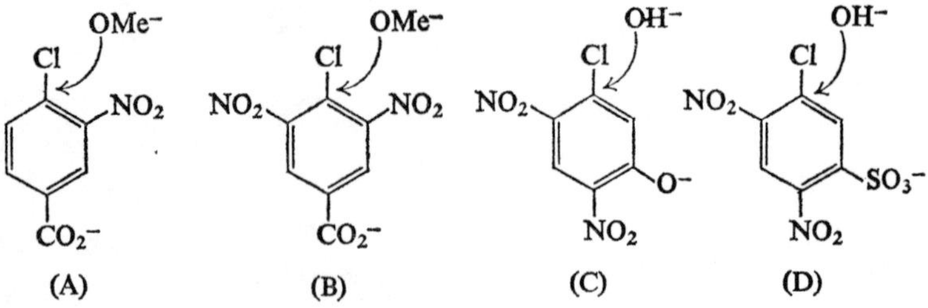

Fig. 105. Some anion–anion aromatic S_N reactions possessing moderately low values of entropy of activation ($\Delta S^{\ddagger}$).

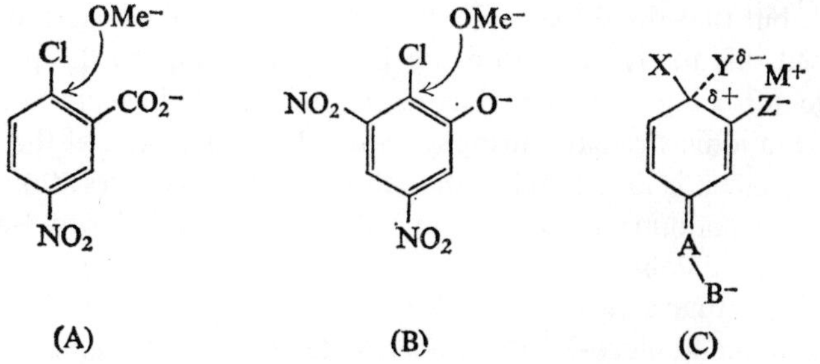

Fig. 106. Some anion–anion reactions with *high* values of the entropy of activation ($\Delta S^{\ddagger}$), (A) and (B); and a suggested general structure for transition states in such cases (C).

and sodium 2-chloro-5-nitrocinnamate *[46b,c]*. The negatively charged group is distant from the ring in both cases, and in *both* the values of $\Delta S^{\ddagger}$ are moderately low, *viz.* about –15 e.u. at 50°.

Apart from these gross effects on the entropy of activation, the Brønsted *[47]* theory of ionic-activity–rate relationships, discussed further by La Mer *[48]* and Davies *[49]* among others, is concerned with changes of activity of ions of charge Z_A and Z_B in forming the "activated complex" of charge $Z_A + Z_B$, *i.e.* $A^{Z_A} + B^{Z_B} \rightarrow (AB)^{Z_A+Z_B}$. In terms of the simple Debye–Hückel theory, as applied to dilute aqueous solutions, the Brønsted equation becomes

$$\log k = \log k_0 + 2CZ_A Z_B \mu^{1/2}$$

where k and k_0 are rate constants at experimental and zero ionic strength (μ); C is the Debye constant; and Z_A and Z_B are the charges of the reagent and the reactant respectively. In less ideal conditions some modified and more complex function of μ is required. When Z_A and Z_B have like sign, their product is positive, and the rate constant increases with ionic strength, *i.e.* $k_2{}^0 < k_2$; whereas with unlike signs the product is negative and the rate constant decreases with ionic strength, *i.e.* $k_2{}^0 > k_2$.

These general predictions have been confirmed for aromatic S_N reactions by Miller and his co-workers *[43–46]*. They showed for example that in the reaction of methoxide ion in methanol with dimethyl-*p*-nitrophenyl sulphonium ion, the plot of $\log_{10} k_2$ against $\mu^{1/2}$ is linear in the range considered, *viz.*, up to $\mu = 0.085$ moles·

References p. 361

litre^{-1}, but that the slope is less than the theoretical slope; and they ascribed this to (*i*) the cation not being spherical, (*ii*) the positive charge not being at the reaction centre, and (*iii*) the occurrence of a change in ionic strength during reaction. The magnitude of the ionic strength effect is illustrated in the following two examples: (*a*) in the reaction of sodium methoxide with *N,N,N*-trimethyl-4-chloro-3-nitroanilinium iodide in methanol at 25.6°, the value of k_2 at $\mu = 0$ is 3.4 times *larger* than at $\mu = 0.0130$; (*b*) in the reaction of sodium methoxide with sodium 4-chloro-3,5-dinitrobenzoate at 25.0°, the value of k_2 at $\mu = 0$ is 2.48 times *smaller* than at $\mu = 0.0147$.

These are gross effects in reactions between ions or in reactions between neutral species which form ions. Ionic strength effects in reactions between an ion and a neutral substrate are relatively small in bimolecular reactions. Ingold *[50a]* has discussed this in relation to aliphatic nucleophilic substitution but there has been little work in this area of aromatic S_N reactions.

Reinheimer and his co-workers *[51]* investigated the reactions of lithium, sodium and potassium methoxide with 1-chloro-2,4-dinitrobenzene in methanol in the presence of added lithium, sodium, and potassium salts. They showed that salt effects were indeed small. The most obvious effects are that lithium salts reduce the rates of sodium and potassium methoxide reactions, and that potassium salts increase the rates of lithium and sodium methoxide reactions. At the same time, in the absence of added salts, the reactivity order is KOMe > NaOMe > LiOMe. They ascribed these facts to the existence of limited ion-pairing in these conditions in the order, $Li^+OMe^- > Na^+OMe^- > K^+OMe^-$, so that reactivity due to "free" methoxide ion is in the reverse order. The salt effects are thus due to the added cation. The effects are less clear when salts with the same cation are added. However there are indications that those added salts, the ions of which are most likely to be completely solvent separated, *e.g.* perchlorates, exert a depressive common-ion (cation) effect on the separation of metal alkoxide ion-pairs. With other salts there are indications of a mild drift towards rate enhancement, and this might be caused by the addition of the salts enhancing the ionizing power of the medium sufficiently to overcome the common-ion effect. There is nevertheless an element of speculation in this explanation, and a good deal more information on such salt effects would be valuable. Marked reduction in availability of a reagent due to strong ion interactions with a highly charged cation are suggested by the results of Lam and Miller *[52]*, which demonstrate that addition of equimolar amounts of thorium nitrate (0.02*M*) in reactions of sodium azide with 1-fluoro-2,4-dinitrobenzene in methanol at 100° reduces rates some 1000-fold.

Any ion-pairing effects will be more potent in solvents of smaller ionising power, with dielectric constant an important feature. Thus Miller and Parker *[7a]* showed that in the reaction of potassium thiocyanate with 1-iodo-2,4-dinitrobenzene in acetone, the reaction is faster in more dilute solutions and this suggests a normal concentration effect on the ionisation equilibrium. Reaction is about 20% faster in 0.024*M* than a 0.045*M* solution, whereas there is little effect of concentration in DMF. A greater difference, though known only qualitatively, is illustrated by the ability of potassium, rubidium, and cesium, but not lithium and sodium, fluoride to replace heavy halogens in suitably activated aromatic substrates in aprotic solvents *[2]*.

In solvents which do not solvate ions well, neutral-reagent–neutral-substrate reactions which form ions, should be facilitated by added salts, and even by the reactants, so that higher order terms may occur in the rate equation. The product ions formed will however act in this way, so that effects of added salts are obscured. No systematic investigation of such salt effects in aromatic S_N reactions systems has been reported, though clearly desirable. The existence of large salt effects in ionisation reactions (aliphatic S_N1) in less polar solvents has been demonstrated, for example, by Winstein and his co-workers *[53]*.

3. Catalytic Effects

(*a*) *Base catalysis*

One form of base catalysis, which leads to a rate enhancement in solvolytic reactions, involves only the reagent and not the substrate. In protic solvents (Solv–H), formation of Ar–Solv compounds by reaction of the solvent acting as a nucleophile with suitable aromatic substrates is facilitated by such reagents as hydroxide and fluoride ions, and triethylamine *[7d,15,54–58]*. With alcohols as examples

References p. 361

of the solvent, interactions are (*i*) $OH^- + ROH \rightleftharpoons H_2O + OR^-$; (*ii*) $2F^- + ROH \rightleftharpoons HF_2^- + OR^-$; (*iii*) $Et_3N + ROH \rightleftharpoons Et_3NH^+ + OR^-$.

In each case, the catalyst acting as a base forms the lyate ion ($Solv^-$), but does not compete with it to any important extent as a nucleophile, these catalysts being more strongly basic than nucleophilic. Relevant factors were discussed on pages 188 and 204; see also ref. *7d* for consideration of hydroxide and fluoride ions. With triethylamine the low reactivity also involves steric factors, less potent in acid-base than electrophile–nucleophile reactions *[59]*, and this is supported for these reactions by Whalley's report *[57b]* of the unsatisfactory nature of pyridine as compared with triethylamine as catalyst. There is a good deal of evidence for another type of base catalysis, which involves interactions with the substrate or intermediates formed from it; either involving an additional mode of action of the reagent, acting as a base, thus leading to a third order term in the kinetic equation, or by the use of added bases *[38,60–65]*.

Where catalytic effects of bases are large, the evidence supports the view that these result from kinetically effective proton removal from intermediates of the type $H\text{–}\overset{+}{Y}\text{–}Ar\text{–}X$. This has been discussed in Chapter 6, and illustrated in Fig. 71 (p. 227). For such interactions to be kinetically effective, the formation of the second transition state (the bond-breaking step) should be rate-limiting. Evidence on this point was supplied by Bunnett and his co-workers *[61]* in the reactions of *N*-methylaniline with 1-halogeno-2,4-dinitrobenzenes, base catalysis being observed only with the fluoro compound. The theoretical calculations, Table 52, and Figs. 67–68 (Chapter 6, p. 220) are also relevant.

Bunnett and his co-workers *[66]* have recently discussed further these catalytic effects, which conform in general to the kinetic law

$$k_A = k' + k''[B]$$

where k_A is the observed second-order rate coefficient; k' and k'' are respectively second- and third-order coefficients, and [B] is the concentration of base. The ratio k''/k' at 1 *M* base concentration may be used as a measure of the acceleration due to base *[36,62]*. They distinguish between one group of reactions with k''/k' between about 0.2 and 5; and another with values >50. Although the mild accelerations (first group) follow the mathematical form appropriate to base catalysis, Bunnett *et al.* do not regard such reactions as exhibiting this, pointing out that in these cases strong bases such as hydroxide ion are no more effective than weak bases, and that some non-basic substances behave similarly. Possibly these mild catalytic effects involve stabilisation of transition states by hydrogen-bond interactions *[62]*, the catalysts acting as hydrogen-bond acceptors. They also point out that with 1-fluoro-2,4-dinitrobenzene there is a very large difference in the extent of catalysis by hydroxide ion of the reaction with aniline ($k''/k' = 2$) and *N*-methylaniline ($k''/k' = 350$), which is inexplicable if *both* are base-catalysed, and involve elimination of hydrogen fluoride from very similar intermediates. They believe that this is the case with *N*-methylaniline, but not aniline, and their view is supported by the calculations and discussions in Chapter 6 (p. 220). They enlarge further on the subject in terms of good and poor leaving groups, but it seems clearer to do so in terms of the relative energy levels of T.St.1 and T.St.2, and the relationship of these to C–X bond dissociation energies. Their more recent work *[66]* adds evidence for base catalysis in expulsion of OMe and OPh, but not OAr groups (where Ar is electron-withdrawing compared to Ph).

The results of Zollinger and his co-workers *[65]* amply demonstrate base catalysis in reactions of 1-fluoro-2,4-dinitrobenzene and also in the reaction of 1-chloro-2,4-dinitrobenzene with *p*-anisidine in benzene. They also demonstrated in aminolysis markedly accelerating effects of added methanol and dimethyl sulphoxide, compared with reactions in pure benzene, and they classify these as medium effects.

(*b*) *Acid and hydrogen-bonding catalysis*

Acid catalysis is well-known in reactions of heterocyclic compounds *[e.g. 37,67,68]* and has been discussed in Chapter 7 (p. 239). It is a consequence of the conversion of neutral heterocyclic nitrogen, which is moderately activating, into quaternary nitrogen, which is more activating *[44b,69]*. As has been pointed out, this intrinsically simple effect may be complex in practice, for many nucleophiles are also bases and compete for acid-catalytic species, while a substituent may have opposite effects on activating an aromatic ring and on the basicity of the ring, thus affecting the proportion of more reactive protonated forms. As well as catalysis due to added acidic species, it should be recalled that reactions of neutral nucleophiles of the type Ÿ–H form acidic products, and this may result in acidic autocatalysis. This and other complicating factors were also discussed in Chapter 7.

Acid catalysis in another form may occur by coordination between a suitable leaving group and a proton, and for this to be kinetically effective to a major degree, it is necessary that the bond-breaking step (T.St.2 formation) be rate-limiting. Reactions of fluoro compounds are especially likely to show this. The concept of intramolecular hydrogen-bonding between hydrogen and fluorine in transition states has been suggested by Chapman and Parker *[70]* in reactions of primary amines with fluoronitroaromatic compounds, and also more recently by Pietra and Fava *[64]*, and Bernasconi and Zollinger *[65d]*. The last authors considered the possibility of the inclusion of an extra mole of reagent in a hydrogen-bonded transition state involving the bond-breaking step. These are illustrated in Fig. 107. However no quantitative significance has been allotted to the effect of such hydrogen-bonding and it is difficult to do so. The effect is in any case likely to be obscured in protic solvents. Bevan and his co-workers *[71]* suggest that hydrogen-bonding, possible in the transition states of reactions of primary and secondary but not tertiary amines, and particularly favouring fluorine displacement, is responsible for its high mobility in such cases. In their report, the relative mobility in acetone is not in fact very high and is greater than unity with pyridine as well as aniline. Since the relative energy levels of T.St.1 and T.St.2, possibly affecting the F/Cl mobility ratio, may differ for other reasons, the evidence is not very conclusive. In the reaction of *N*-methylaniline

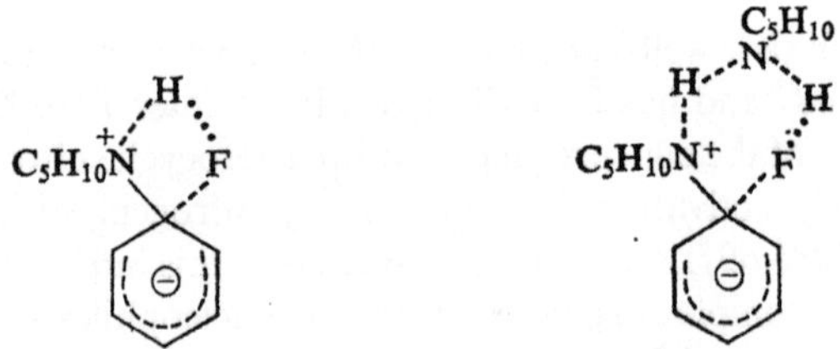

Fig. 107. Catalysis by hydrogen-bonding in the transition state.

with 1-fluoro-2,4-dinitrobenzene this factor is no doubt involved, but also obscured in the protic solvent, ethanol *[61]*. A weak acid, acetic acid, showed no catalytic effect. In nitrobenzene *[33]* the formation of T.St.2 is apparently the rate-limiting step despite such interactions.

To obtain direct and clear evidence of this type of acid catalysis, it is necessary to choose a system in which the reactivity of the nucleophile is essentially unaffected by the presence of acid, *i.e.* one in which the reagent is a kinetic nucleophile of reasonable strength, but very weak as a thermodynamic base; and in which T.St.2 formation is rate-limiting. According to Miller's calculations *[7c,d]* the reaction of iodide ion with 1-fluoro-2,4-dinitrobenzene is such a system. This is the subject of a preliminary study by Lam and Miller *[52a]*; a very large catalytic effect was demonstrated in comparing reactivity of potassium iodide and hydriodic acid in methanol with this substrate. The rate ratio (HI/KI) is approximately $5 \cdot 10^5$ at 145°, for equimolar concentrations (0.02*M*) of reagent and substrate.

Negative catalysis will of course occur in reactions of nucleophiles, X^-, on addition of acid, if the species HX is not a strong acid in the conditions of the reaction. It is of interest therefore that such negative catalysis occurs in the reaction of azide ion with 1-fluoro-2,4-dinitrobenzene *[52b]*, a reaction in which formation of T.St.1 is calculated to be rate-limiting; whereas positive catalysis by acids is found in the reaction of thiocyanate ion, in which formation of T.St.2 is calculated to be rate-limiting. If the difference in behaviour were simply one of relative acid strength, thiocyanic acid > hydrazoic acid *[72,73]*, then one would simply observe less negative catalysis.

Hydrogen-bond (donating) catalysis is similar in character to acid catalysis, and such effects in stabilising transition states by intramolecular hydrogen-bonding and favourable electrostatic interactions have been illustrated by the higher reactivity of *ortho-* than *para-*halogenonitro compounds with primary and secondary amines, but not tertiary amines and alkoxides *[36,38]*, and similarly with the reactivity of halogen *ortho-* and *para-* to heterocyclic nitrogen, discussed above and illustrated in Fig. 103.

Shepherd and Fedrick *[74]* report many examples of hydrogen-bonding and other electrophilic catalysis, with supporting evidence. The latter include, for example, substances providing electrophilic acyl

References p. 361

groups forming species such as $=\overset{|}{N}{}^{+}\text{–SOCl}$ and $=\overset{|}{N}{}^{+}\text{–COR}$. They also point out that azinones and azinethiones have zwitterionic characteristics, with the ring electron-deficient, and that this may be enhanced by protonation, just as with *N*-oxides, for example. Chan and Miller *[75]* have referred to the catalytic effect of added *p*-nitrophenol in reactions of *p*-nitrophenoxide ion with monochloro-diazabenzenes.

(*c*) *Bifunctional catalysis*

A compound with suitable structure, *viz.*, an amphoteric compound H-Acid–$\dot{X}$-Base, with the appropriate geometrical arrangement of its component functional groups, ought to be able to combine base and acid catalytic functions.

Zollinger and his co-workers *[65]* have demonstrated this with, for example, α- but not γ-pyridone. Shepherd and Fedrick *[74]* have also suggested a similar role for ethanolamine in heterocyclic systems and a similar explanation for high reactivity with *o*-chloronitrobenzene has been given *[76]*. This type of catalysis has been discussed in Chapter 7, and also illustrated by Fig. 103 (p. 324).

(*d*) *Metal ion catalysis*

(*i*) *Catalysis by thorium ion*

Electrophilic (including acid) catalysis of S_N reactions by coordination between a leaving group and an electrophile is well-known in aliphatic S_N reactions and is regarded as conferring S_N1-like character on the reaction *[50b]*. In terms of the addition–elimination mechanism of aromatic S_N2 reactions however, this should be found only in those cases in which the formation of the second transition state (the bond-breaking step) is rate-limiting. On the basis of Miller's calculations *[7]* this is likely to be generally observed, for example, in reactions of heavy anionic nucleophiles with aromatic fluoro compounds. Its occurrence in the specific case of reaction of iodide ion with 1-fluoro-2,4-dinitrobenzene, with hydrogen ion as the electrophilic catalyst *[52a]*, has already been mentioned. Lam and Miller have also considered another example, with thorium ion, as it exists in methanol solution, as the electrophilic catalyst, *viz.* in the reaction of thiocyanate ion with 1-fluoro-2,4-dinitrobenzene, which according to Miller's calculations *[7c,d,e]* is also a reaction in which formation of T.St.2 is rate-limiting. In accord with this, the uncatalysed reaction in methanol is found to be very slow *[cf. 7a,43c]* and accompanied by solvolysis and side-reactions. In the presence of equimolar amounts (0.02*M*) of thorium ion however, the rate is increased by about 2000-fold at 100°, with concurrent reduction in the value of $\Delta E^{\ddagger}$. There is also an increase in the value of $\Delta S^{\ddagger}$. They also considered the corresponding reaction with azide ion not only because the calculations indicated that it would *not* be subject to electrophilic catalysis (since formation of T.St.1 is rate-limiting), but because this reagent and thiocyanate ion are alike in many respects, both being regarded as pseudo-halogens *[77–81]*, thus having some resemblances also to iodide ion. Further, as kinetic nucleophiles towards aromatic carbon, their reactivity falls smoothly in the order $N_3^- > SCN^- > I^-$ *[7c,d,e]*. In accordance with this prediction, the rate is in fact not enhanced by the addition of thorium ion. On the contrary the reactivity of azide ion is markedly reduced. It is suggested that this is due to the effects of ion interactions and perhaps even bond-formation, and this is supported by conductivity measurements. These results give further strong support for the addition–elimination mechanism.

(*ii*) *Catalysis by copper compounds*

Such catalysis in aromatic S_N reactions has been longest known in Sandmeyer *[82]* and related reactions which take place in aqueous solution, though other solvents have been utilised. These have been discussed at length with reference to the mechanism by a number of other workers *[e.g. 83–90]*. These reactions involve cuprous-salt catalysed replacements, shown in the most general form as:

$$ArN_2^+ + X^- \xrightarrow{Cu^I} ArX + N_2$$

Numerous reports of catalytic activity by other transition metal compounds have appeared, but none of these matches cuprous ion

References p. 361

in the range of reactions catalysed, or generally in the mildness of the conditions in which reactions may occur.

In the absence of catalysts most of these reactions do not proceed at all readily. They have been included in the discussion of S_N1 reactions in Chapter 2 (p. 31). It is noteworthy *[91–93]* that, more specifically, the reactions of fluoride, chloride and bromide leading to replacement of the diazonium group do not proceed readily without catalysts, but those of iodide ion do. This supports the concept that the reactions are nucleophilic substitutions, since the kinetic nucleophilicity of iodide ion in protic solvents is substantially greater than that of bromide, chloride and fluoride ion *[7d]*. Similar reactions with other nucleophiles are also well-known *[82,94]*. Even now the mechanism of this reaction is obscure, though it is generally recognised that cuprous complexes with the aromatic substrate are formed, and have been isolated in a number of cases.

Hodgson and his co-workers *[84]* suggested a mechanism in which the role of the cuprous salt is to increase the nucleophilic power of halide ion, reaction being attack on the electron-deficient carbon atom to which the diazonium group is attached. This is essentially the S_N1 reaction of diazonium salts (Chapter 2). The halide ion is carried in a complex copper salt and there is interchange with halide ions in solution, so that the evidence on product formation in mixed chloride–bromide systems is largely evidence for the reactivity order $Br^- > Cl^-$, the kinetic nucleophilicity order in protic solvents *[7c,d,e,]*.

Waters *[87]* suggested that the role of the cuprous salt is to reduce the aryl diazonium salt to an aryl radical with elimination of nitrogen. Some relevant equations are:

$$ArN_2^+ + Cu^I \rightarrow Ar\cdot + N_2 + Cu^{II}$$

$$Ar\cdot + Cl^- + Cu^{II} \rightarrow ArCl + Cu^I$$

$$Ar\cdot + ArN_2^+ + Cu^I \rightarrow Ar{-}N{=}N{-}Ar + Cu^{II}$$

$$Ar\cdot + Ar\cdot \rightarrow Ar{-}Ar$$

This mechanism seems particularly appropriate for reductive side-reactions, but not so appropriate for the displacement reaction.

Cowdrey and Davies *[88]* investigated these reactions in buffered acid solution, and showed that the rate of reaction is proportional to the concentration of diazonium ion and $[Cu^ICl_2]^-$, but that $[Cu^ICl_4]^{3-}$ is inactive. They also pointed out a close parallelism in substituent effects, and even Arrhenius parameters (especially the low values of $\log_{10} B$), in these and diazo coupling reactions *[93–95]*. Their suggested mechanism, while in accord with the kinetic data for the Sandmeyer reactions, seems inconsistent with the parallelism demonstrated with the coupling reactions. They suggest the following scheme (in which the charge on atoms within a complex is not shown):

$$[ArN_2]^+ + [Cu^ICl_2]^- \longrightarrow [Ar\text{–}N{\equiv}N{\rightarrow}Cu^ICl_2]^0$$

$[Ar\text{–}N{\equiv}N{\rightarrow}Cu^ICl_2]^0$ $\xrightarrow{ArN_2^+}$ $\left[\begin{matrix} Ar\text{–}N{\equiv}N \searrow \\ \qquad\qquad Cu^ICl_2 \\ Ar\text{–}N{\equiv}N \nearrow \end{matrix}\right]^+ \longrightarrow ArCl + Cu^ICl + N_2\ (+ArN_2^+)$

$[Ar\text{–}N{\equiv}N{\rightarrow}Cu^ICl_2]^0 \longrightarrow ArCl + Cu^ICl + N_2$

$$\left[\begin{matrix} Ar\text{–}N{\equiv}N \\ Ar\text{–}N{\equiv}N \end{matrix} Cu^ICl_2\right]^+ \xrightarrow{Cu^ICl_2^-} Ar\text{–}N{=}N\text{–}Ar + 2Cu^{++} + 4Cl^- + N_2$$

$$\left[\begin{matrix} Ar\text{–}N{\equiv}N \\ Ar\text{–}N{\equiv}N \end{matrix} Cu^ICl_2\right]^+ \xrightarrow{Cu^ICl_2^-} ArH + 2Cu^{++} + 4Cl^- + N_2 + ArN_2^+$$

The yield of azo compound is low when the concentration of ArN_2^+ is low, in accord with their mechanism.

Since the coupling reaction involves coordination by a nucleophilic carbon atom of an aromatic ring activated for aromatic electrophilic substitution, to the terminal nitrogen as acceptor, as follows, it is

$$X\text{–}C_6H_4\text{–}\overset{+}{N}{\equiv}N \quad C_6H_5\text{–}\ddot{Y} \longrightarrow X\text{–}C_6H_4\text{–}N{=}N\text{–}C_6H_4\text{–}\ddot{Y}$$

difficult to see the parallel with coordination *by* the terminal nitrogen *to* the copper as suggested by Cowdrey and Davies. Indeed such coordination seems unlikely on several grounds, *viz.*, (*i*) the unfavourable formation of adjacent charge; (*ii*) the improbability of nucleophilic reactivity of $\text{–}\overset{+}{N}{\equiv}N{:}$, when even neutral $:N{\equiv}N:$ lacks it; (*iii*) the unlikelihood of finding parallel substituent effects with standard types of aromatic substituent in reactions of opposite type.

A mechanism now suggested involves nucleophilic attack by the halogen of $[Cu^{I}Cl_2]^-$ on the terminal nitrogen, to form a complex with covalently bonded chlorine coordinated to a copper(I) salt (*cf.* catalysis in Friedel–Crafts reactions), *e.g.*

$$[ArN_2]^+ + [Cu^{I}Cl_2]^- \rightarrow [Ar{-}N{=}N{-}Cl \rightarrow Cu^{I}{-}Cl]^0.$$

This resembles the suggestion of Hodgson and his co-workers except that they postulated initial attack on the ring carbon atom.

There are possible variations in the following suggested sequence, *e.g.* the timing of the expulsion of copper(I) chloride, but it is nevertheless in accord with the available kinetic data. It also suggests a *cis*-configuration of the azo group, as the reactive form.

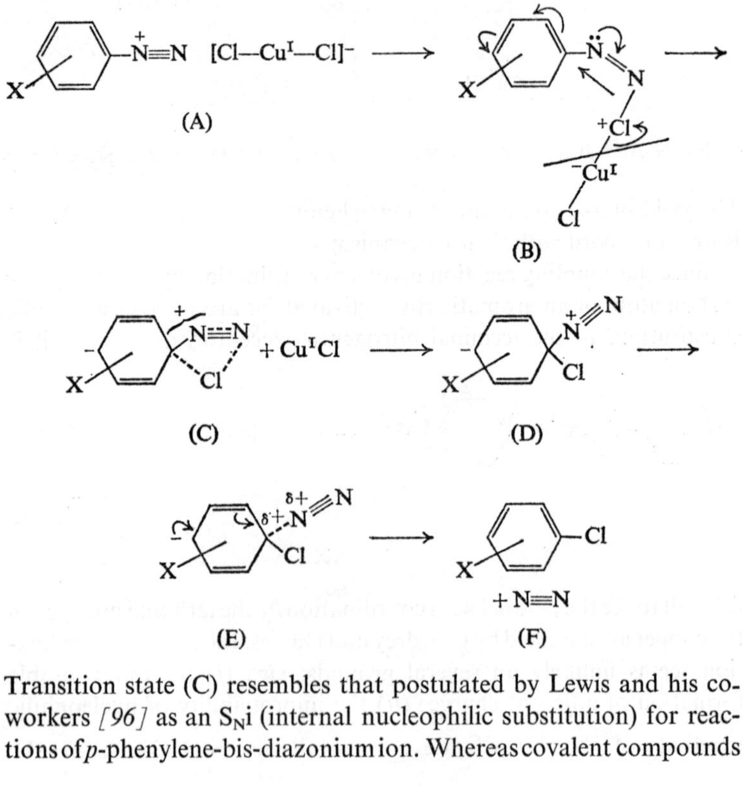

(A) (B)

(C) (D)

(E) (F)

Transition state (C) resembles that postulated by Lewis and his co-workers *[96]* as an S_Ni (internal nucleophilic substitution) for reactions of *p*-phenylene-bis-diazonium ion. Whereas covalent compounds

Ar–N_2–Hal have low stability, those of structure Ar–N_2–Hal → Cu^{I}Hal may be more stable. Since there is some steric strain in (C) it is likely that the N–Cl bond is very tenuous; this suggests that the C–Cl bond is close to a full bond, which would correspond to the usual pattern. It does not seem necessary to postulate any rupture of the C–N bond at this stage of the reaction.

The catalytic effect of cuprous salts depends on (*i*) the thermodynamic stability but kinetic lability of copper(I) salts *[97]*, thus ensuring ready availability of a cuprous complex which can coordinate then release halogen, as shown; (*ii*) the presumed much lower solvation energy of $[Cu^{I}Cl_2]^-$ than Cl^-, thus effectively causing a marked increase in nucleophilic power; and (*iii*) the greater stability of covalent structures Ar–N=N–Hal→Cu^{I}–Hal than of Ar–N=N–Hal; (*iv*) the suitable geometry of a *cis*-benzeneazo–cuprochloride complex for nucleophilic attack on aromatic carbon by halogen linking nitrogen and copper.

Another example of copper(I) catalysis is the Ullman reaction, copper(I)-catalysed halogen exchange. As discussed in earlier chapters, simple halogenobenzenes and derivatives containing electron-releasing groups are very unreactive via the addition–elimination S_N2 mechanism. Facile reaction via the elimination–addition S_N2 mechanism requires the use of very powerful bases, such as amide ion. S_N1 heterolysis is not favoured for reasons given in Chapter 2, section, 1, p. 29.

In these substrates which are very unreactive towards nucleophiles, catalysis of halide exchange by copper(I) compounds in aprotic solvents has been demonstrated by Bacon and Hill *[97]*; the catalysis is specific to copper(I), as far as is known. The reactions are characterised by having only slight substituent effects and the mobility order I > Br > Cl ≫ F. The kinetic form of the reactions was also shown to be first order in aromatic substrate, and in copper(I) compound. In accord particularly with the halogen mobility order, Bacon and Hill, though suggesting alternative mechanisms, envisage the key initial step of coordination by aromatic halogen to copper(I). This results in a marked increase in mobility. It is relevant that Bolto and Miller *[43b]* have demonstrated the reactivity order $Ar\overset{+}{X} > Ar\overset{+}{X}–\overset{-}{Y} > ArX$, and the coordinating power of halogens is well-known to be in the order I > Br > Cl ≫ F. In the mechanisms postulated by Bacon and

Hill, solvent or other suitable ligand molecules (L) are included to give tetrahedrally coordinated copper(I) species. Using ArBr and $Cu^{I}Cl$ as the example, suggested details are:

$$\mathrm{ArBr + Cu^{I}L_3Cl \rightleftharpoons Ar\text{–}\overset{+}{Br}\text{–}\underset{\substack{|\\ Cl(+L)}}{Cu^{I}L_2}}$$

(A)

thence

$$(i)\quad \mathrm{(A)} \longrightarrow \mathrm{Ar}\begin{smallmatrix}\cdot\cdot Br \searrow \\ \cdot\cdot Cl \nearrow\end{smallmatrix}\mathrm{Cu^{I}L_2} \xrightarrow{(+L)} \mathrm{ArCl + Cu^{I}L_3Br}$$

$$\text{or}\quad (ii)\quad \mathrm{(A)} \xrightarrow{(+L)} \mathrm{Ar\text{—}\overset{+}{Br}\text{—}Cu^{I}L_3} \quad \mathrm{Cl^-} \longrightarrow \mathrm{ArCl + Cu^{I}L_3Br}$$

$$\text{or}\quad (iii)\quad \mathrm{(A)} \longrightarrow \mathrm{Ar^{\bullet} + Br\underset{\substack{|\\ Cl}}{Cu^{II}}L_3} \longrightarrow \mathrm{ArCl + Cu^{I}L_3Br}$$

In (*i*) one could similarly postulate part-bonds from copper(I) to halogen and full bonds to Ar in a cyclohexadienide type complex.

Mechanism (*ii*) seems more probable for several reasons, (*a*) because in this the aromatic species in which halogen exchange occurs is ArX^+, a more reactive type than $\mathrm{Ar\text{–}\overset{+}{X}\text{–}\overset{-}{Y}}$; (*b*) because the reactivity order of halogen in the reagent is Cl > Br > I, and this fits the order of kinetic nucleophilicity of the anionic forms in aprotic solvents (see earlier, p. 318, however, the relative kinetic lability of $X\text{–}Cu^{I}\text{–}Y$ species must also be involved here and no doubt is responsible for the order Hal > CN); (*c*) the special catalytic activity of Cu^{I} salts associated with the known kinetic lability of $Cu^{I}XY$ species; (*d*) the adverse kinetic effect of added cations which are likely to form $M^{+}Hal^{-}$ ion pairs. There seem no forceful reasons from the available data to postulate a homolytic mechanism, and the occurrence of the reaction with iodo compounds, though species $I\text{–}Cu^{II}\text{–}Hal$ are unlikely, militates against it.

The generally slight substituent effects do not favour any particular mechanism, except that of the preliminary coordination step ($Ar\text{–}Hal \rightarrow Cu^{I}\text{–}$), since conjugation between heavy halogens and the ring is slight, and thus the coordination would be little affected by substituents. The special efficacy of the *o*-nitro group may be presumed to be correlated with the favourable electrostatic interaction with positively charged halogen, and this is similar to the effects

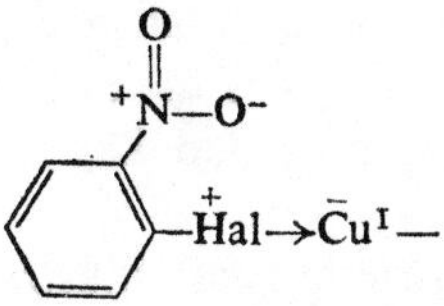

$$\overset{+}{N}(=O)-O^{-} \quad \text{—} \overset{+}{\text{Hal}} \rightarrow \overset{-}{\text{Cu}}{}^{\text{I}}\text{—}$$

suggested as contributing to the higher reactivity of some neutral reagents with *ortho-* than with *para*-nitroaromatic substrates (see earlier) in normal S_N2 reactions. The special efficacy of copper(I) salts also supports the ionic mechanism (*ii*), the factors involved being the thermodynamic stability and kinetic lability of these compounds already referred to *[97a]*.

Similar kinetic investigations have recently been made by Stephen and Castro *[98]*, and Weingarten *[99]*. The former, by using cuprous acetylide, suggested a mechanism essentially the same as alternative (*i*) of Bacon and Hill. Weingarten used various copper(I) salts in diglyme with phenoxide reagents and suggested an alternative mechanism which involves coordination by the π-electrons of the ring to copper(I). This, however, seems to require much larger substituent effects, and also to be less specific in catalyst requirements.

A number of related reactions have been discussed in the recent review by Bacon and Hill *[97a]*.

(*e*) *Photocatalysis*

There have been a number of reports in recent years of photocatalysed aromatic S_N reactions. The light and dark reactions may be the same, but are usually different.

Papers by Havinga *[100]*, Zimmerman *[101]*, Letsinger *[102]* and their co-workers, deal with photocatalysed reactions which either do not occur in the dark or for which dark reactions are different.

In essence, they showed that in the photocatalysed reactions an electron-withdrawing group, the nitro group, is activating for nucleo-

References p. 361

philic substitutions from the *meta*- but not (or only mildly) from the *para*-position; and an electron-releasing group, the methoxyl group, is *activating* as a *para*- but not a *meta*-substituent. It may be recalled (Chapter 4, pp. 77, 88) that the nitro group in normal S_N2 reactions is powerfully activating in *ortho*- and *para*-positions, and much less so in the *meta*-position, whereas the methoxyl group is deactivating in *ortho*- and *para*-positions, and has a smaller effect from the *meta*-position.

The photocatalysed reactions are illustrated in Figs. 108 and 109 in which photo-excited states are shown in the forms suggested by Zimmerman *[101]*.

The explanation of photo-activation by the *para*-methoxyl group is also in accord with M.O. calculations which suggest that (*i*) an electron-acceptor group *para* to an electron-donor group in a benzene ring acquires negative charge, primarily at the expense of the aromatic ring, when raised to the first excited state *[103]*; and (*ii*) that the

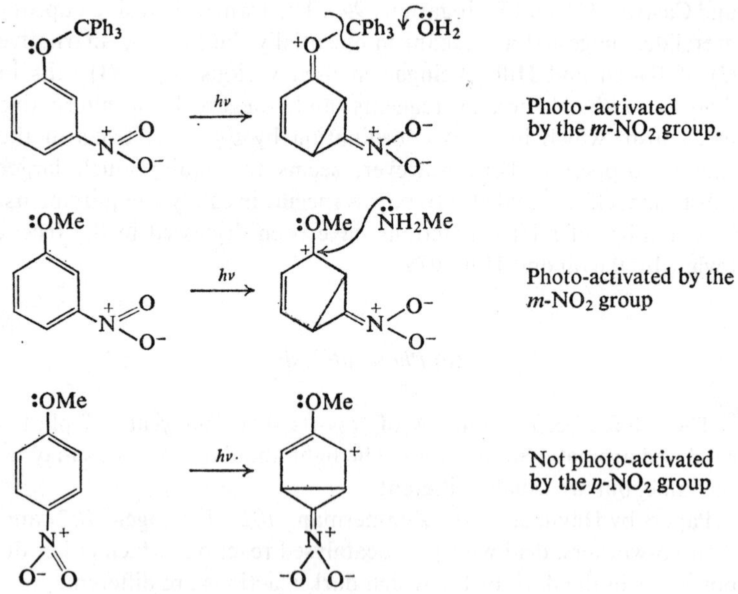

Photo-activated by the *m*-NO_2 group.

Photo-activated by the *m*-NO_2 group

Not photo-activated by the *p*-NO_2 group

Fig. 108. Comparison of the effect of *meta*- and *para*-nitro groups in photo-activated nucleophilic displacement in both side-chain and ring.

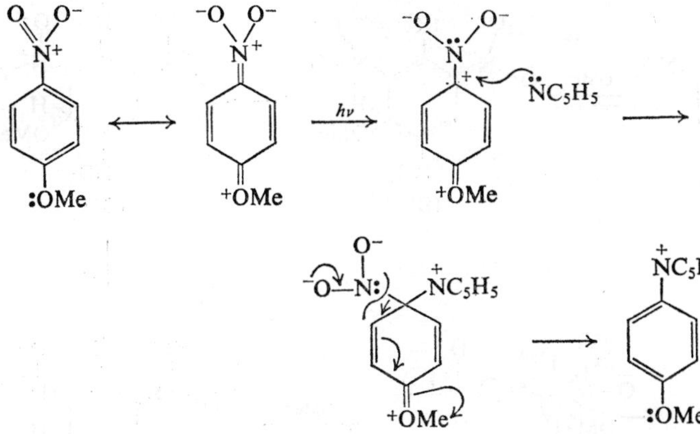

Fig. 109. Activation by a *para*-methoxyl group in the photocatalysed displacement of a nitro group by pyridine.

para-position in anisole in the first-excited state is electron deficient relative to both the *para*- and *meta*-positions in the ground state, and the *meta*-position in the excited state *[104]*.

The photo-activated displacement of a nitro group reported by Gold and Rochester *[105]* in reactions of some picryl compounds with two moles of methoxide ion may be similarly explained by a scheme involving (*a*) normal (dark) activated addition of a methoxyl group to positions *ortho*- and *para*- to nitro groups; followed by (*b*) a photoactivated displacement of nitro groups in positions *ortho*- and *para*- to methoxyl by a second molecule of reagent. The scheme is illustrated in Fig. 110.*

As illustrated, the nitro group *ortho* to X is displaced, but the scheme could equally account for displacement of the nitro group *para* to X. Gold and Rochester do not state which isomer is obtained. Also relative stabilities may be such that in some cases the photo-activated form may be obtained from a species as in B1 of Fig. 110, though this will lack stabilisation by electrostatic interaction between X^+ and the negative oxygens of *ortho*-nitro groups. It is illustrated in Fig. 111 and is presumably the case for X = H in reaction with aqueous alkali.

* Addition of OMe^- can also take place at C-1 in a dark reaction leading to normal displacement of X by OMe^-.

References p. 361

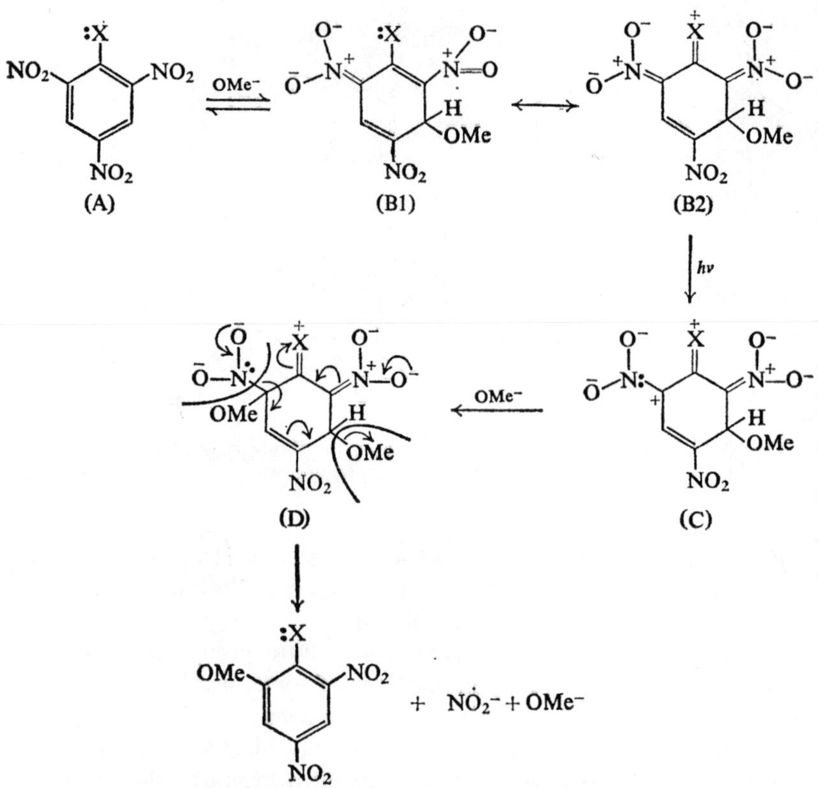

Fig. 110. Scheme for the reaction of picryl-X with two mols. of OMe^- in an overall photo-activated displacement involving X.

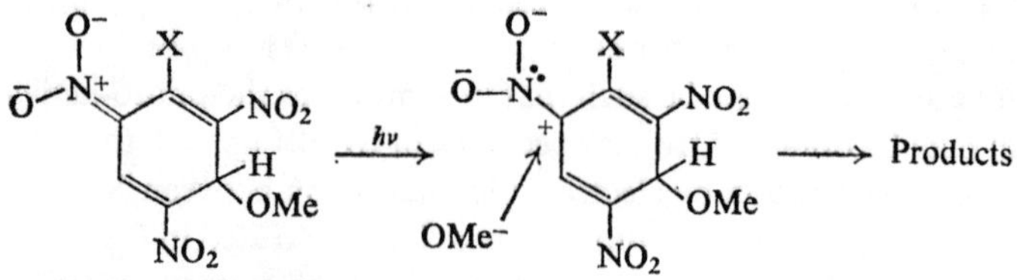

Fig. 111. Scheme for the reaction of picryl-X with two mols. of OMe^- in an overall photo-activated displacement not involving X.

In contrast to these reactions, Johnson and Rees *[106]* report a photocatalysed reaction which gives the *same* products as in the dark reaction, *viz.* the reaction of 4-nitropyridine-1-oxide with piperidine forming 4-piperidinopyridine-1-oxide and piperidinium nitrite.

The aromatic substrate is a type with a very powerful activating group *[69]* and a highly mobile displaced group *[43b]*. The key is however that the *N*-oxide group is one of a number of groups which, being able to supply or withdraw electrons as required by the reagent, can activate *both* aromatic S_E and S_N reactions *[107,108]* and we may correspondingly expect generally that such substrates will give the same products in dark and photo-activated reactions, in which electron-attracting groups activate in the former case, and electron-releasing groups activate in the same position in the latter case. The parallel dark and light reactions are illustrated in Fig. 112.

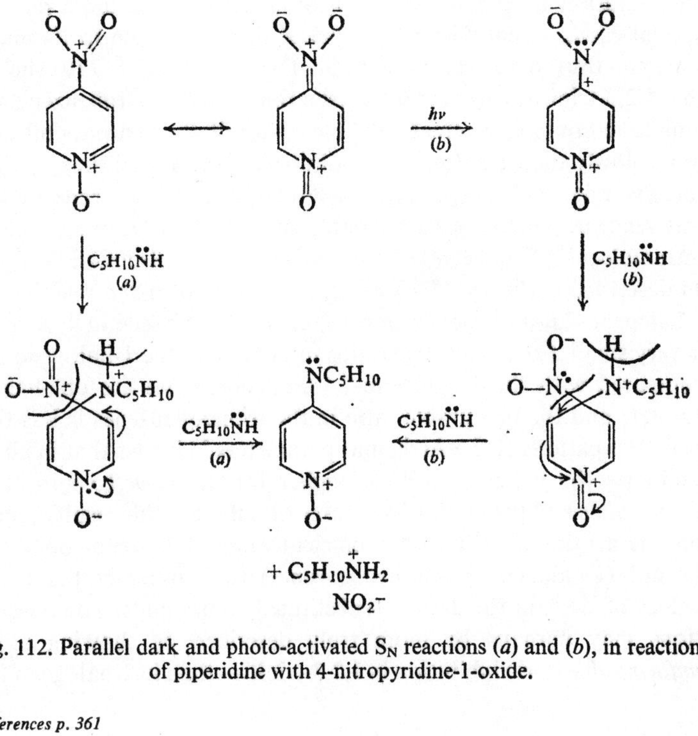

Fig. 112. Parallel dark and photo-activated S_N reactions (*a*) and (*b*), in reactions of piperidine with 4-nitropyridine-1-oxide.

References p. 361

4. Steric Effects

Consideration of steric effects has arisen earlier in the context of several of the chapters. Although some special S_N1 reactions have been described, which, as dissociative processes, are intrinsically less susceptible to steric effects *[109]*, aromatic S_N reactions are almost exclusively bimolecular in character.

In elimination–addition S_N2 reactions, an essential requirement is the absence of a substituent in one position *ortho* to the displaced group, and this means that there is rarely any important steric hindrance to the subsequent associative process, *viz.* addition of a nucleophile to a benzyne, though no doubt a system could be found to demonstrate some measure of steric hindrance.

The associative process in the addition–elimination S_N2 reactions does not generally lead to large steric effects, because formation of the transition state involves a change from a planar to a tetrahedral configuration at the reaction centre. Indeed this change may, in special cases, actually lead to a release of steric strain in the initial state and thus to a steric acceleration. Parker and Read *[110]* showed that 1,2,3-trinitrobenzene reacts faster than 1,2,4-trinitrobenzene with aniline in ethanol, whereas 1-halogeno-2,6-dinitrobenzenes all react more slowly than 1-halogeno-2,4-dinitrobenzenes; and they ascribed this to steric acceleration in the reaction of 1,2,3-trinitrobenzene with this reagent. This seems reasonable since the favourable effects of intramolecular hydrogen-bonding, solvation and electrostatic forces as described earlier (p. 324) are applicable also in the reactions of 1-halogeno-2,6-dinitrobenzenes. The effect, though seemingly definite is very small, for the adverse ratios in the case of the 1-halogeno-2,6- compared with the 1-halogeno-2,4-dinitrobenzenes are one-half or greater; and the favourable ratio in the trinitrobenzenes is less than two. The pattern of the Arrhenius parameters indicates that in all the examples the value of $\Delta E^{\ddagger}$ is higher for the di-*ortho*-substituted compounds and the mildly favourable or unfavourable results depend on whether the value of $\Delta S^{\ddagger}$ is sufficiently high (1,2,3-trinitrobenzene) or not (1-halogeno-2,6-dinitrobenzenes) to counteract the higher values of $\Delta E^{\ddagger}$ in the di-*ortho*-substituted compounds. The observations may perhaps be more truly described as showing a *less unfavourable* steric inhibition in 1,2,3-trinitro- than in 1-halogeno-2,6-

dinitro-benzenes; the favourable intramolecular factors then being sufficient to overcome it in the trinitro compound but not the others. A comparison with methoxide reactions might throw light on this suggestion, while further investigation of possible cases of steric acceleration is clearly desirable.

An early but comprehensive survey by Brady and Cropper *[60, cf. 111]* of the reactions of a range of amines with 1-chloro-2,4-dinitrobenzene in ethanol shows very clearly the onset of steric effects with sterically-demanding branched amines. Some of their results are given in Table 66, in which primary and secondary amines are shown separately to avoid introducing into comparisons possible reactivity differences between the classes of amines (see Chapter 6, p. 216). Ammonia is excluded because of fundamental differences from primary and secondary amines, apart from steric factors (Chapter 6, p. 224).

Although they do not regard their values of the Arrhenius parameters as more than approximate, the available data, which do not however include values for the least reactive amines, suggest that the steric effect is mainly associated with an increase in $\Delta E^{\ddagger}$.

Miller and his co-workers *[44,46,112]* have investigated a series of *ortho*-substituted aromatic substrates in reactions with methoxide ion in methanol. Most of their results are comparisons of 2-substituted 1-chloro-4-nitrobenzenes with 4-substituted 1-chloro-2-nitrobenzenes; and of 6-substituted 1-chloro-2,4-dinitrobenzenes with 4-substituted 1-chloro-2,6-dinitrobenzenes. The group substituent effects (substitutent rate factors, S.R.F. or f) *[112a]* are the rate ratios for the substituted compared with the parent compound (hydrogen instead of the substituent). The positional effects are then the ratios of the f values which correspond to the partial rate factor (P.R.F. or f) in aromatic S_E reactions. What was then called a Steric Index but now an Ortho Index, is the ratio, S.R.F. *p*-X/S.R.F. *o*-X, or f_p^X/f_o^X in the notation used for aromatic substitution generally. Values of the Index less than unity correspond to an *ortho*-accelerative, and more than unity to an *ortho*-decelerative effect, the magnitude of which is given by the divergence from unity. The implicit assumption in the use of these Indices is simply that the susceptibility of any pair of reactions to substitutent effects, measured conveniently by the Hammett reaction constant (ρ), is the same. To a close approximation this is in accord

TABLE 66

REACTIVITY OF AMINES WITH 1-CHLORO-2,4-DINITROBENZENE IN ETHANOL AT 25° *[60]**

Primary amines	Rate constant: $10^3 k_2$ ($l \cdot mole^{-1} \cdot sec^{-1}$)	Secondary amines	Rate constant: $10^3 k_2$ ($l \cdot mole^{-1} \cdot sec^{-1}$)
$MeNH_2$	31.6	Me_2NH	3.55
$EtNH_2$	9.2	[piperidine ring] NH	153
n-$PrNH_2$	9.6	Et_2NH	1.9
n-$BuNH_2$	10.0	(n-Pr)(Et)NH	1.7
$Me_2CHCH_2NH_2$	6.8	$(Me_2CHCH_2)_2NH$	0.58
(Et)(Me)CH—NH_2	0.91	[2-methylpiperidine ring] NH, Me	0.22
tert-$BuNH_2$	0.038	$(Me_2CH)(Et)NH$	*ca.* 0.01
		$(Me_2CH)_2NH$	*ca.* 0.01
		[2,6-dimethylpiperidine ring] NH, Me, Me	*ca.* 0.01
		[tetrahydroquinoline ring] N-H	*ca.* 0.01
		[2-methyltetrahydroquinoline ring] N-H, Me	*ca.* 0.003

* The values for low reactivity species are unreliable because of concurrent reaction with the lyate ion.

with the body of data on substituent effects in aromatic S_N reactions. Even the change from the mono- to the di-nitro series causes only a small change ρ, being smaller in the dinitro series *[107a]*. The ρ-value is also known to be largely independent of the displaced group. In addition a few rate data are available for reactions of substituted fluoro- or chloro-benzenes *[28,112,113]*. These are shown as Table 67, while data of Miller and his co-workers for the mono- and di-nitro series are given in Tables 68 and 69 (see also Chapter 4, Section 6, p. 95).

As a general conclusion, with methoxide and hydroxide ions, which are reagents of low steric requirements, *ortho* effects are very small in the substituted halogenobenzenes and halogenonitrobenzenes. Noticeable but not gross decreases of rate appear for the substituted halogenodinitrobenzenes, *i.e.* when *both* positions *ortho* to the point of substitution are occupied. These seem to involve field effects and effects due to inhibition of conjugation, as well as primary steric effects *[50c]*.

The few results for the substituted halogenobenzenes demonstrate *ortho*-acceleration for the diazonium group. This is expected for reaction with an anion. Since in fact the rate measured is for the

TABLE 67

COMPARISON OF THE REACTIONS OF *ortho*- AND *para*-SUBSTITUTED HALOGENOBENZENES

Substituent X	Leaving group	Substituent Rate Factor (S.R.F. or f)		Ortho Index (f_p^X/f_o^X)
H	Hal	1	1	1
N_2^+[a]	F	—	—	0.10
NO_2[b]	F	$1.33 \cdot 10^9$	$8.54 \cdot 10^8$	1.46
NO_2[b]	Cl	$7.07 \cdot 10^{10}$	$2.10 \cdot 10^{10}$	3.37
NO_2[b]	Br	—	—	3.87
NO_2[b]	I	—	—	3.78
NO_2[c]	F	—	—	0.220
NO_2[c]	Cl	—	—	0.363
NO_2[c]	Br	—	—	0.288
NO_2[c]	I	—	—	0.17
Cl[b]	Cl	—	—	0.71

[a] With OH^- in H_2O at 0°. [b] With OMe^- in MeOH at 50°. [c] With piperidine in EtOH at 90°.

TABLE 68

COMPARISON OF *ortho*- AND *para*- SUBSTITUENT EFFECTS IN REACTIONS OF SUBSTITUTED CHLORONITROBENZENES WITH OMe^- IN MeOH AT 50° ($\rho = 3.90$)

Substituent X	Substituent Rate Factor (S.R.F. or f)		Ortho Index
	f_p^X	f_o^X	(f_p^X/f_o^X)
H	1	1	1
CO_2^- [a, b]	7.12	0.373	19.1
$CH{=}CH{-}CO_2^-$ [a]	16.7	2.10	7.95
SO_2^- [a, c]	5.44	0.512	10.6
NMe_3^+	21,300	28,900	0.738
$CONH_2$	262	462	0.567
CO_2Me	1560	174	8.97
COMe	1990	246	8.06
COPh	2660	21.6	123
CHO [d]	4500	570	*ca.* 8
CN [d]	7800	5830	*ca.* 1.3
NO_2	114,000	33,900	3.37
SO_2Me	12,850	2310	5.58
CH_3	0.119	0.211	0.564
CF_3	813	155	5.25
F	0.895	*ca.* 20	*ca.* 0.02
Cl	14.0	12.0	1.17

[a] At approximately the same ionic strength. [b] At zero μ, $f_p^{CO_2^-} = 3.37$. [c] With hydroxide ion in water and estimated for $\mu = 0$. [d] Estimated approximately by use of data obtained for the reactions of SCN^- with iodonitrobenzenes in MeOH.

reaction with a small amount of diazonium ion in equilibrium with a diazohydroxy compound, the value of the Ortho Index (= 0.1) should be taken as very approximate. The comparisons of *o*- and *p*-halogeno-nitrobenzenes clearly show no steric effect depending on the size of halogen with either methoxide or piperidine as reagent. With the latter the *ortho*-nitro is more activating than the *para*-nitro group, and this is due to intramolecular hydrogen-bonding and favourable electrostatic interactions in the transition-state *[36]* between the piperidino and nitro group, as discussed earlier (p. 324).

The Ortho Index for the carboxylate group is 19.1 in the mononitro

TABLE 69

COMPARISON OF *ortho*- AND *para*-SUBSTITUENT EFFECTS IN REACTIONS OF SUBSTITUTED CHLORODINITROBENZENES WITH OMe^- IN MeOH AT 0° ($\rho = 3.80$)

Substituent X	Substituent Rate Factor (S.R.F. or f)		Ortho Index
	f_p^X	f_o^X	(f_p^X/f_o^X)
H	1	1	1
CO_2^- [a]	11.5	0.00702	1640
$CONH_2$	240	27.9	8.60
$CONMe_2$	64.7	0.188	344
CO_2Me	728	5.57	131
COPh	821	0.613	1340
CH_3	0.106	0.0598	1.78
Ph	1.28	0.0788	16.2
Cl	8.49	3.72	2.28
OMe	0.00544	0.0906	0.0600
NH_2	0.000518	0.00212	0.244

[a] At approximately the same ionic strength.

series (Table 68), and 1640 in the dinitro series (Table 69). In contrast the methyl group has values close to unity (see below). It appears therefore that the effect of the carboxylate group is an unfavourable field effect, for reaction with an anion, noticeable even in the mononitro series and becoming large in the dinitro series. There a less unfavourable direction of approach, available in the mononitro series, is blocked in the dinitro series by the second *ortho*-substitutent. Since the *para*-carboxylate group is only mildly activating (at zero ionic strength, μ, the $f_p^{CO_2^-}$ value is only 3.37, as compared with the reported value of 7.12 at $\mu = 0.138$), the *ortho*-carboxylate group is deactivating. In the absence of the unfavourable field effect, as with the neutral reagent piperidine, Bunnett and his co-workers subsequently showed *[3b]* that the *ortho*-carboxylate group, like the *ortho*-nitro, is more activating than when it is in the *para* position, due to factors already described. Further support is given by the smaller Ortho Index for the cinnamate group (mononitro series), especially as in this case the value of 7.95 may be ascribed in the main to a polar substituent effect, *viz.* a $-T$ effect operating more effectively from the *para*- than the *ortho*-position. Like the carboxylate group, the sulphonate group, with an Ortho Index of 10.6 (mononitro series), is activating from the *para*- and deactivating from the *ortho*-position, with an anionic reagent (hydroxide ion).

Among a series of neutral COX groups, the amide group is the least activating from the *para*-position yet the most activating from the *ortho*-position *[112,114]*. This relatively high reactivity is shared by the CONHPh group but not the $CONMe_2$ group. The explanation offered is that the phenomenon is due to favourable Cl---H–N bonding in the transition state when the *ortho*-group is a primary or secondary amide. In the mononitro series the Ortho Index is actually less than unity, *viz.* 0.567, but in the dinitro series, incursion of a steric effect raises the value to 8.60. Nevertheless this is less than for any of the other COX groups.

The methoxycarbonyl and acetyl groups have Ortho Indices between 8 and 9 in the mononitro series. This may be a combination of a weak primary steric effect and inhibition of conjugation by forced rotation out of the ring plane. The large Ortho Index of 123 for the benzoyl group shows a substantially larger effect of similar origin. The additional steric factor in the dinitro series raises Ortho Indices for electrically neutral groups by a factor of about 10. The Ortho Indices there are: $CONH_2$, 8.60; CO_2Me, 131; COPh, 1340.

In the dinitro series data are available for the dimethylamido group. It is unlike the amido group, and has large steric requirements, similar to those of the benzoyl group, and the Ortho Index is nearly as large, *viz.* 344. This is assumed to include inhibition of conjugation with the ring, and also within the amide group. In the mononitro series results were vitiated, at the higher temperatures necessary, by facile hydrolysis of the $CONMe_2$ group.

The methyl group is seen to have virtually no steric effect, with the reagents considered, and little polar effect, but a group such as the tert-butyl group would be expected to exhibit a substantial steric effect. Bevan and his co-workers *[115]* compared the *ortho*- and *para*-tert-butyl groups in reactions of substituted *m*-fluoronitrobenzenes with methoxide ion in methanol. Their data lead to an Ortho Index of about 54 at 130.4°. An inhibiting secondary steric effect between the *para*-tert-butyl and *meta*-nitro group may be assumed, so that the "true" value is greater than 54. Fierens and Halleux *[116]* have made similar measurements with iodide ion in acetone in the dinitro series. Their results give an Ortho Index of 0.555 for the methyl group, but 91.9 for the tert-butyl group. This is a clear primary steric effect shown also with this reagent.

It is convenient at this point to consider the trimethylammonio group since its geometric steric effect may be taken as approximately equal to that of the tert-butyl group. However both σ-inductive and field effects operate more effectively from the *ortho* position, and the electrostatic effect is particularly favourable with an anionic reagent. In view of the presumed approximate equivalence of the geometric steric effect of the tert-butyl and the trimethylammonio group, the Ortho Index of 0.738 of the trimethylammonio group in the mononitro series represents a favourable polar rate effect of the order of 100 or more for the attached cationic group. The direct comparison of reactions of *N*-methyl-2-chloro- and *N*-methyl-4-chloropyridinium ions with *p*-nitrophenoxide ion in methanol *[44b,107]* gives an Ortho Index of 0.0439 for a cationic ring atom. This is not as low as the value for the trimethylammonio group in the absence of steric effects, *viz.* about 0.01×0.738, and this is ascribed to a more effective relay of electronic effects by the π-system of the ring favouring the *para*-group, in the former case.

The trifluoromethyl group is similar in size to the methyl group, but the Ortho Index is 5.25, as compared with 0.564 for the methyl group. The effect is presumably polar in origin, and may involve hyperconjugative activation by the trifluoromethyl group, more effective from the *para* position. A possible contributory cause is repulsive forces between the electronegative fluorine atoms and the reagent.

The steric requirements of a benzene ring are somewhat greater than those of a methyl group and its polar effect ($+T$) is very weak *[40b,c, 107]*, being only slightly activating for S_N reactions. Any steric inhibition will therefore show up in a change from activation to deactivation. In the dinitro series this is confirmed *[40b]* and the Ortho Index is 16.2. This shows the effect clearly, though it is by no means a large one.

The nitro and methylsulphonyl group are mildly less activating as *ortho*- than *para*-substituents, with Ortho Indices of 3.7 and 5.88 in the mononitro series *[40,112]*. A comparison of the reactivity of 1-chloro-2,4- and -2,6-dinitrobenzene with that of *o*-chloronitrobenzene measures the activating power of an *o*-nitro group when the

References p. 361

other *ortho* position is occupied by a nitro group, increasing the Index by a factor of about 12 when *both ortho*-positions are occupied. It reflects the constraint of the direction of approach of the reagent when one *ortho* position is occupied by the variable substituent, and the other by the standard substituent, the nitro group.

Groups which are attached by hetero atoms, and which are not bulky, are expected to have *ortho*-accelerative effects of polar origin. As $-I+M$ groups (see Chapter 4, p. 107) they have larger σ-inductive effects from the *ortho*- than the *para*-positions, and smaller electron-releasing conjugative effects. Both favour higher reactivity of the *ortho*- than the *para*-substituted compounds. To the extent that the activating σ-inductive component is the greater, the *ortho*-acceleration is in the order $F > Cl \sim OR > NR_2$. The conjugative component contribution is in the order $NR_2 > OR > F > Cl$.

In the mononitro series the Ortho Index for fluorine is about 0.05, whereas for chlorine it is 1.17. For chlorine it is 2.26 in the dinitro series, and 0.71 in the unsubstituted series *[112]*. Fluorine thus demonstrates the polar effect without evidence of steric or field effects. The chlorine values suggest counteracting weaker polar effects and very slight steric or field effects. It is relevant that X-ray data show that hexafluorobenzene is planar; but that in hexachlorobenzene the chlorine atoms are staggered and about 12° out of the ring plane *[117,118]*.

The methoxyl group gives a clear example of *ortho*-acceleration also, with an Ortho Index, in the dinitro series, equal to 0.06 at 0°. The amino group is a further example. Though the Ortho Index has the higher value of 0.24, its value is only approximate because the results for the *para*-compound are imprecise, being affected by side-reactions at the necessarily elevated temperatures used. The higher value for amino than for methoxyl, may however, reflect a greater contribution of the σ-inductive polar effect to the *ortho* substituent effect in the case of the methoxyl group.

Results for aldehyde and nitrile groups in reactions with methoxide ion suffer from competition by the electrophilic carbonyl and nitrile carbon for the nucleophile, and this is affected by steric factors. Interactions are the reversible formation of hemiacetal and imino-ether conjugate bases, *viz.*

$$-\mathrm{CH{=}O} + \mathrm{\dot{O}Me^-} \rightleftharpoons -\mathrm{CH}\langle^{\mathrm{O^-}}_{\mathrm{OMe}}$$

$$-\mathrm{C{\equiv}N} + \mathrm{OMe^-} \rightleftharpoons -\mathrm{C}\langle^{\mathrm{{=}N^-}}_{\mathrm{OMe}}$$

The relevance of these interactions has been discussed and further investigated by Miller and his co-workers *[112]*.

The small linear nitrile group has virtually no adverse steric requirement, but the iminoether has. This inhibits the side-chain addition whenever there is an *ortho*-group other than hydrogen. As a result 2-chloro-5-nitrobenzonitrile reacts smoothly and cleanly with methoxide ion to displace chlorine. The results for 4-chloro-3-nitrobenzonitrile are affected by the side reaction, but from estimates based on the reaction of thiocyanate ion with 4-iodo-3-nitrobenzonitrile *[119]*, a "true" value of f_p^{CN} may be obtained and leads to an Ortho Index of about 1.3. This confirms the virtual absence of steric factors, since a value of about 1.3 is expected for conjugative activating groups (see Chapter 4, p. 97) in the absence of steric effects.

The steric requirements of the formyl group with an *ortho*-substituent probably cause it to rotate out of the plane of the ring so that there is no "coplanarity barrier" to the side-chain addition. Thus for both *ortho*- and *para*-formyl groups, estimates of the "true" value of f^{CHO} were based on results from reactions of thiocyanate with iodo compounds. The Ortho Index thus obtained has a value of about 8, closely similar to values for COMe and CO_2Me, and similarly implies weak steric factors only.

Brieux and Deulofeu *[29]* and Sbarbarti *[120]* have made similar sets of measurements with piperidine in benzene as reagent. The fuller set of results at 75° of the latter author are shown in Table 70. These results, although with a reagent of greater steric requirements than methoxide ion, are in some cases similar to those already discussed, but in others specific factors operate. It should be noted that for the 1-chloro-2-nitro-6-X-benzenes, which are compared with 1-chloro-2-nitro-4-X-benzenes, both positions *ortho* to the point of substitution are occupied so that this corresponds to the dinitro series of Miller and his co-workers.

The carboxyl group, as already mentioned above, behaves quite

TABLE 70

COMPARISON OF REACTIONS OF *ortho*- AND *para*-SUBSTITUTED 1-CHLORO-2-NITROBENZENES WITH PIPERIDINE IN BENZENE AT 75° ($\rho \sim 4$)

Substituent X	Substituent Rate Factor (S.R.F. or f)		Ortho Index
	f_p^X	f_o^X	(f_p^X/f_o^X)
H	1	1	1
CO_2H[a]	21.8	14.7	1.48
CO_2Et	1010	42.6	23.7
NO_2	42,200	1820	23.2
CH_3	0.145	0.0049	29.6
Cl	4.64	1.38	3.36
Br	7.19	1.14	6.31
OMe	0.032	0.150	0.213
NH_2	0.0021	0.0033	0.638

[a] Assumed completely neutralised and therefore as CO_2^-.

differently since with a neutral reagent there are favourable instead of unfavourable electrostatic plus hydrogen-bonding interactions. The Ortho Index is 1.48 at 75°, in contrast to the value for reaction with methoxide ion in the dinitro series of 1640 at 0° (the value at 100° is only 61). The Ortho Index for the ethoxycarbonyl group is 23.7 at 75°, instead of 131 at 0° (84 at 100°) in the dinitro series for the methoxycarbonyl group (methoxide ion reaction), and this suggests a polar effect resembling that with the carboxylate ion, but much less marked, due to the carbonyl oxygen bearing a fractional negative charge.

The Ortho Index for the nitro group is 23.2 at 75° as compared with 40.2 at 0° (38 at 100°) for reaction with methoxide ion. The difference is ascribed to favourable effects resembling those described for the carboxylate group (and alkoxycarbonyl groups). These have been discussed at some length by Bunnett and his co-workers *[36]*. Chlorine has only a slightly higher Ortho Index, 3.36, than in reaction with methoxide ion, 2.28 at 0°, confirming the very slight character of its steric effect. This is somewhat enhanced with bromine, Ortho Index, 6.31 at 75°, in the reaction with piperidine.

The *ortho* effects of methoxyl and amino groups have been classified as polar in origin and this is confirmed by similar results with piperidine. Values of Ortho Indices are 0.213 and 0.638, as compared with 0.06 and 0.24, at 0° (0.217 and 0.396 at 100°) in the reactions with methoxide ion.

The methyl group shows definite evidence in piperidine of a steric effect, whereas little or none is evident with methoxide ion. Ortho Indices are 29.6 with the former and 1.78 with the latter at 0° (1.91 at 100°). In view of the demonstration by Fierens and Halleux *[116]* of the incursion of a marked steric effect of *ortho*-tert-butyl, but not *ortho*-methyl, -ethyl or -isopropyl groups, in reactions with iodide ion, it would be of interest to make measurements in this alkyl series with piperidine.

Substituent effects at more distant sites may demonstrate secondary steric effects, such as by inhibition of conjugation, and cross-conjugation. Evidence for this is available from comparison of *meta*- and *para*-substituent effects (see Chapter 4, p. 114), as in reactions of 1-chloro-2,4-dinitro-5-X- and 1-chloro-2,6-dinitro-4-X-benzenes *[112c,121]*. In a manner similar to the previous comparison, ratios of *para*- and *meta*-S.R.F. or f values, f_p^X/f_m^X, are treated as an Index, previously described as the Conjugative Index but now the Meta Index. In the main such values reflect the large polar, especially conjugative, differences between effects from the *para*- and *meta*-positions, but some steric effects are seen. In the dinitro series, for example, the Meta Index for chlorine is 1.17 at 0° whereas a direct comparison of *m*- and *p*-dichlorobenzene gives an Index of 0.28 *[113]*. The difference though small is ascribable to mild interference by the 5-chlorine with coplanarity of the 4-nitro group.

The sulphonate group is activating at the *para*-position but deactivating at the *meta*-position with a Meta Index of 279 at 50°. This is ascribed to very marked steric inhibition of coplanarity of the 4-nitro group by the 5-sulphonate group. Similarly the carboxylate group has a Meta Index of 55 at 50°. Miller and his co-workers have attempted to assess divergences from a Hammett plot in terms of steric and cross-conjugation factors, with some success. The results, though speculative, support the view that such effects are present in other less obvious cases.

Brieux and his co-workers *[34c]* have compared *meta*- and *para*-substituent effects with piperidine as reagent. Marked differences in polar effects are obvious from their work but since there is no substituent in the 4-position in their compounds only cross-conjugation effects would be expected.

Steric effects should also be noticeable when the steric requirements of entering and leaving groups are substantial. There is evidence for this in the heavy nucleophile interactions described by Miller and his co-workers *[7]*, where both entering and leaving groups are heavy atoms or attached to the reaction centre by heavy atoms. It leads to relatively low values of k_2 (and high values of $\Delta G^{\ddagger}$), consequent on a substantial reduction of $\Delta S^{\ddagger}$ of the order of 10 e.u., though somewhat counteracted by lower values of $\Delta E^{\ddagger}$.

The low values of $\Delta S^{\ddagger}$ are appropriate for steric interactions not involving any major degree of compression. The cause of the lower values of $\Delta E^{\ddagger}$ is obscure. Such reductions are not found, for example, in saturated aliphatic S_N2 exchanges of halogen (Finkelstein reactions) in acetone *[109]*, though in the aromatic S_N2 reactions the angle between entering and leaving groups is lower, *viz.* about 109° instead of 180°. It might be thought therefore that the favourable mutual polarisability factor (London forces) discussed by Bunnett *[122]* might in fact be operative *[cf. 7e]*, as a counterbalancing rate factor, though not leading to a rate increase. However the $\Delta E^{\ddagger}$ values for the highly polarisable thiomethoxide and thiophenoxide groups fit a pattern based on expected changes in the relative levels of transition states 1 and 2, without inclusion of any polarisability factors. The reasons thus remain obscure and further investigation is called for.

Reactions of amines with 1-chloro-2,4-dinitrobenzene are known (see above) to demonstrate primary steric effects. There appear however to be no investigations of changes in this pattern as the steric requirements of the replaced group are varied, though some simple halogen replacement patterns are known. However, these appear to be normal with piperidine and aniline for example, while abnormalities with *N*-methylaniline are due to changes in relative energy levels of T.St.1 and T.St.2 (see Chapter 6, p. 225), and differences so far found with heavy neutral nucleophiles *[123,124]* seem to be of similar origin also, so that evidence for such a change is still lacking.

In summary, there is clear evidence of steric effects in aromatic S_N reactions and although some of these are quite substantial, very large effects, such as are found in many other classes of reactions, have not so far been found and are not expected.

REFERENCES

1 S. Winstein, D. Darwish and N. J. Holness, *J. Am. Chem. Soc.*, 78 (1956) 2916.

2*a* G. C. Finger and C. W. Kruse, *J. Am. Chem. Soc.*, 78 (1956) 6034; *b* G. C. Finger and L. D. Starr, *J. Am. Chem. Soc.*, 81 (1959) 2674; *c* G. C. Finger, C. D. Starr, D. R. Dickerson and J. Harner, *J. Org. Chem.*, 28 (1963) 1666; *d* G. C. Finger, D. R. Dickerson, T. Adl and T. Hodgins, *Chem. Commun.*, (1965) 430.

3 N. Kornblum, R. K. Blackwood and J. W. Powers, *J. Am. Chem. Soc.*, 79 (1957) 2507.

4 R. T. Major and H. J. Hess, *J. Org. Chem.*, 23 (1958) 1563.

5 H. Zook and T. J. Russo, *J. Am. Chem. Soc.*, 82 (1960) 1258.

6*a* H. E. Zaugg, B. W. Horrom and S. Borgwardt, *J. Am. Chem. Soc.*, 82 (1960) 2895; *b* H. E. Zaugg, *J. Am. Chem. Soc.*, 83 (1961) 837.

7*a* J. Miller and A. J. Parker, *J. Am. Chem. Soc.*, 83 (1961) 117; *b* A. J. Parker, *J. Chem. Soc.*, (1961) 1328; *Quart. Rev. (London)*, 16 (1962) 163; *c* J. Miller, *J. Am. Chem. Soc.*, 85 (1963) 1628; *d* D. L. Hill, K. C. Ho and J. Miller, *J. Chem. Soc., B*, (1966) 299; *e* K. C. Ho, J. Miller and K. W. Wong, *J. Chem. Soc., B*, (1966) 310.

8 L. Clark, *J. Phys. Chem.*, 65 (1961) 1651.

9 R. Fuchs, G. McCrary and J. J. Bloomfield, *J. Am. Chem. Soc.*, 83 (1961) 4281.

10 D. J. Cram, R. Rickborn, C. A. Kingsbury and P. Haberfield, *J. Am. Chem. Soc.*, 83 (1961) 5835.

11 C. A. Kingsbury, *J. Org. Chem.*, 29 (1964) 3262.

12 J. E. Gordon, *J. Am. Chem. Soc.*, (a) 86 (1964)4492; (b) 87 (1965) 1499.

13 H. K. Schläfer and W. Schaffernicht, *Angew. Chem.*, 17 (1960) 618.

14*a* J. Murto and A. M. Hiiro, *Suomen Kemistilehti*, B37 (1964) 177; *b* J. Murto, *Suomen Kemistilehti*, B34 (1961) 92.

15 R. Bolton and J. Miller, unpublished work.

16 W. M. Weaver and J. D. Hutchinson, *J. Am. Chem. Soc.*, 86 (1964) 261.

17 S. Winstein, L. G. Savedoff, S. G. Smith, I. D. H. Stevens and J. S. Gale, *Tetrahedron Letters*, (1960) 24.

18 R. L. Heppolette and J. Miller, *Australian J. Chem.*, 9 (1956) 293.

19 G. P. Briner and J. Miller, *J. Chem. Soc.*, (1954) 4682.

20 K. C. Ho and J. Miller, *Australian J. Chem.*, 19 (1966) 423.

21*a* J. Delpuech, *Tetrahedron Letters*, (1965) 211; *b Angew Chem. (Intern. Ed.)*, 6 (1967) 1046.

22 D. J. Cram, B. Rickborn and G. R. Knox, *J. Am. Chem. Soc.*, 82 (1960) 6412.

23 J. Murto, *Acta Chem. Scand.*, 18 (1964) 1029.

24 J. Miller, *Rev. Pure Appl. Chem. (Australia)*, 1 (1951) 171.

25 A. J. Parker, private communication.

26 R. Bolton, J. Miller and A. J. Parker, *Chem. and Ind.*, (1963) 492.

27 H. Suhr, *Chem. Ber.*, 97 (1964) 3277.

28 N. B. Chapman, R. E. Parker and P. W. Soanes, *J. Chem. Soc.*, (1954) 2109.

29 J. A. Brieux and V. Deulofeu, *J. Chem. Soc.*, (1954) 2519.

30 M. Freymann and R. Freymann, *Arch. Sci. (Geneva)*, 11 (1958) Sp. No. 233; *Chem. Abstr.*, 53 (1959) 21173[c].

31 M. Okuda, *Nippon Kagashu Zasshi*, 82 (1961) 1118; *Chem. Abstr.*, 56 (1962) 2089[i].

32 H. Fritzsche, *Z. Naturforsch.*, 19a (1964) 1132.
33 G. S. Hammond and L. R. Parks, *J. Am. Chem. Soc.*, 77 (1955) 340.
34*a* N. E. Sbarbarti, T. H. Suarez and J. A. Brieux, *Chem. and Ind.*, (1964) 1754; *b* W. Greizerstein and J. A. Brieux, *J. Am. Chem. Soc.*, 84 (1962) 1032; *c* W. Greizerstein, R. A. Bonelli and J. A. Brieux, *J. Am. Chem. Soc.*, 84 (1962) 1026.
35 F. Pietra, *Tetrahedron Letters*, (1965) 745.
36*a* J. F. Bunnett and R. J. Morath, *J. Am. Chem. Soc.*, 77 (1955) 5051; *b* J. F. Bunnett, R. J. Morath and T. Okamoto, *J. Am. Chem. Soc.*, 77 (1955) 5085.
37*a* R. R. Bishop, E. A. S. Cavell and N. B. Chapman, *J. Chem. Soc.*, (1952) 437; *b* E. A. S. Cavell and N. B. Chapman, *J. Chem. Soc.*, (1953) 3392; *c* N. B. Chapman and C. W. Rees, *J. Chem. Soc.*, (1954) 1190; *d* N. B. Chapman and D. Q. Russell-Hill, *J. Chem. Soc.*, (1956) 1563.
38*a* S. D. Ross in S. G. Cohen, A. Streitwieser and R. W. Taft, (Eds.), *Progress in Physical Organic Chemistry*, Vol. 1, Interscience, New York, 1963, p. 31; *b* S. D. Ross and M. Finkelstein, *J. Am. Chem. Soc.*, 85 (1963) 2603.
39 G. Illuminati and G. Marino, *Chem. and Ind.*, (1963) 1287.
40*a* J. Miller and V. A. Williams, *J. Am. Chem. Soc.*, 76 (1954) 5482; *b* J. Miller and V. A. Williams, unpublished work.
41 E. Berliner and L. C. Monack, *J. Am. Chem. Soc.*, 74 (1952) 1574.
42 A. A. Frost and R. G. Pearson, *Kinetics and Mechanism*, Wiley, New York, 1961, Chapter 7.
43*a* B. A. Bolto and J. Miller, *J. Org. Chem.*, 20 (1955) 558; *b* B. A. Bolto and J. Miller, *Australian J. Chem.*, 9 (1956) 74, 304; *c* B. A. Bolto and J. Miller, unpublished work.
44*a* B. A. Bolto, M. Liveris and J. Miller, *J. Chem. Soc.*, (1956) 750; *b* M. Liveris and J. Miller, *Australian J. Chem.*, 11 (1958) 297.
45*a* J. Miller, *J. Am. Chem. Soc.*, 76 (1954) 448; *b* J. Miller, *J. Am. Chem. Soc.*, 77 (1955) 180; *c* M. Liveris, P. G. Lutz and J. Miller, *J. Am. Chem. Soc.*, 78 (1956) 3375.
46*a* C. P. Leung and J. Miller, unpublished work; *b* R. L. Heppolette, J. Miller and A. J. Parker, *Chem. and Ind.*, (1954) 904; *c* R. L. Heppolette, J. Miller and V. A. Williams, unpublished work.
47 J. N. Brønsted, *Z. Phys. Chem.*, 102 (1922) 169, and later papers.
48 V. K. La Mer, *Chem. Rev.*, 10 (1932) 185, 192, and later papers.
49*a* C. W. Davies, *J. Chem. Soc.*, (1938) 2092; *b* C. W. Davies in G. Porter (Ed.) *Progress in Reaction Kinetics*, Vol. 1, Pergamon Press, London, 1961, p. 161–186.
50 C. K. Ingold, *Structure and Mechanism in Organic Chemistry*, Cornell Univ. Press, Ithaca, N.Y., (1953), (a) p. 362; (b) p. 357–360; (c) p. 60, 400–403.
51 J. D. Reinheimer, W. F. Kieffer, S. W. Frey, J. C. Cochran and E. W. Barr, *J. Am. Chem. Soc.*, 80 (1958) 164.
52*a* K. B. Lam and J. Miller, *Chem. Commun.*, (1966) 642; *b* K. B. Lam and J. Miller, unpublished work.
53 S. Winstein, S. Smith and D. Darwish, *J. Am. Chem. Soc.*, 81 (1959) 5511.
54 G. Schiemann, *Ber.*, 62 (1929) 1794.
55 M. J. Rarik, R. Q. Brewster and F. B. Dains, *J. Am. Chem. Soc.*, 55 (1933) 1289.
56 R. W. Bost and F. Nicholson, *J. Am. Chem. Soc.*, 57 (1935) 2368.

57a J. H. Brown, C. W. Suckling and W. B. Whalley, *J. Chem. Soc.*, (1949) S95; *b* W. B. Whalley, *J. Chem. Soc.*, (1950) 2241.
58 N. N. Vorozhtsov and G. G. Yakobson, *Zh. Obshch. Khim.*, 28 (1958) 40.
59 H. C. Brown, H. Bartholomay and M. D. Taylor, *J. Am. Chem. Soc.*, 66 (1944) 435 and later papers.
60 O. L. Brady and F. R. Cropper, *J. Chem. Soc.*, (1950) 507.
61a J. F. Bunnett and K. M. Pruitt, *J. Elisha Mitchell Sci. Soc.*, 73 (1957) 297; *Chem Abstr.*, 52 (1958) 10077e; *b* J. F. Bunnett and J. J. Randall, *J. Am. Chem. Soc.*, 80 (1958) 6020.
62 S. D. Ross and M. Finkelstein, *J. Am. Chem. Soc.*, 79 (1957) 6547 and later papers.
63 H. Suhr, *Ber. Bunsenges. Phys. Chem.*, 67 (1963) 893.
64 F. Pietra and A. Fava, *Tetrahedron Letters*, (1963) 1535.
65a B. Bitter and H. Zollinger, *Angew .Chem.*, 70 (1958) 246; *b* B. Bitter and H. Zollinger, *Helv. Chim. Acta*, 44 (1961) 812; *c* H. Zollinger, *Angew. Chem.*, 73 (1961) 1257; *d* C. Bernasconi and H. Zollinger, *Tetrahedron Letters*, (1965) 1083; *Helv. Chim. Acta*, 49 (1966) 103, 2570; 50 (1967) 3; *e* C. F. Bernasconi, M. Kaufmann and H. Zollinger, *Helv. Chim. Acta*, 49 (1966) 2563; *f* G. Becker, C. F. Bernasconi and H. Zollinger, *Helv. Chim. Acta*, 50 (1967) 10.
66a J. F. Bunnett and R. H. Garst, *J. Am. Chem. Soc.*, 87 (1965) 3875, 3879; *b* J. F. Bunnett and C. Bernasconi, *J. Am. Chem. Soc.*, 87 (1965) 5209.
67 C. K. Banks, *J. Am. Chem. Soc.*, 66 (1944) 1127.
68 J. D. Reinheimer, J. T. Greig, R. Garst and R. Schrier, *J. Am. Chem. Soc.*, 84 (1962) 2770.
69 M. Liveris and J. Miller, *J. Chem. Soc.*, (1963) 3486.
70a N. B. Chapman and R. E. Parker, *Chem. and Ind.*, (1951) 248; *b* N. B. Chapman and R. E. Parker, *J. Chem. Soc.*, (1951) 330.
71 T. O. Bamkole, C. W. L. Bevan and J. Hirst, *Chem. and Ind.*, (1963) 119.
72 T. Susuki and H. Hagisawa, *Bull. Inst. Phys. Chem. Res.* (*Tokyo*), 21 (1942) 597.
73 E. A. Burns and F. D. Chang, *J. Phys. Chem.*, 63 (1959) 1314.
74 R. G. Shepherd and L. F. Fedrick in A. R. Katritzky, A. J. Boulton and J. M. Lagowski (Eds.), *Advances in Heterocyclic Chemistry*, *Vol. 4*, Academic Press, New York, p. 145–423.
75 T. L. Chan and J. Miller, *Australian J. Chem.*, 20 (1967) 1595.
76 G. E. Ficken and J. R. Kendall, *J. Chem. Soc.*, (1959) 3988.
77 A. W. Browne, A. B. Hoel, G. B. L. Smith and F. H. Swezey, *J. Am. Chem. Soc.*, 45 (1923) 2541.
78 L. Birckenback and K. Kellermann, *Ber.*, 58 (1925) 786, 2377.
79 P. Walden and L. F. Audrieth, *Chem. Revs.*, 5 (1928) 339.
80 T. Moeller, *Inorganic Chemistry*, Wiley, New York, 1952, p. 463–480.
81 B. E. Douglas and D. H. McDaniel, *Concepts and Models in Inorganic Chemistry*, Blaisdell, New York, 1965, p. 248–250.
82 T. Sandmeyer, *Ber.*, 17 (1884) 1633, 2650 and later papers.
83 A. R. Hantzsch, *Ber.*, 28 (1895) 1751 and later papers.
84a H. Hodgson, S. Birtwell and J. Walker, *J. Chem. Soc.*, (1941) 770; (1944) 18; *b* H. Hodgson and E. Marsden, *J. Chem. Soc.*, (1944) 22; *c* H. Hodgson, *Chem. Rev.*, 40 (1947) 251.
85 P. Waentig and J. Thomas, *Ber.*, 46 (1913) 3923.
86 E. Pfeil and O. Velten, *Ann.*, 561 (1949) 220; 562 (1949) 165; 565 (1949) 183.

87 W. A. Waters, *J. Chem. Soc.*, (1942) 266.
88*a* W. A. Cowdrey and D. S. Davies, *J. Chem. Soc.*, (1949) 549; *b* W. A. Cowdrey and D. S. Davies, *Quart. Rev.*, 6 (1952) 358.
89 K. H. Saunders, *The Aromatic Diazo Compounds and their Technical Applications*, 2nd Ed., Longmans Green and Co., London, 1948. (A 3rd Edition is in preparation.)
90 J. F. Bunnett and R. E. Zahler, *Chem. Rev.*, 49 (1951) 273.
91*a* P. Griess, *Ann.*, 137 (1866) 89; *b* P. Griess, *Ber.*, 1 (1868) 190; *c* P. Griess, *Ber.*, 18 (1885) 961.
92 A. Hantzsch and R. Vock, *Ber.*, 36 (1903) 2060.
93 F. J. Buchmann and C. S. Hamilton, *J. Am. Chem. Soc.*, 64 (1942) 1359.
94 T. Sandmeyer, *Ber.*, 20 (1887) 1494.
95 J. B. Conant and W. D. Peterson, *J. Am. Chem. Soc.*, 52 (1930) 1220.
96 E. S. Lewis and W. H. Hinds, *J. Am. Chem. Soc.*, 74 (1952) 304.
97*a* R. G. R. Bacon and H. A. O. Hill, *Quart. Rev.*, 19 (1965) 95; *b* R. G. R. Bacon and H. A. O. Hill, *Proc. Chem. Soc.*, (1962) 113; *c* R. G. R. Bacon and H. A. O. Hill, *J. Chem. Soc.*, (1964) 1097, 1108.
98 R. D. Stephen and C. E. Castro, *J. Org. Chem.*, 28 (1963) 3313.
99 H. Weingarten, *J. Org. Chem.*, 29 (1964) 977, 3624.
100*a* E. Havinga, R. O. de Jongh and W. Dorst, *Rec. Trav. Chim.*, 75 (1956) 378; *b* E. Havinga and R. O. de Jongh, *Bull. Soc. Chim. Belges.*, 71 (1962) 803; D. F. Nijhoff and E. Havinga, *Tetrahedron Letters*, (1965) 4199; *d* M. E. Kronenberg, A. van der Hayden and E. Havinga, *Rec. Trav. Chim.*, 85 (1966) 56; *e* R. O. de Jongh and E. Havinga, *Rec. Trav. Chim.*, 85 (1966) 275; *f Heterocyclic Photosubstitution Reactions in Aromatic Compounds*, Conseil de Chimie Solvay at University of Brussels, 1965.
101*a* H. E. Zimmerman in W. A. Noyes, G. S. Hammond and J. N. Pitts (Eds.), *Advances in Photochemistry*, Vol. 1, Interscience, New York, 1963; *b* H. E. Zimmerman, *Tetrahedron*, 19 (1963) Suppl. 393; *c* H. E. Zimmerman and S. Somasekhara, *J. Am. Chem. Soc.*, 85 (1963) 922.
102*a* R. O. Letsinger and O. B. Ramsay, *J. Am. Chem. Soc.*, 86 (1964) 1447; *b* R. O. Letsinger, O. B. Ramsay and J. H. McCain, *J. Am. Chem. Soc.*, 87 (1965) 2945.
103 R. Grinter and E. Heilbronner, *Helv. Chim. Acta*, 45 (1962) 2496.
104 H. E. Zimmerman and V. R. Sandel, *J. Am. Chem. Soc.*, 85 (1963) 915.
105*a* V. Gold and C. H. Rochester, *Proc. Chem. Soc.*, (1960) 403; *b* V. Gold and C. H. Rochester, *J. Chem. Soc.*, (1964) 1704.
106 R. M. Johnson and C. W. Rees, *Proc. Chem. Soc.*, (1964) 213.
107*a* J. Miller, *Austral. J. Chem.*, 9 (1956) 61; *b* J. Miller and A. J. Parker, *Austral. J. Chem.*, 11 (1958) 302.
108 K. Ahmed, M. Hassan, J. Miller, D. B. Paul and L. Y. Wong, unpublished work.
109 C. K. Ingold, *Quart. Rev.*, 11 (1957) 1.
110 R. E. Parker and T. O. Read, *J. Chem. Soc.*, (1962) 3149.
111 N. B. Chapman and R. E. Parker, *J. Chem. Soc.*, (1951) 3301.
112*a* J. Miller, *J. Chem. Soc.*, (1952) 3550; *b* J. Miller and V. A. Williams, *J. Chem. Soc.*, (1953) 1475; *c* R. L. Heppolette, M. Liveris, P. G. Lutz, J. Miller and V. A. Williams, *Australian J. Chem.*, 8 (1955) 454; *d* B. A. Bolto, J. Miller and V. A. Williams, *J. Chem. Soc.*, (1955) 2926; *e* R. L. Heppolette, J. Miller and V. A. Williams, *J. Chem. Soc.*, (1955) 2929; *f* B. A. Bolto, J. Liveris and J. Miller, *J. Am. Chem. Soc.*, 78 (1956) 1975;

g N. S. Bayliss, R. L. Heppolette, L. H. Little and J. Miller, *J. Am. Chem. Soc.*, 78 (1956) 1978; *h* H. W. Leung and J. Miller, unpublished work.
113 J. Miller and J. M. Wrightson, *Abstr. 112*th *Meeting, Am. Chem. Soc.*, (1947) 16J.
114 J. Miller, *J. Am. Chem. Soc.*, 76 (1954) 448; 77 (1955) 182.
115 C. W. L. Bevan, T. O. Fayiga and J. Hirst, *J. Chem. Soc.*, (1956) 4284.
116 P. J. C. Fierens and A. Halleux, *Bull. Soc. Chim. Belges.*, 64 (1955) 696; 704; 709; 717.
117 G. Ferguson and J. M. Robertson, in V. Gold (Ed.) *Advances in Physical Organic Chemistry*, Vol. 1, Academic Press, New York, 1963, p. 233.
118 C. A. Coulson and D. Stocker, *Mol. Phys.*, 2 (1959) 397.
119 B. A. Bolto, J. Miller and A. J. Parker, *J. Am. Chem. Soc.*, 79 (1957) 5969.
120 N. E. Sbarbarti, *J. Org. Chem.*, 30 (1965) 3365.
121*a* M. Liveris, P. G. Lutz and J. Miller, *Chem. and Ind.*, (1952) 1222; *b* M. Liveris, P. G. Lutz and J. Miller, *J. Am. Chem. Soc.*, 78 (1956) 3371.
122*a* J. F. Bunnett, *J. Am. Chem. Soc.*, 79 (1957) 5969; *b* J. F. Bunnett and J. D. Reinheimer, *J. Am. Chem. Soc.*, 81 (1959) 315; 84 (1962) 3284.
123 A. J. Parker in *Organic Sulfur Compounds*, Pergamon Press, London, 1961, p. 103–111.
124 J. Miller and H. W. Yeung, unpublished work.

Chapter 9

SOME LESS COMMON AND RELATED AROMATIC S_N REACTIONS

1. Introduction

In the course of the preceding chapters the following mechanisms have been discussed: (*i*) the special S_N1 reactions of diazonium ions, and S_N1-like copper-catalysed reactions; (*ii*) elimination–addition (benzyne) S_N2 reactions; (*iii*) activated and photo-catalysed addition–elimination S_N2 reactions. Reactions of some related systems were also considered in Chapter 7.

There are some aromatic S_N reactions of limited scope not covered by these categories; and many related reactions, some of which are now considered briefly.

2. Less Common Aromatic S_N Reactions

(*a*) *1,2-Addition–elimination reactions*

Addition–elimination reactions already described are of the 1,1 type, *i.e.* the elimination takes place from the same carbon as that to which the nucleophile adds.

Bordwell and his co-workers *[1]* have shown that the substitution of 2-bromothianaphthene-1,1-dioxide (2-bromobenzo[*b*]thiophen-1,1-dioxide) by piperidine and some other nucleophiles give exclusively 3-piperidinobenzo[*b*]thiophen-1,1-dioxide or the corresponding product. They showed that, to a major extent at least, the reaction does not take place via 2-bromo-3-(1-piperidino)-2,3-dihydrobenzo[*b*]-thiophen-1,1-dioxide. Alternative modes of formation are illustrated in Figs. 113 and 114. The former illustrates reaction via a dihydro derivative. An alternative 1,2 addition–elimination mechanism, shown in Fig. 114, is similar to, but not identical with, that suggested by Bordwell and his co-workers. Both involve initial addition at the 3-

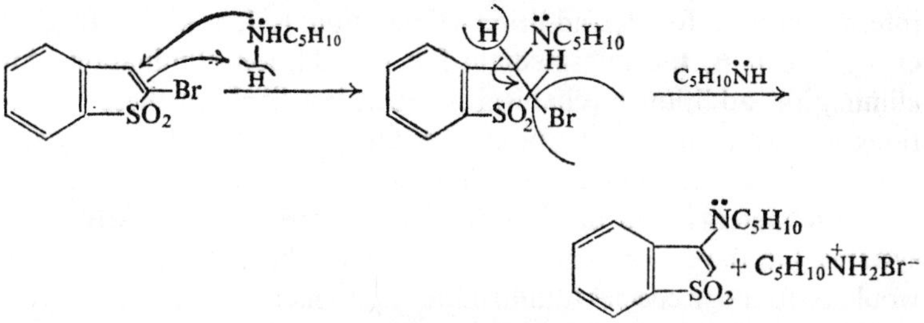

Fig. 113. Formation of 3-piperidinobenzo[*b*]thiophen-1,1-dioxide from 2-bromobenzo[*b*]thiophen-1,1-dioxide via a dihydro derivative.

position. This in itself is not an unusual feature since in reactions of activated compounds the initial addition takes place at points of similar activation irrespective of the presence of a readily replaceable group, though if there is hydrogen at any of these points addition is not commonly followed by elimination to give overall substitution unless a suitable oxidising agent is present.

Features of this reaction are (*i*) the quasi-aromatic character of the thiophen ring; (*ii*) the preference for intermediates in which the carbocyclic ring remains benzenoid; (*iii*) the presence of a powerfully activating ring group (dipolar $>\overset{+}{S}(=O)O^-$). Thus, as in Fig. 115, the

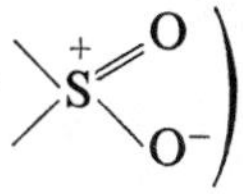

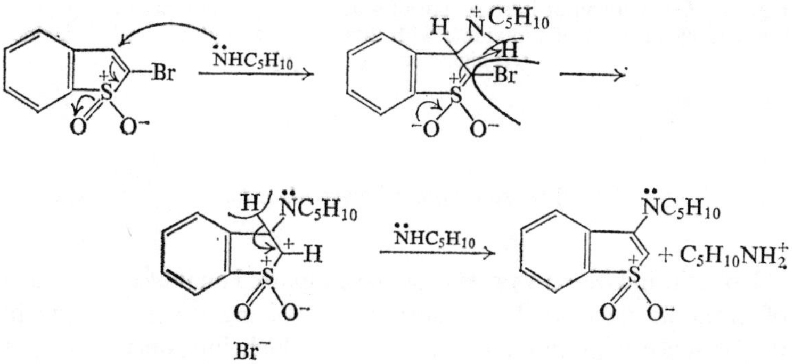

Fig. 114. Formation of 3-piperidinobenzo[*b*]thiophen-1,1-dioxide from 2-bromobenzo[*b*]thiophen-1,1-dioxide by a 1,2-addition–elimination mechanism.

intermediate A for 1,1-addition–elimination is less stable (higher energy) than B, the intermediate for 1,2-addition–elimination. An elimination–addition mechanism is much less likely in these conditions, and would in any case be expected to give a mixture of products.

1,2-Addition–elimination reactions are unlikely to be found in fully aromatic systems, and are less facile in monocyclic quasi-aromatic systems. On the other hand, there is little doubt that such reactions would be found in species fulfilling the requirements mentioned above, and the suggestion of a 1,2-addition–elimination mechanism in some similar species has been made more recently by Kauffmann *et al.* [2].

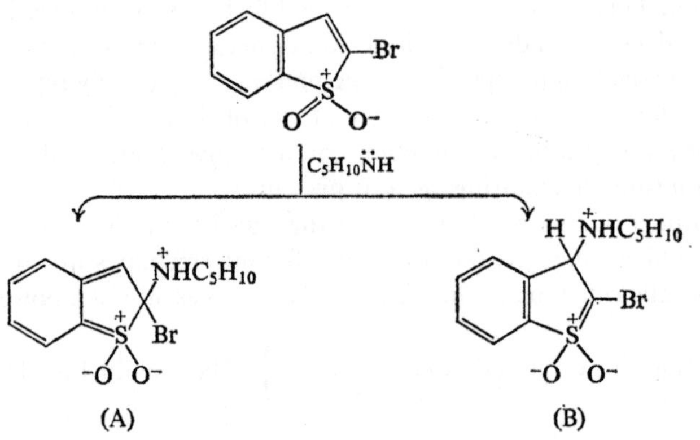

Fig. 115. Comparison of intermediates for 1,1- and 1,2-addition–elimination of bromine by reaction of piperidine with 2-bromobenzo[*b*]thiophen-1,1-dioxide: (A) higher energy than (B).

(*b*) S_N *reactions of benzenium ions*

Kovacic and his co-workers [3] have reported an interesting mode of aromatic nucleophilic substitution, involving trichloroamine in the presence of aluminium chloride, in which introduction of the nucleophile into species unactivated in the normal sense is accompanied by introduction of chlorine. The key is the prior formation of an intermediate, *viz.* the chlorobenzenium ion, by *electrophilic* substitution:

$$Cl^+ + C_6H_5\text{-}\ddot{X} \longrightarrow \text{(4-chloro-4H-benzenium)}{=}\overset{+}{X}$$

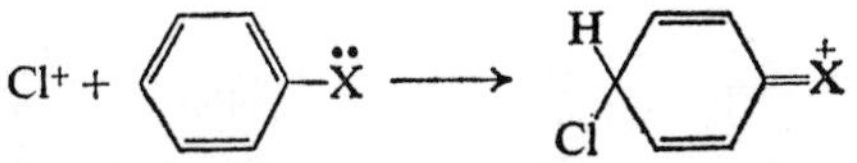

As an electron deficient species, this is susceptible to nucleophilic attack. The major products are *o*- and *p*-chloroanilines, together with *m*-X-aniline *[cf. 4]* as well as considerable amounts of *p*-chloro-X-benzene.

A detailed mechanism now suggested, and accounting for all products, is given below and involves attack by an anionic nucleophile $[NCl_2\text{–}AlCl_3]^-$ on activated positions of a benzenium ion, with alternative 1,1- and 1,2-addition–elimination modes, followed by electrophilic substitution on nitrogen. The benzenium ions may be regarded as positively charged quinomethanes (*cf.* Chapter 7, p. 303).

The reactions of trichloroamine-aluminium chloride with a species PhX are the following:

(*i*) Friedel–Crafts type activation of NCl_3:

$$NCl_3 + AlCl_3 \rightleftharpoons \overset{+}{N}Cl_3\text{–}\overset{-}{Al}Cl_3 \rightleftharpoons Cl^+[NCl_2\text{–}AlCl_3]^-$$

(*ii*) Electrophilic attack on PhX by Cl^+:

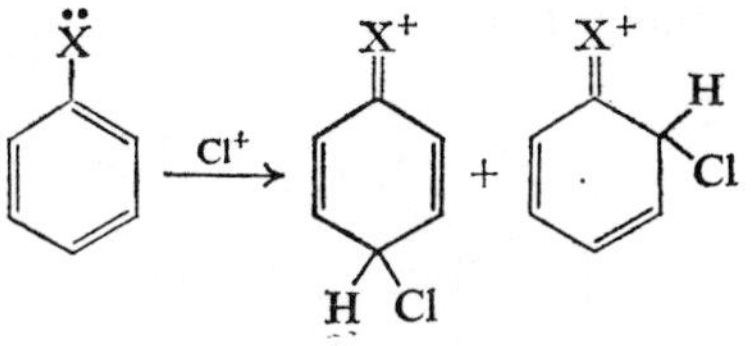

$$C_6H_5\ddot{X} \xrightarrow{Cl^+} \text{(A)} + \text{(B)}$$

(A) (B)

(*iii*) Nucleophilic attack by $[NCl_2\text{–}AlCl_3]^-$ on chlorobenzenium ions, *e.g.* (*ii*) (A), followed by elimination of HX (1,1-addition–elimination):

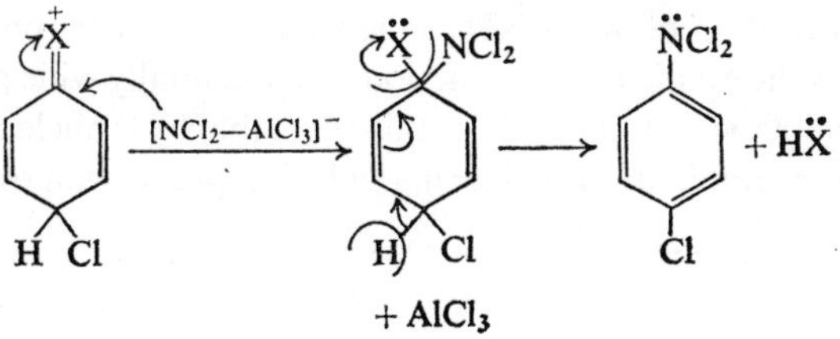

$$\text{(A)} \xrightarrow{[NCl_2\text{—}AlCl_3]^-} \text{intermediate} + AlCl_3 \longrightarrow p\text{-}ClC_6H_4\ddot{N}Cl_2 + H\ddot{X}$$

References p. 388

(*iv*) Electrophilic substitution of 2 moles of PhX by p-$ClC_6H_4NCl_2$, resulting in electrophilic exchange on nitrogen of the *N*-chloroamine:

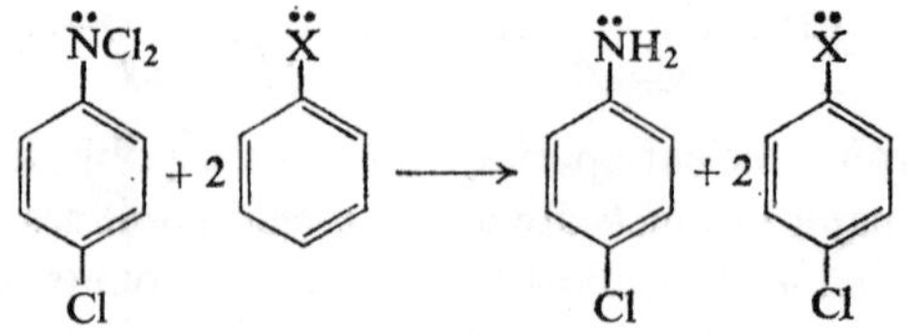

(*v*) and (*vi*) Similar reactions to (iii) and (iv) with species (B).

(*vii*) Alternative nucleophilic attack on benzenium ions, *e.g.* (A), followed by elimination of HCl (1,2-addition–elimination):

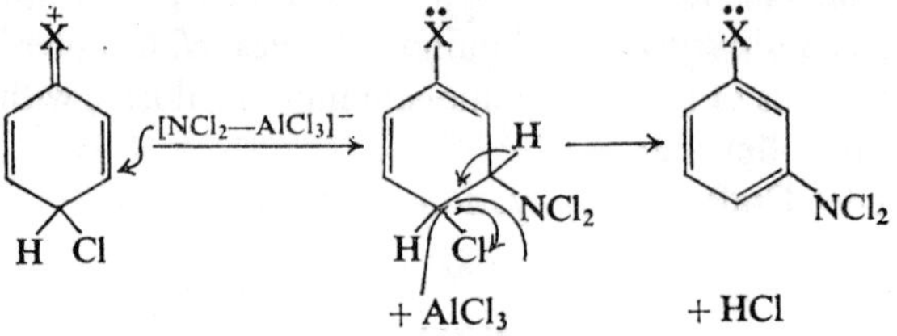

(*viii*) Reactions similar to (*iv*):

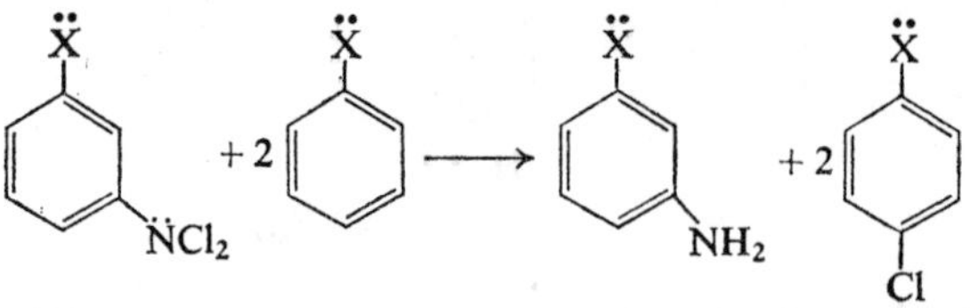

(*ix*) and (*x*) Similar reactions to (*vii*) and (*viii*) with species (B).

Blocking the positions *meta* to X increases the yield of the nucleophilic substitution of X by NH_2 as Kovacic *et al.* have pointed out. Further, as they indicate, the reaction has potentially wide generality. Alternatively, with a group X not displaceable by a nucleophile, the 1,2-addition–elimination displacing chlorine predominates.

(c) *Alkaline hydrolysis of* p-*nitroso*-N,N-*dimethylaniline*

The low mobility of the dimethylamino group was discussed in Chapter 5 (p. 168), reference being made specifically to the very low reactivity of *p*-nitroso-*N*,*N*-dimethylaniline with methoxide ion in methanol reported by Bevan and his co-workers *[5]*. They also compared this with its relatively facile reaction with hydroxide ion in water. The mechanism suggested draws an analogy with the Mannich reaction. In view of the preceding section, and the discussion of quinone reactions in Chapter 7, a broader analogy is with S_N reactions of quinones.

The detailed mechanism is given as Fig. 116. In this the nitrosoamine is written in quinonoid form which probably represents the actual structure more truly *[6–9]*.

The hydroxide ion is able to displace the dimethylamino group while methoxide does not, although the latter is more nucleophilic (Chapter 6, p. 201), because the reaction requires proton removal from the added group to drive the reaction forward. The 3-position relative to the dimethylamino group is also activated and facile displacement of suitable groups should be observable.

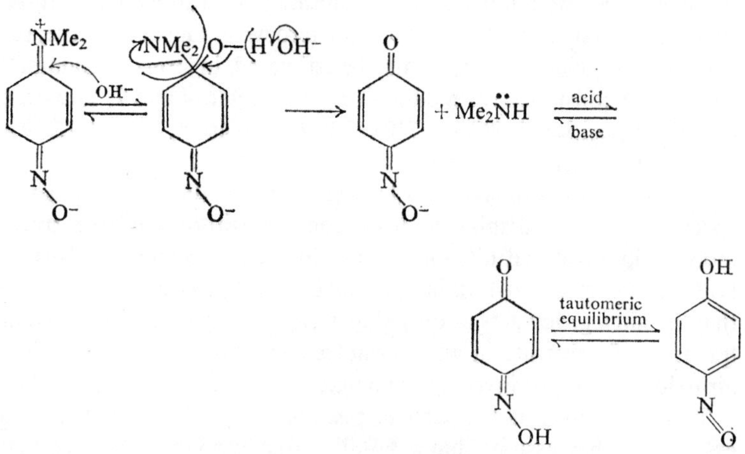

Fig. 116. Alkaline hydrolysis of *p*-nitroso-*N*,*N*-dimethylaniline.

References p. 388

(*d*) *Some reactions at methine carbon atoms*

Situations frequently arise where the position most susceptible to nucleophilic attack has no group replaceable by nucleophilic displacement. Hydrogen is in this category except that it may be removed by oxidation by the air, an added oxidising agent, or the aromatic substrate itself. In specific conditions it may be displaced as hydride ion. In some other activated substrates replaceable groups may be present but are in less activated positions, and reversible addition at the most activated positions accompanies the overall displacement reaction. Additions of nucleophiles to polynitro compounds, producing coloured intermediates, are well-known examples of addition without subsequent elimination, so that the addition is readily reversed. The Janovsky reaction *[10–17]* is an example, and such reactions were referred to in Chapter 5 (pp.161, 173).

There are a number of cases in which hydrogen is reported as displaced from activated positions in preference to displacement of a group X, of a type normally preferentially displaced, at a less activated position. A recent interesting example, with the substrate as oxidising agent, is the report by Richter and Rustad *[18]* of the reaction of diethyl sodiomalonate with 4-nitroquinoline-1-oxide to give a low yield of diethyl sodio(4-nitro-3-quinolyl)malonate-1-oxide.

Apart from the relatively weak additional activation by the fused ring, which activates all points in the heterocyclic ring in any case, the activation at positions 2- to 4- may be estimated, in terms of Hammett substituent constants, from data given in Chapters 4 and 7. These are (at 50°): 2-position, $1.502 + 0.710 = 2.212$; 3-position, $1.178 + 1.219 = 2.397$; 4-position $= 1.526$.

The 3-position is thus the most activated but, in the absence of some special means of displacing hydrogen, substitution of the much more easily displaced nitro group would occur. Richter and Rustad however reported that 4-aminoquinoline-1-oxide was also obtained, so that the nitro group acts as the oxidising agent, apparently of a further quantity of substrate in an intermolecular redox reaction. Possibly air oxidation is also involved. The reaction is illustrated in Fig. 117.

This mechanism is the same as postulated by Richter and Rustad except that they assume that a 3,4-dihydro compound is formed first by further addition of a proton from a molecule of solvent (ethanol).

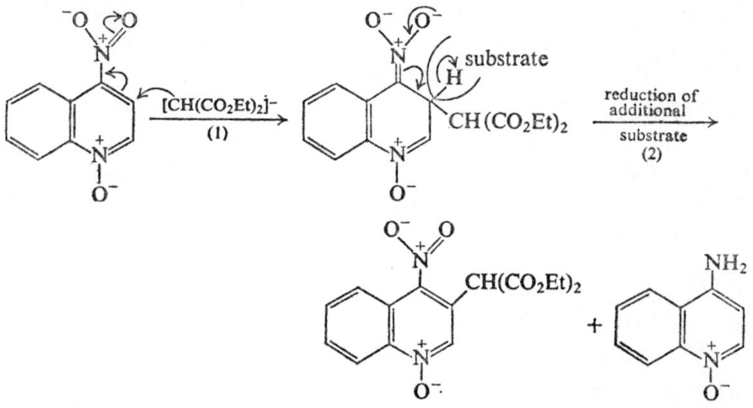

Fig. 117. Nucleophilic substitution of hydrogen in 4-nitroquinoline-1-oxide. Stage (1), nucleophilic addition; stage (2), elimination of hydride by reduction of additional substrate.

Such reactions nevertheless suggest that when a methine carbon atom is the most activated point, there is a special factor which favours attack by a carbanion at that point, as compared with attack by other nucleophiles, in almost all of which the nucleophilic or bond-forming atom is a hetero atom. Most probably, the factor involved is the ease of oxidation, involving reaction with an electrophile, of the methine hydrogen with its bonding electrons, which is adversely affected by electron-withdrawing groups attached to the methine carbon, as with hetero-atom nucleophiles. Significantly, and in support, in the Von Richter reaction (see below), the carbanionic reagent, unusually, is powerfully electron-withdrawing as an attached group (cyano group) and in this case the methine hydrogen is removed as a proton instead of being oxidised.

In contrast, and as expected, it has been reported *[19]* that n-butyl mercaptide ion reacts with 4-nitroquinoline-1-oxide in a normal displacement reaction of the nitro group.

(*e*) *The Von Richter reaction*

Another reaction in which the initial addition takes place other than at the point where a replaceable group is present is the Von Richter reaction. The reagent is also a carbanion, *viz.* the cyanide ion. The reaction is the conversion of a range of aromatic nitro compounds to carboxylic acids by reaction with cyanide ion in aqueous alcoholic medium. It was first described by Von Richter *[20]* and it was later shown that the carboxyl group is *ortho* to the position vacated by the nitro group *[20c,21,22]*. The *overall* reaction thus corresponds to a 1,2-addition–elimination process.

Little further work was done on this reaction until a series of studies by Bunnett and his co-workers *[23]* led to elucidation of most of its salient features, and to the postulation of a reaction mechanism. Further investigation by Rosenblum *[24]* has led to a modification of the mechanism, although the essential features are unchanged, and the modified mechanism has received support from the report by Ullman and Bartkus *[25]* that a postulated intermediate, *viz.* 3-indazolone, gives the required carboxylic acid with elimination of nitrogen.

The essential features of the reaction are:

(*i*) the carboxyl group is *ortho* to the position vacated by the nitro group;

(*ii*) hydrogen from the hydroxylic solvent is incorporated;

(*iii*) the carboxyl group is not formed from the corresponding nitrile or amide by hydrolysis;

(*iv*) substituents *ortho* to the nitro group inhibit but do not necessarily preclude reaction;

(*v*) some activating substituent(s) should be present but if too activating alternative reactions occur;

(*vi*) nitrogen gas is evolved in a yield corresponding to that of carboxylic acid, one of the nitrogen atoms in each molecule of nitrogen gas comes from the nitro group (shown by use of ^{15}N), and the other nitrogen comes from cyanide ion by a path not involving its hydrolysis to ammonia;

(*vii*) by use of $H_2{}^{18}O$ enriched solvent it was shown *[26]* that the atom % of ^{18}O in the carboxyl group is half that of the solvent, indicating that one oxygen comes from the solvent and the other comes from the nitro group.

The mechanism in accord with all these factors and the supporting evidence of Ullman and Bartkus *[25]*, is illustrated in Fig. 118.

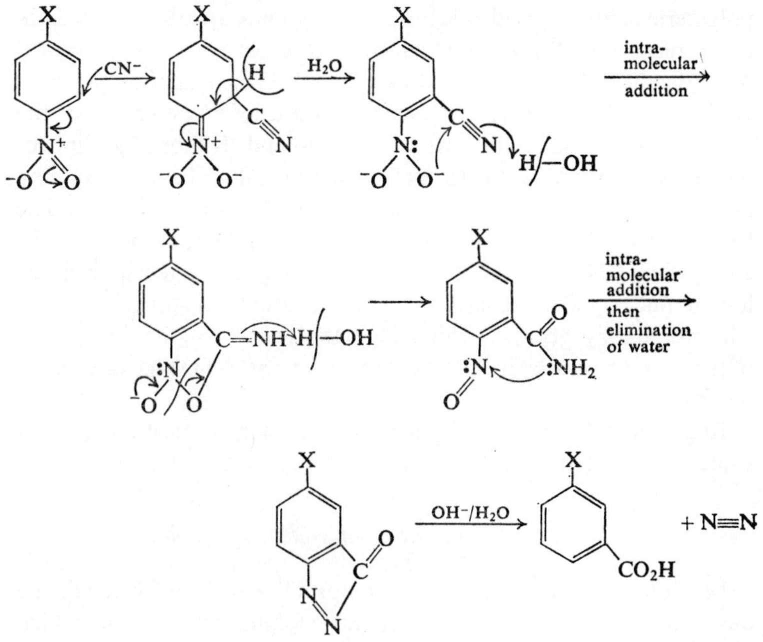

Fig. 118. Mechanism of the Von Richter reaction.

The ability of this reaction to compete with normal substitution derives from (*i*) the combined activating power of NO_2 and X, at the position *ortho* to the nitro group, sufficiently exceeding that *para* to the nitro group; (*ii*) the relatively high acidity of the

$$({}^{-}O)_2\overset{+}{N}{=}C{-}\overset{H}{\underset{|}{\overset{|}{C}}}{-}C{\equiv}N$$

system, which permits facile replacement of hydrogen as a *proton*, its bonding electrons being involved instead in subsequent steps; (*iii*) the availability of subsequent intramolecular redox reactions involving the nitro group as oxidant and nitrile group as reductant, leading then to intramolecular side-chain nucleophilic addition reactions; (*iv*) the facile hydrolysis of the indazolone thus formed.

Bunnett and his co-workers *[23]* showed that what seem to be

References p. 388

polymeric acidic, as well as neutral, compounds are also obtained in these reactions. With *p*-nitroanisole as substrate, they identified 4,4′-dimethoxy-2,2′-dicarbamylazoxybenzene as a neutral product. Beckwith and Miller *[27a,b]* carried out a brief investigation of the reactions of cyanide ion with 1-chloro- and 1-fluoro-2,4-dinitrobenzenes and recorded the formation of a red oil probably containing an azo compound, and also isolated a 2,4-dinitrophenyl ether. The formation of the latter suggests that alkoxide ions, formed by the equilibrium $CN^- + ROH \rightleftharpoons HCN + OR^-$, can compete with cyanide ion in nucleophilic displacement of the usual substituent groups, whereas alkoxy groups, which lack the acid-strengthening power of the cyano group, cannot compete in reactions of the type now being described.

Rogers and Ulbricht *[27c]* have shown that the reaction takes a somewhat different and more complex course in DMSO.

(*f*) *Reactions of 3-nitrobenzylidene compounds*

Loudon and Smith *[28]* have reported that 3-nitrobenzylidene dichlorides undergo an abnormal nucleophilic substitution, which they suggest is initiated by an attack on methine carbon, and that the hydrogen at that point is removed as a proton, following elimination of chloride from the benzylidene carbon. In their reactions the substrates possess also a tolylsulphonyl group *ortho* to the dichloromethyl group, but they pointed out that their reactions are analogous to those of 3-nitrobenzylidene acetals reported by Kliegl and Hölle *[29]*.

The essence of these reactions is that the methine position *ortho* to the nitro group is the most powerfully activated and that elimination of a group X with its bonding electrons from the side-chain allows the necessary internal electron redistribution, including removal of the methine hydrogen as a proton.

The σ^--value for the *ortho*-nitro group has been quoted already, *viz.* 1.219. SO_2X groups, like the nitro group, have very substantial inductive components in their polar effects *[25]* and since the σ^--value for the *p*-SO_2Ph group is 1.117 *[30,31a]*, similar to the value for the *p*-NO_2 group, the value for the *m*-SO_2 tolyl group may be estimated as $\nless 0.7$, similar to the value for the *m*-NO_2 group. From results of

Miller and his co-workers *[30b,32]* the σ^--values for $CHCl_2$ and $CH(OR)_2$ groups may be estimated as about 0.3. The total activation is thus large, corresponding to a $\Sigma\sigma^-$ of about 2.2 for the reactions investigated by Loudon and Smith, and about 1.5 for those investigated by Kliegl and Hölle.

The details of the suggested mechanism, shown in a rather general form, are given in Fig. 119. In this an extra (final) stage is shown in which a normal side-chain substitution of the benzyl compound takes place.

Presumably the position *para* rather than *ortho* to the benzylidene group, being both *ortho* to the nitro group and *meta* to R, is preferred for steric reasons.

Similar reactions should be found wherever corresponding structures occur, and as indicated by comparison with the work of Kliegl and Hölle, the R group *ortho* to the benzylidene group is not essential but acts purely as an additional activating group. As far as the polar effect is concerned, an activating group *meta* to the benzyli-

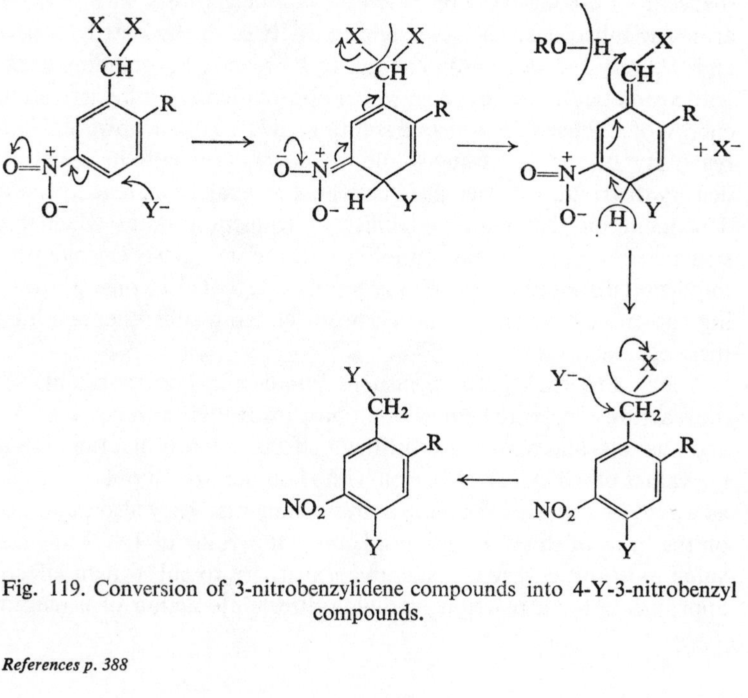

Fig. 119. Conversion of 3-nitrobenzylidene compounds into 4-Y-3-nitrobenzyl compounds.

References p. 388

dene group, and thus *ortho* or *para* to methine carbon atoms at which the nucleophile might react, would be more effective; but this is likely to be counteracted in part by steric factors, since each methine position would then have *two ortho* groups.

(*g*) *Reactions with the solvated electron*

Considerable interest in reactions of the solvated electron has developed in recent years *[e.g. 33,34]*. It can be regarded as the simplest base and nucleophile and is known to be very reactive.

Hart and his co-workers *[35]* have shown that the hydrated electron (e^-_{aq}) is a powerful nucleophile which reacts with aromatic substrates to form carbanions as primary products. Secondary reactions include decomposition to a radical and an anion, and proton abstraction from the solvent. In forming the carbanion no new bond is formed, but the process bears some resemblance to the formation of cyclohexadienide (benzenide) σ-complexes by reactions of nucleophiles with reactive aromatic substrates. The very high reactivity of the hydrated electron and the very low activation energy for its reactions, including those with aromatic substrates *[36]*, which approximates to the activation energy of diffusion in aqueous solution *[37]* suggest, however, that reactivity of the substrate would be correlated with its electron-density distribution, whereas in activated aromatic S_N2 reactions it is dependent on the relative stability of transition states which are structurally close to the cyclohexadienide adducts. Despite this important distinction it remains generally true that electron-withdrawing substituents facilitate and electron-releasing substituents inhibit these e^-_{aq} reactions.

Anbar and Hart *[35b]* measured bimolecular rate constants for the reaction of e^-_{aq} with a range of mono-substituted benzenes, toluenes and phenols, and reported substituent effects in terms of a function η, the values of which correlate well with Hammett σ- but not σ^--values, as would be expected from the above comments. They also estimated on the basis of this correlation a Hammett ρ-value of 4.8. While this value indicates a very considerable sensitivity to substituent effects, appropriate for a reaction involving direct interaction of a reagent with an aromatic ring, it should be recalled that aromatic S_N2 reactions of species ZArX, in which there is a substituent Z and a leaving group X, which proceed by the addition–elimination mechanism via σ-complexes, have much higher values of ρ in the range 8–9 *[31]* (*cf.* discussion in Chapter 4, p. 65).

3. Some Other Nucleophilic Reactions

There are many reactions closely related to substitution reactions on aromatic carbon, which should be referred to very briefly, though any full discussion is beyond the scope of this book. They include nucleophilic additions to aromatic carbon, nucleophilic ring fission, nucleophilic condensation by side-chain α-CH groups, and nucleophilic substitution on atoms attached directly to the ring (α-atoms).

(*a*) *Nucleophilic additions; nucleophilic ring fission; nucleophilic condensations by α-CH groups*

Examples of the first three are the following:
(*i*) covalent hydration.

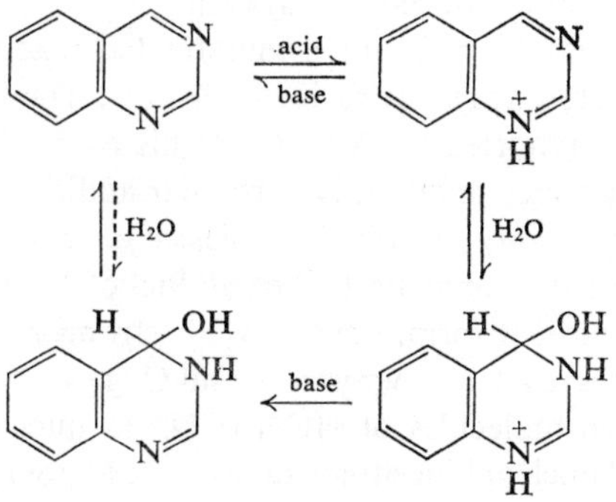

acid
base
+
N
H
H_2O
H_2O
H OH
NH
N
base
H OH
NH
+
N
H

(*ii*) fission of a pyridinium ring (X is an electron-withdrawing group):

$$\text{Pyridinium-}\overset{+}{N}\text{-X} \xrightarrow{Ph\ddot{N}H_2} Ph\overset{+}{N}H{=}(CH)_5{-}NHX$$

$$\downarrow PhNH_2$$

$$Ph\overset{+}{N}H{=}(CH_5){-}NHPh + XNH_2$$

(*iii*) formation of stilbenes by condensation

$$NO_2{-}C_6H_3(NO_2){-}CH_2{-}H \;(:Base)\;\; R{-}CH{=}O \;\; H{-}Solv. \longrightarrow NO_2{-}C_6H_3(NO_2){-}CH_2{-}CHOH{-}R$$

$$\downarrow$$

$$NO_2{-}C_6H_3(NO_2){-}CH{=}CH{-}R + H_2O$$

Because the nucleophilic addition reactions form reduced aromatic systems, they are essentially confined, among suitably activated systems, to polycyclic systems in which the delocalisation energy *per ring* is less than in a single benzene ring, and which form conjugated benz-derivatives of a reduced ring system.

Covalent hydration, is a much investigated reaction of this type which has featured in recent reviews *[38–40]*. The reaction used as illustration was reported by Albert and his co-workers *[41]*. Since water is a weak nucleophile this particular addition reaction occurs only with very highly activated species, *e.g.* with the quinazoline cation but not quinazoline itself. The adduct of the latter is however known as a transient form, see (*i*). With stronger nucleophiles less activated species react, *e.g.* Bergstrom and Ogg *[42]* have reported the addition to two molecules of HCN (CN^- as nucleophile) to quinoxaline. Initial nucleophilic attack takes place at points most activated for substitution by the nucleophile concerned.

Nucleophilic fission of an aromatic ring requires a very highly activated system, and the formation of products with considerable delocalisation energy such as unit-charged open-chain polymethine anils: reactions forming these have been known for a long time. The reaction used as an illustration was reported by Zincke and his co-workers *[43]*. Reaction is initiated by nucleophilic attack as in aromatic S_N reactions.

Side-chain (α-CH) nucleophilic condensations are less closely related to aromatic S_N reactions. The reaction is initiated by a nucleophilic addition as in aldol-type reactions. In the example given *[43]* 2,4-dinitrotoluene acts as donor molecule and an aldehyde as acceptor molecule.

Activation of the α-CH group requires electron-withdrawal as in S_N reactions, but essentially only conjugative withdrawal is effective. Substituent effects do not therefore entirely parallel those in aromatic S_N reactions in which inductive activation is also important. Such differences in activation were referred to by Miller and his co-workers *[30,45]*. A feature of reactions in which substituent effects are largely conjugative, as compared with those where inductive effects are also important, is a relative enhancement of reactivity of groups such as COR and depression of reactivity of groups such as SO_2R. There is a similar relative enhancement of NO as against the NO_2 group, but the comparison is complicated by the ability of nitroso compounds to act also as acceptors in addition reactions.

(*b*) *Nucleophilic substitution on α-atoms*

Another class of reactions of interest is that in which the nucleophile attacks an atom attached directly to the ring and thus displaces the whole aryl group with its bonding electrons. This requires catalysis by protic solvent or other proton source, which must be of sufficiently low electrophilic power not to react with the nucleophile. The overall reaction, omitting charge, is illustrated in Fig. 120.

Such reactions are facilitated by electron-withdrawing groups, but unlike the side-chain condensations, which they resemble to some extent, inductive as well as conjugative electron-withdrawal is effective.

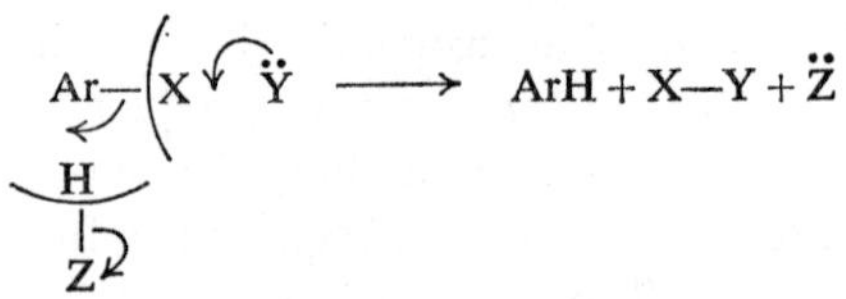

$$\text{Ar—X} \; \ddot{\text{Y}} \longrightarrow \text{ArH} + \text{X—Y} + \ddot{\text{Z}}$$
$$\begin{array}{c}\text{H}\\ |\\ \text{Z}\end{array}$$

Fig. 120. Nucleophilic substitution on X attached directly to an aromatic ring.

In this they are much more like ordinary aromatic S_N reactions, and as in these, the nitro group is highly effective.

Blatt and Tristram *[46]* have thus shown that picryl chloride reacts with iodide ion in formic or acetic acid at steam-bath or boiling-point temperature, and in acetone at room temperature to form 1,3,5-trinitrobenzene. They suggest that reaction proceeds by initial halide exchange (ordinary S_N reaction) to form picryl iodide which then reacts with iodide ion to form 1,3,5-trinitrobenzene and iodine. An alternative is a direct displacement on chlorine followed by reaction of iodine monochloride with iodide ion. Both are illustrated in Fig. 121.

It is relevant that Miller's calculations *[47,48]*, which are, for example, in reasonable agreement with experimental observations of chloride exchange of picryl chloride, iodide exchange of 1-iodo-2,4-dinitrobenzene, and reaction of iodide ion with 1-bromo-2,4-dinitrobenzene, suggest that the reactivity of picryl chloride with iodide ion in a protic solvent such as an alcohol is low, *viz.* $k_2 \sim 5 \cdot 10^{-4}$ at 100°. No doubt the carboxylic acid solvents used have somewhat different solvent effects, but not such as to lead to an enhancement of reactivity as in aprotic solvents. On the other hand one would expect the substitution on halogen to proceed more readily on iodine than chlorine, so that some detailed kinetic investigations are warranted. In either case one may assume that S_N2 halide exchange, especially I–I exchange, accompanies displacement on halogen.

The faster reaction in acetone, as mentioned above, is characteristic of the solvent effect on reactivity of anionic nucleophiles (see Chapter 8, p. 315).

Reactions of this type have been found useful in preparation of parent heterocycles from halogeno derivatives. For preparative convenience phosphorus is used with hydriodic acid.

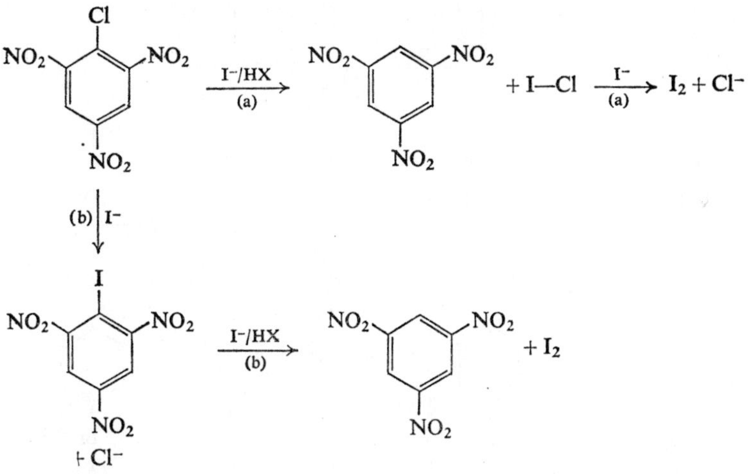

Cl

NO_2 NO_2 $\xrightarrow[(a)]{I^-/HX}$ NO_2 NO_2 $+ I{-}Cl \xrightarrow[(a)]{I^-} I_2 + Cl^-$

NO_2 NO_2

(b) I^-

I

NO_2 NO_2 $\xrightarrow[(b)]{I^-/HX}$ NO_2 NO_2 $+ I_2$

NO_2 NO_2

$+ Cl^-$

Fig. 121. Alternative modes of formation of 1,3,5-trinitrobenzene by reaction of iodide ion with picryl chloride.

(*c*) *Some rearrangements*

There are many rearrangements of nucleophilic character, and a few of these are considered briefly.

(*i*) *Base-catalysed isomerisation*

Wotiz and Huba *[49]* reported the isomerisation of 1,2,4- to 1,3,5-tribromobenzene by the action of sodium amide in liquid ammonia. This has been confirmed and extended by Bunnett and Mayer *[50]*, using also potassium anilide.

Demonstrating the rearrangement by potassium anilide of 1-iodo-2,4-dibromobenzene mainly to 1-iodo-3,5-dibromo-, 1,3,5-tribromo-, 1,2,4-tribromo- and 1-bromo-2,4-diiodobenzene; the ineffectiveness of external halide ions; and the failure of 1,2,4-trichloro- and -triiodobenzene to rearrange, they suggest a mechanism involving proton removal to form aryl anions which then react with the aryl halides by nucleophilic substitution on halogen, as in section (*b*) above. The facility of the latter reaction is in the order $I > Br > Cl$, whereas that of the formation of the aryl anions occurs more readily

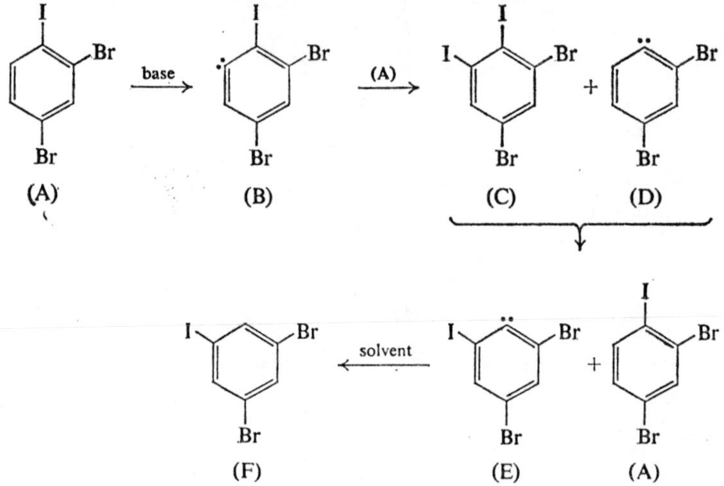

Fig. 122. A sequence of base-catalysed isomerisation of 1-iodo-2,4-dibromo benzene.

in the reverse order. The combination of these factors leads to the observed pattern.

There are many possible reactions which are illustrated by one sequence as Fig. 122.

(*ii*) *Smiles rearrangement*

The Smiles rearrangement *[51]* has been discussed by Watson *[52]* and at considerable length by Bunnett and Zahler *[53]* in their review of aromatic S_N reactions.

These reactions are essentially base-catalysed intramolecular aromatic S_N reactions, which may be generally represented as in Fig. 123.

Since X bears the side-chain, which may alternatively be another ring system, univalent groups such as the halogens are excluded, and since the reaction involves the mobility of X as a displaced group, the relevant factors discussed in Chapter 6 may be applied. The reactions are thus largely confined to the Group VI elements oxygen and sulphur, and more particularly to the more mobile higher valency, and thus more electronegative, forms of sulphur such as SO_2R

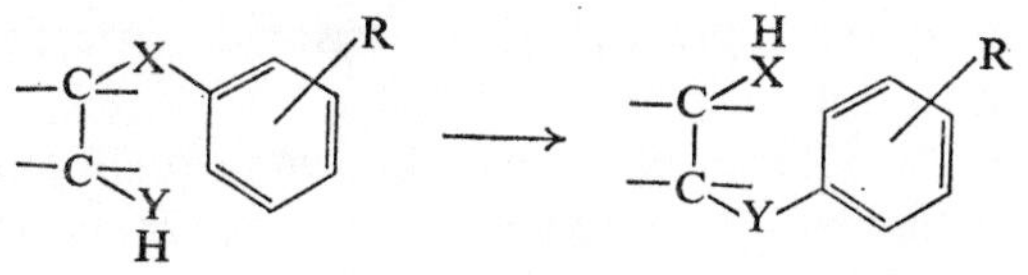

Fig. 123. General representation of the Smiles rearrangement.

groups. This is especially the case if there is no activating group in the ring to which X is attached.

In considering the nature of YH, the factors discussed in Chapter 5 are relevant. For example in electrically neutral forms the order is N > O. The situation may however be complex, for at some pH values oxygen but not nitrogen nucleophilic groups may be sufficiently ionised to be more reactive. High pH may on the other hand be an adverse factor if it converts an activating group from a neutral into a negatively charged and thus less activating form. Another interesting feature is that while NHX groups, where X is electron-withdrawing, are less nucleophilic than NHR groups (R = alkyl or H), at high pH values an NX^- but not an NR^- group may be formed, with the reactivity order $NX^- > NHR$.

If the YH group is also attached to an aromatic ring as, for example, in suitable diaryl sulphones, there will be substituent effects in that ring also. To the extent that such substituents affect the YH group, they are by and large opposite to those in the ring to which X is attached, but substituents favourable in this way are contrariwise likely to exert an unfavourable effect on the linking group (sulphone commonly). However the situation is not as complex as it might appear, since a substituent will be *meta* to one and *para* to the other. These various features are straightforward in origin, though complex in interaction.

As an intramolecular reaction the ability to take up the necessary steric configuration is a prerequisite. A favourable conformation is commonly one of several, the others being unfavourable. Bunnett and Okamoto *[54]* have shown that the introduction of some 6-substituents into 2-hydroxy-2′-nitrodiphenyl sulphone causes a very large increase in rate, irrespective of the electronic effects of the groups but dependent on size. They suggest that the 6-substituent raises the energy of unfavourable conformations but not of the

References p. 388

favourable conformation. They demonstrated steric accelerations approaching 10^6-fold increase in rate.

Reverse Smiles rearrangements are known in conditions in which XH (the structure on the right hand side of Fig. 123) is ionised as X^- (*e.g.* SO_2H at moderately low pH values) but YH (structure on left hand side of Fig. 123) is in the neutral form.

There is no specific reason why the atoms linking X and Y (Fig. 123) should necessarily be carbon, but there has been little work on other systems. Nevertheless Bunnett and Zahler *[53]* draw an analogy between the Smiles rearrangement and those of nitrophenylsulphonylguanidines via unstable *N*-sulphinic acids.

(*iii*) *Hauser rearrangement*

The Hauser Rearrangement is another intramolecular S_N reaction *[55]* and has also been referred to in a review by Bunnett *[56]*. The reaction requires strongly basic conditions and involves removal of a proton from a methyl group α to quaternary nitrogen. The carbon atom thus rendered strongly nucleophilic then attacks the *ortho* carbon, mildly activated by a $—CH_2\overset{+}{N}{<}$ group. It is illustrated in Fig. 124.

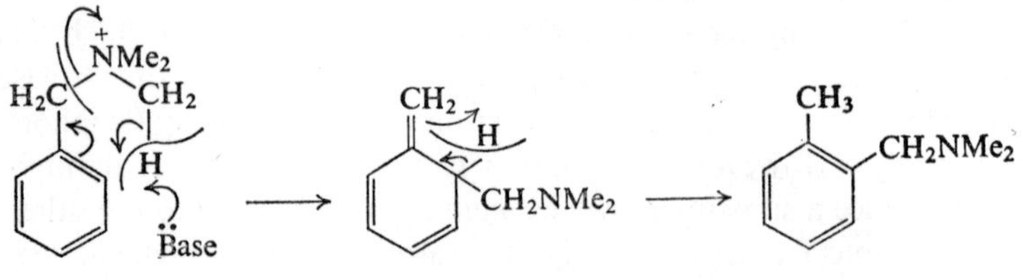

Fig. 124. The Hauser rearrangement of the benzyltrimethylammonium ion.

Hauser and his co-workers *[55]* showed that corresponding reactions occur with sulphonium ions and even sulphides (thioethers) but not ethers, for which an alternative rearrangement which does not involve substitution into the aromatic ring occurs. They also showed that the rearrangement of benzylammonium ions could be repeated with the products after requaternisation.

(*d*) *Intramolecular* S_N *reactions*

In general terms these introduce no features of novel theoretical interest. They have stringent steric requirements for reaction to take place. Some of these reactions have just been referred to in connection with rearrangements. They may also for example lead to cyclisations and there are numerous examples of this type.

A simple example to illustrate the type of reaction is given in Fig. 125. In this example Mathur and Sarbhai *[57]* have shown that a lactone attached to a biphenyl system is formed by a carboxylate group as nucleophile, and nitro as the leaving group.

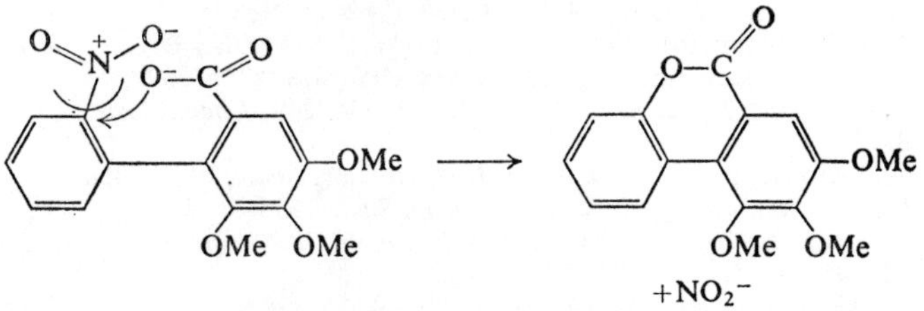

$+NO_2^-$

Fig. 125. An example of an intramolecular aromatic S_N reaction.

(*e*) *Reductions by hydride ion*

Reductions are related to nucleophilic substitution. When they involve direct displacement of a group by hydride, they are in fact straightforward aromatic S_N reactions. Such reactions are quite common.

Karabatsos *et al.* *[58]* have shown that in 8-bromonaphthoic acid, for example, hydride ion (from lithium aluminium hydride) not only reduces the carboxyl to the methylol group but also replaces bromine by hydrogen. By using the deuteride they showed that deuterium appeared only attached to oxygen and in the 8-position. There is thus no indication of a benzyne reaction. They considered a variety of similar reactions and demonstrated that the replacement of hydride ion is activated in the normal manner by electron-withdrawing groups.

References p. 388

REFERENCES

1 F. G. Bordwell, B. B. Lampert and W. H. McKellin, *J. Am. Chem. Soc.*, 71 (1949) 1702.
2 T. Kauffmann, A. Risberg, J. Schultz and R. Weber, *Tetrahedron Letters*, (1964) 3563.
3 P. Kovacic, J. J. Hiller, J. F. Gormish and J. A. Levisky, *Chem. Commun.*, (1965) 580.
4 P. Kovacic, C. T. Goralski, J. J. Hiller, J. A. Levisky and R. M. Lange, *J. Am. Chem. Soc.*, 87 (1965) 1262.
5 C. W. L. Bevan, J. Hirst and A. J. Foley, *J. Chem. Soc.*, (1960) 4543.
6 B. G. Gowenlock and W. Lüttke, *Quart. Revs.*, 12 (1958) 321.
7 A. Schors, A. Kraaijeveld and E. Havinga, *Rec. Trav. Chim.*, 74 (1955) 1243.
8 H. Jaffé, *J. Am. Chem. Soc.*, 77 (1955) 4448.
9 D. Hadži, *J. Chem. Soc.*, (1956) 2725.
10 J. V. Janovsky, *Ber.*, (*a*) 19 (1886) 2158; (*b*) 24 (1891) 971.
11 N. V. Sidgwick, in T. W. J. Taylor and W. Baker (Eds.), *Organic Chemistry of Nitrogen*, Rev. Ed., Clarendon Press, Oxford, 1942, p. 259.
12 T. Canbäck, *Farm. Revy*, 48 (1949) 217, 234, 249; *Chem. Abstr.*, 43 (1949) 6175a.
13 M. Ishidate and T. Sukaguchi, *J. Pharm. Soc. Japan*, 70 (1950) 444.
14 M. J. Newlands and F. Wild, *J. Chem. Soc.*, (1956) 3686.
15 M. Kimura and M. Thoma, *Yakugaku Zasshi*, 78 (1958) 1401; *Chem. Abstr.*, 53 (1959) 8056d.
16 M. Akatsuka, *Yakugaku Zasshi*, 80 (1960) 375, 378; *Chem. Abstr.*, 54 (1960) 18407b.
17 S. S. Gitis *et al.*, (*a*) *Zh. Obshch. Khim.*, 27 (1957) 1894; (*b*) *ibid.*, 29 (1959) 2646, 2648; (*c*) *ibid.*, 30 (1960) 3810; (*d*) *Dokl. Akad. Nauk, S.S.S.R.*, 144 (1962) 785.
18 H. J. Richter and N. E. Rustad, *J. Org. Chem.*, 29 (1964) 3381.
19 L. Bauer and T. E. Dickerhofe, *J. Org. Chem.*, 31 (1966) 939.
20 V. von Richter, *Ber.*, (*a*) 4 (1871) 21, 459, 553; (*b*) 7 (1874) 1145; (*c*) 8 (1875) 144.
21 P. Griess, *Ber.*, 5 (1872) 192.
22 W. Korner, *Gazz. Chim. Ital.*, 4 (1874) 306.
23*a* J. F. Bunnett, J. F. Cormack and F. C. McKay, *J. Org. Chem.*, 15 (1950) 481; *b* J. F. Bunnett, M. M. Rauhut, D. Knutson and G. E. Burrell *J. Am. Chem. Soc.*, 76 (1954) 5755; *c* J. F. Bunnett and M. M. Rauhut, *J. Org. Chem.*, 21 (1956) 934, 939, 944.
24 M. Rosenblum, *J. Am. Chem. Soc.*, 82 (1960) 3797.
25 E. F. Ullman and E. A. Bartkus, *Chem. and Ind.*, (1962) 93.
26 D. Samuel, *J. Chem. Soc.*, (1960) 1318.
27*a* A. L. Beckwith, B.Sc. Hons., *Thesis*, University of W. Australia, 1951; *b* A. L. Beckwith and J. Miller, unpublished work; G. T. Rogers and T. L. V. Ulbricht, *Tetrahedron Letters*, (1968) 1029.
28 J. D. Loudon and D. M. Smith, *J. Chem. Soc.*, (1964) 2806.
29 A. Kliegl and W. Hölle, *Ber.*, 59 (1926) 901.
30*a* R. L. Heppolette and J. Miller, *J. Chem. Soc.*, (1956) 2329; *b* R. L. Heppolette and J. Miller, unpublished work.
31*a* J. Miller, *Australian J. Chem.*, 9 (1956) 61; *b* J. Miller and K. Y. Wan, *J. Chem. Soc.*, (1963) 3492.

32 R. L. Heppolette, J. Miller and V. A. Williams, *J. Chem. Soc.*, (1955) 2929.
33 *Solvated Electron*, Advances in Chemistry Series, No. 50, Am. Chem. Soc., Columbus, Ohio, 1965.
34 D. C. Walker, *Quart. Rev.*, 21 (1967) 79.
35*a* S. Gordon, E. J. Hart and J. K. Thomas, *J. Phys. Chem.*, 68 (1964) 1262, 1271; *b* M. Anbar and E. J. Hart, *J. Am. Chem. Soc.*, 86 (1964) 5633; *c* S. Gordon, E. J. Hart, J. K. Thomas and A. Szutka, *J. Phys. Chem.*, 69 (1965) 289.
36 M. Anbar, Z. B. Alfass and H. Bregman-Reisler, *J. Am. Chem. Soc.*, 89 (1967) 1263.
37 H. S. Taylor, *J. Chem. Phys.*, 6 (1938) 331.
38 W. L. F. Armarego in A. R. Katritzky (Ed.), *Advances in Heterocyclic Chemistry*, Vol. 1, Academic Press, New York, 1963, p. 253–309.
39 A. Albert and W. L. F. Armarego in A. R. Katritzky, A. J. Boulton and J. M. Lagowski (Eds.), *Advances in Heterocyclic Chemistry*, Vol. 4, Academic Press, New York, 1965, p. 1–42.
40 D. D. Perrin in *Advances in Heterocyclic Chemistry*, Vol. 4, Academic Press, New York, 1965, p. 43–73.
41 A. Albert, D. J. Brown and H. C. S. Wood, *J. Chem. Soc.*, (1954) 3832.
42 F. W. Bergstrom and R. A. Ogg, *J. Am. Chem. Soc.*, 53 (1931) 245.
43*a* T. Zincke, *Ann.*, 330 (1904) 361; *b* T. Zincke, G. Heuser and W. Möller, *Ann.*, 333 (1904) 296.
44 J. Thiele and R. Escoles, *Ber.*, 34 (1901) 2842.
45 J. Miller and A. J. Parker, *Australian J. Chem.*, 11 (1958) 302.
46 A. H. Blatt and E. W. Tristram, *J. Am. Chem. Soc.*, 74 (1952) 6273.
47*a* J. Miller, *J. Am. Chem. Soc.*, 85 (1963) 1628; *b* D. L. Hill, K. C. Ho and J. Miller, *J. Chem. Soc.*, *B*, (1966) 299.
48 F. H. Kendall and J. Miller, *J. Chem. Soc.*, *B*, (1967) 119.
49 J. H. Wotiz and F. Huba, *J. Org. Chem.*, 24 (1959) 595.
50 J. F. Bunnett and C. E. Mayer, *J. Am. Chem. Soc.*, 85 (1963) 1891.
51 L. A. Warren and S. Smiles, *J. Chem. Soc.*, (1931) 914, and subsequent papers.
52 H. B. Watson, *Ann. Repts. Progr. Chem.* (Chem. Soc. London), 36 (1939) 197.
53 J. F. Bunnett and R. E. Zahler, *Chem. Rev.*, 49 (1951) 273.
54 J. F. Bunnett and T. Okamoto, *J. Am. Chem. Soc.*, 78 (1956) 5363.
55*a* C. R. Hauser and S. W. Kantor, *J. Am. Chem. Soc.*, 73 (1951) 4122; *b* W. B. Brasen, C. R. Hauser and S. W. Kantor, *J. Am. Chem. Soc.*, 75 (1953) 2660; *c* D. N. van Eenam and C. R. Hauser, *J. Am. Chem. Soc.*, 78 (1956) 5698.
56 J. F. Bunnett, *Quart. Rev.*, 12 (1958) 1.
57 K. B. L. Mathur and K. P. Sarbhai, *Tetrahedron Letters*, (1964) 1743.
58 G. J. Karabatsos, R. L. Shone and S. E. Scheppele ,*Tetrahedron Letters*, (1964) 2113.

SUBJECT INDEX

Printed in Great Britain by Spottiswoode, Ballantyne & Co. Ltd., London and Colchester